Contributions to Indian Geography . . . XIII

READINGS IN POLITICAL GEOGRAPHY

CONTRIBUTIONS TO INDIAN GEOGRAPHY

Series editor: R.P. Misra

Published Volumes

Concepts and Approaches
Geomorphology
General Climatology
Applied Climatology
Resource Geography
Population Geography
Urban Geography
Agricultural Geography
Rural Geography
Regional Geography
Environmental Geography
Geography of the Mountains
Readings in Political Geography

CONTRIBUTIONS
TO INDIAN GEOGRAPHY . . . XIII

READINGS IN POLITICAL GEOGRAPHY

Editor
CHANDRA PAL SINGH
Department of Geography
University of Delhi, Delhi 110007

HERITAGE PUBLISHERS, New Delhi
INDIA **1994**

ISBN 81-7026-062-0 (set)
81-7026-194-5

Published by B.R. Chawla for Heritage Publishers, 5 Ansari Road, Daryaganj, New Delhi 110002
Lasertypeset by Calligraphics, 13-D, Kamla Nagar, Delhi-7 and Printed by Efficient Offset Press, 215, Shahjada Bagh, Industrial Complex, Phase II, Delhi-110 035.

Foreword

Indian geography has made notable progress during the last twenty-five years or so. It is now an integral part of the school curriculum and almost every university in India has a full- fledged geography department with graduate as well as postgraduate course. In terms of books and research articles published annually, it is not far behind. A descriptive discipline of yesteryears, it is fast evolving as an analytical one with a good theoretical and empirical base and capable of contributing to local regional and national planning and development.

Its growth and evolution are, however, not smooth. There are many teething problems which not only have to be properly understood but also resolved soon. These problems are touched upon in some detail in the introductory chapter. It may be mentioned here that the progress of the discipline lies with geographers themselves. They must further build on the foundation already laid, even if it is weak. It is their responsibility to strengthen it to construct a new edifice.

The purpose of this series on *Contributions to Indian Geography* is to bring together the writing of Indian geographers on various subfields and thus to enable our young friends in colleges and universities to appreciate the strengths and weaknesses of the discipline better and to enrich themselves by the work done by their own colleagues.

The series is presently planned to be of fifteen volumes:

This is the thirteenth in the series. The other volumes already published are:

Volume	I	Concepts and Approaches
Volume	II	Geomorphology
Volume	III	General Climatology
Volume	IV	Applied Climatology
Volume	V	Resource Geography
Volume	VI	Population Geography
Volume	VII	Urban Geography
Volume	VIII	Agricultural Geography
Volume	IX	Rural Geography
Volume	X	Regional Development
Volume	XI	Environmental Geography
Volume	XII	Mountain Geography
Volume	XIII	Readings in Political Geography

More volumes may be added later on

Volume	XIV	Cultural Geography
Volume	XV	Research Methods

I am grateful to all the editors of various volumes for taking extraordinary pains to collect relevant materials and to edit the manuscript for publication. I must express may thanks to the authors of various articles included in the volumes for permitting us to publish them.

I am grateful to Professor Chandra Pal Singh for agreeing to edit this volume.

New Delhi
1 February 1994

R.P. Misra

Acknowledgements

The editor expresses his deep gratitude to Professor R.P. Misra for giving encouragement to bring out this volume. Thought provoking discussions with him compelled the Editor to have a fresh and closer look at the works of K.M. Panikkar and C.S. Venkatachar and to include them here. Sincere thanks to all the authors and publishers whose articles have been included in this volume. Proper acknowledgement for each one of them has been made at the beginning of their respective contribution. But in a few cases it has proved impossible to trace the copyright holder. The Editor hopes that such authors/publishers would realize the academic value of inclusion of their article and under the circumstances would accord their permission ipso facto.

CONTRIBUTORS

Sudeepta Adhikari	Department of Geography, Patna University, Patna, Bihar, India.
Walter K. Andersen	Bureau of Intelligence and Research, Department of State, Washington D.C., USA.
S.S. Bhattacharya	Institute for Defence Studies and Analyses, Sapru House, Barakhamba Road, New Delhi, India.
M.V. Bratersky	Institute of the USA and Canada, Academy of Sciences, Moscow, Russia.
C.D. Deshpande	Formerly Professor of Geography, University of Bombay and Jawaharlal Nehru University, New Delhi, India.
S.A.H. Haqqi	Formerly Professor, Department of Political Science, Aligarh Muslim University, Aligarh, U.P., India.
Ashok Kapur	Department of Political Science, University of Waterloo, Waterloo, Ontario, Canada.
S.I. Lunyov	Institute of Oriental Studies, Academy of Sciences, Moscow, Russia.
K.P. Misra	Centre for International Politics, Organization and Disarmament, School of International Studies, Jawaharlal Nehru University, New Delhi, India.
S.D. Muni	South Asian Studies Division, School of International Studies, Jawaharlal Nehru University, New Delhi, India.
Iqbal Narain	Senior Fellow, Institute of Development Studies, 8 B, Jhalana Institutional Area, Jaipur, Rajasthan, India.
K.M. Panikkar	Renowned historian and diplomat. Formerly Vice-Chancellor, University of Mysore, Mysore, Karnataka, India.
Niharranjan Ray	Formerly Director, Institute of Advanced Studies, Simla, Himachal Pradesh, India.
B.K. Roy Burman	Formerly Deputy Register General, Census of India, Block I, West Block, R.K. Puram, New Delhi, India.
Lloyd I. Rudolph	Department of Political Science, University of Chicago, Chicago, Illinois, USA.
Susanne H. Rudolph	Department of Political Science, University of Chicago, Chicago, Illinois, USA.

Vijay K. Seth	Faculty of Management Studies, University of Delhi, Delhi, India.
Chandra Pal Singh	Department of Geography, Delhi School of Economics, University of Delhi, Delhi, India.
B.G. Tamaskar	Noted Historical Geographer, Pune, Maharastra, India.
C.S. Venkatachar	Civil servant and diplomat, formerly Secretary to the Government of India.
B. Vivekanandan	West European Studies Division, School of International Studies, Jawaharlal Nehru University, New Delhi, India.

Contents

I | Introduction

CHANDRA PAL SINGH

Political geography as a discipline is about half a century old in India. Basu (1944), Pithawala (1944), Mookerjee (1945, 1946, 1947), Chatterjee (1947), Chaudhuri (1947, 1949), Spate (1948) and Bose (1949) wrote articles on political themes concerning problems faced by India during the mid-forties. Their writings covered subjects like the evolution of national frontiers, partition of Bengal as well as of India, ceasefire line in Kashmir, geography and the world war, etc. These reflected the immediate political concerns of the nation and the world. Besides, all regional texts of geography pertaining to the Indian subcontinent had a section on politico-administrative regions. Some political aspects were also dealt with in accounts of explorations of well known travellers earlier. Their accounts of the Himalayan and trans-Himalayan regions contain many valuable aspects of terrain and geography which can be part of the country's strategy. But most of these were not conscious efforts to contribute towards the systematic study of political geography of India.

In fact, the beginning of geography in India as a modern discipline could not have been at a worse time than during the inter-war years. The first department of geography in India at post-graduate level in any university came into existence at Aligarh in 1931, which was followed by Calcutta ten years later. At that time geography had just come out of the fire of environmental determinism versus possibilism debate in the West and it was cooking in the frying pan of chronological-ideographic paradigm. It was in this kind of theoretical debate in the West which was not yet decided either way that Indian geography finds its origins. Those Indian geographers who went abroad for higher degrees, and there was a bigger movement to the United Kingdom, during the forties and fifties, imported the knowledge and prevalent concepts and methodology of the subject to India. The course contents of geography in Indian universities in the earlier years reflected this inheritance. There was heavy dependence universally on teaching of physical, regional and human geography and surveying and cartography. There was no inclusion of political geography in any of the universities of India at that time. It was also because that was the darkest period for political geography, particularly in the West where the subject had become "too hot to handle" for its association with the then despised concept of geopolitics. If there were any writings of Indian geographers that can be called as political geography at that time, it was because they were handling a political subject, without a debate about the concepts, tools and techniques of political geography, except occasionally in the writings of Mookerjee and Chaudhari. The upheavals of the partition of India, conflict with Pakistan over Kashmir, and the impact of the world war were natural concerns of the time, also for geographers who wrote on such political subjects.

Ramesh D. Dikshit has been the foremost contributor from the Indian subcontinent to the theoretical and conceptual growth of political geography as a sub-discipline, particularly since the 1970s. But his focus was mostly global and largely theoretical. Political geography of the state of India received scanty attention from him till recently. Others whose work can be legitimately placed in this category have remained few in number. Only three authors can be cited in this regard — G.S. Singh (1969), Sukhwal (1971, 1986), and Mohiuddin (1987) — who attempted to analyse the interrelationships of the geography of India with its political aspects. There have been less than half a dozen case studies of small parts of the country, some of which have misleading titles masquerading as political geography of India giving an impression that they cover the whole country.

This paucity of literature on political geography of India is mainly due to lack of clarity regarding the conceptal framework required for this purpose. A discussion regarding this is necessary even as a justification for inclusion of contributions in this volume from different social sciences. Very often the student of geography is not clear regarding what is political, and the student of other social and physical sciences thinks, erroneously, that description of the surface of the earth and knowledge of what is where is geography. The compartmentalization of politics and geography and poor understanding of their respective students of each other, as well as their understanding by students of other social and physical sciences, have hindered growth of a political geography which recognizes the inevitability of interrelationships between politics and geography. The role of geography in politics has been recognized for a long time by some discernible scholars, but the role of politics in geography came to be widely recognized mainly after the conceptual revolution in geography in 1950s and 1960s. It was in the wake of redefining of geography in the general systems framework (Ackerman, 1963) that these interlinkages between politics and geography could be understood better. Under the general systems framework, the overriding problem of geography came to be the full understanding of the vast system comprising man and the natural environment (see C.P. Singh, 1981, for further discussion on this theme). Geography's organizing concept became spatial distributions and space relations on the earth's surface of the vast overriding system comprising all subsystems of man and the natural environment. Political geography according to this view is concerned with the interaction of geographical area and political process. Every political process has a geographical area uniquely associated with it and no geographical area escapes some relation with a political process. Such a definition of political geography takes into account the symbiotic and indivisible relationship with those human subsystems where and due to which politics and power are generated — society and economy within the natural environment. Although close interaction of political science, sociology and economics has produced very useful fields of political sociology and political economy improving understanding of how man organizes himself in relation with other man, they fall short of explaining how man, with his socio-economic-political systems, organizes himself in relation with the natural environment. It is this link which is vital to the understanding of relationship between politics and the environment. The environmental and ecological problems of the world today are often the result of unhealthy nexus between the exploiter of resources and politicians and administrators. The ideological arguments supported by the coercive powers of the state often determine what and who will own and exploit natural and human resources within the territorial extent of a country, which is a political entity. The traditional concern of political geography, geopolitics, has lost some relevance in the post-Cold War world. Political geography does not have to remain concerned with geopolitical rivalry and the politics of deterrence any more. It has to include the interaction of man with the environment and human and ecological security in it.

There have been two political geographies till recently: political *geography* (work by human geographers on political topics) and *political* geography (work by those in other disciplines on topics arguably falling within the scope of political geography) (O'Loughlin quoted in Reynolds, 1992, 93). This is indicative of apparent diversity, fragmentation and lack of focus in both the 'fields'. Their integration is possible under the systems framework where social, economic and political subsystems of man interact with the physical and biotic subsystems of environment within the territorial framework created by man. Space is something that cuts across all the subsystems. Space per se can have no general effects and it cannot be viewed as a container somehow separate from material objects. But spatial arrangements of what is human and natural contribute to the processes. Transactions among the spatial structures of human and natural phenomena bind them together and make them inseparable. The study of politics cannot be done without taking account of the society and its economy, as well as its resource base and its spatial extent and context in which they evolve. That is what gives politics its peculiar characteristics. Politics of each country is bound to be different because of the peculiarities of its society, economy, physical environment and the particular historical context within which they have interacted with each within their territorial extent, and with other similar territorial entities.

The focus of political geography therefore cannot remain confined only to human geography as some scholars (Dikshit, 1992, 523) have argued. For attaining disciplinary coherence political geography of today has to take into account the interplay of human and natural systems. It has to become a means of picking up the thread of the complex web of reality from the political end to reach to the environmental end. Without this political geography cannot help to understand the problems of the environment that the world seems to be suddenly facing at the end of the twentieth century. Searching for the political in geography and geographical in politics, geopolitical advantages of particular territories, and politics of place may be insufficient in searching for answers for poverty and famine in Africa, consequences of deforestation in large parts of the world, nuclear waste, industrial and urban pollution, depletion of ozone layer and global climate change. These are isssues which are linked up with the security of man, as well as of all living beings on the planet. The geopolitical rivalry and politics of deterrence is no more valid with the collapse of the USSR and dramatic political changes in Europe and elsewhere. An ecologically sustainable future and a just economic and political order impinge upon the security of man from the local to global scale.

Such a political geography opens up interaction with not only social sciences but also environmental sciences. Concepts and tools developed in these sister disciplines are going to be useful to the study of political geography of a place, an area or region, a country or regional groupings of countries in the world. For instance, a village has a physical structure and spatial extent which is bounded by an administrative boundary. It comprises several socio-economic-religious groups, wherein status of the individual and the group to which he belongs is determined and power and authroity are generated and allocated. The village does not exist in isolation. It has linkages with other villages and towns of different ranks in the settlement hierarchy of the area through its social, economic, political and communication linkages. The politics of caste, religion, class, or any other aspect of the village may spill over its administrative boundary and form an integral part of politics at higher scale. The channels of information and communication of the village become the means of ideas and materials to travel in both directions. A study of political geography of the village cannot remain confined to its politics and geography. It has to understand the social and economic structure of the village, and interrelationships of different socio-economic groups with each other, and their linkages with the world outside the village. Natural resources of the village, their ownership and method of exploitation, and the consequences of the demands made by the local and outside people on the nature within the administrative extent of the village are also of equal interest and concern of political geography of that village. The starting point of the study is political, which should remain in the focus throughout, but it can be studied only when all other human and natural systems existing at that place are put in proper perspective. Such an approach is essentially interdisciplinary.

When this approach is extended to include the entire territorial extent of India then a political geography of India has to include the society, economy and polity and the environment, their regional configurations and manifestations, and the political framework within which they are interlinked with each other. The country so organized internally exists among other similar entities, the nation states, at the world scale. The internal organization of the country determines its level of wealth and knowledge and through them power, which in turn determines its international standing among the comity of nations (C.P. Singh, 1992). Such a view does not abandon the traditional concept of geopolitics in political geography. Geopolitics in the present framework deals with how geopolitical reasoning is integrated into political discourse to maintain social and political relations of dominance within contemporary international politics. It helps articulate the linkages between changing material conditions in the world economy and the creation of a global geopolitics to preserve, maintain, or enhance existing structures of material advantage (Reynolds, 1992, 394).

The contemporary concerns of environmental degradation and the politics of the environment are also essential aspects of political geography being advocated here. The term security, which was traditionally concerned with the state providing protection from threats to social order from external human enemies, has not assumed new meaning with the environmental hazards transcending national boundaries. The accumulation of nuclear and chemical wastes, rising levels of pollution, high energy consumption, related with the high consumerist culture of the North are responsible for ecological damage even in the far away countries of the South. These issues cannot be studied under the conventional geopolitical and state-centric preoccupations of political geography. Political geography of the 1990s has to incorporate linkage between sovereignty, security and environment (see Delby, 1992, 504) and extend its analysis beyond an explicitly spatial focus to examine questions of environmental politics in the context of the emergence of a global civil society. Political geography of India should therefore include a new section on the stage of the environment in India, where a fast increasing and already huge population is making increasing demands on the environment and the growing industrialization and urbanization are contributing rapidly to its degradation. A study of who gets what, where and how much and their linkages with the environment within the country and also with other parts of the world, and environmental politics from the local to global scale would be a significant contribution to the political geography of India. The interaction of man with his environment for survival within a systems framework offers the theoretical foundations of the sub-discipline.

Such an approach to the study of political geography of India leads us to take into account work already done not only by geographers but also by other social and environmental scientists who study the complex reality around us on this planet. These together could form the building blocks of a conceptual framework for political geography of India, in which concepts, tools and techniques of social and environmental sciences are exchanged freely. Such an exchange would not only be beneficial to political geography but would also enrich the sister disciplines. It is with this rationale that contributions from different social sciences have been included in this volume under readings in political geography of India. These are examples for political geographers of the work already done in disciplines working towards a common goal — how man organises himself in relation to other man and collectively to the environment within a territory enclosed by well defined boundaries, and the interrelationships of such a political entity with similar other entities in the world. They are intended to give the student of political geography a glimpse of how other sciences have approached their subject in their study of India. It is for this reason that the original form of the contributions has been retained as far as possible, as abridgement of quoting out of context can often distort the intention or the views of the author and the reader receives them in a light in which they were never intended. Besides, they are also meant to provide direct access to them by students not having adequate library facilities for all the articles published in different national and international journals.

II

The historian-diplomat K.M. Panikkar can be considered as the first political geographer of India who studied the role of geographical factors in Indian history as early as 1955. Despite ritualistic references to his work geographers have not followed the lead given by him. His use of concepts like space, friction of distance, annihilation of space and distance, role of relative location, shape and size of territory was ahead of geographers not only in India but also abroad. The impact of Schaefer's article (1953) on geography, which among others had argued for use of such concepts in geography, was several years away, in fact almost a decade away. There is no evidence to show that Panikkar was aware of the beginnings of the internal debate in geography in the West in the fifties. The fallout of that debate reached India by the second half of 1960s, by which time Panikkar's writings on the subject were more than a decade old. Political geographers in India who should have been the natural inheritors of Panikkar's ideas paid little attention to them. Other social scientists refrained from incorporating his ideas and concepts presumably thinking that they had no jurisdiction over them because of the suffix geography in his title. Since physical facts of geography remain unaltered since Panikkar wrote about their role in India's history most relevant sections of his book are included here under chapters II and III in section I of this volume.

Panikkar argues that both the internal policies of a country and its external relations are governed largely by its unalterable geographical conditions and their relationships in space to other countries. India has to realize her position in the world where geography has placed her in relation to other countries. He points out that the most significant and revolutionary activity of man against geography has been the attempt to conquer space. "The experience of Napolean and Hitler in Russia, of Hidoyeshi in the 17th century in China, of the Moghuls and even the British who took over 100 years to conquer in India, show that space is of the greatest importance in the life of a nation. The dynamics of space is one of the favourite themes of geopoliticians and yet both distance and space have contracted woefully in the last few decades. Technology has helped to make distance and space relative." This quotation from Panikkar is even more relevant to the present than it was at the time of his writing. But how many studies of political geography of India have incorporated it in their analysis of shrinkage of space due to technological change, being accelerated in its speed with each succeeding year. Besides, at a time when *lebensraum*, or living space, became the war cry of European nations during the inter-war years, Panikkar showed that desire for space did not have the same significance in India. She had no necessity to expand from the point of view of security or defence.

In his second contribution to the present volume Panikkar traces the succession of invasions and population movements, in which the role of physiography, distance and direction have been emphasized. He assesses the location of India astriding Indian Ocean and the necessity of controlling the seas because of the indivisibility of sea on the globe. He traces the development how for the first time the subcontinental power of India was subjugated from invasions and penetration from the coast from different directions from Calicut, Bombay, Madras and Calcutta and ruled from a distant country, changing political relationships of the two distant territories. The spatial organization of the vast communication network of subcontinental scale in southern Asia can best be understood by following the routes of invasion from land and sea on India in the past.

A mention of S.P. Chatterjee's article (1982) can be made here for his analysis of the role of geographical factors in the evolution of political history of India, which could not be included in this volume. He noted the role of the alignments of her hills and mountains, plateaus and plains, and drainage system, which made it possible for the Aryans to come into India during the historical past through the natural gaps in the northwestern mountains. Other waves of immigrants and invaders also came during different periods through the northeastern passes. These immigrants preferred to make India their permanent home and merged themselves with the local inhabitants, thereby making this country of diverse people and culture. Location of settlements, selection of occupations, growth of trade and communication linkages were initially developed on the basis of congenial physical environment. Despite recent growth of science and technology, the diverse landforms of India still continue to exert strong influence in shaping the political history of this subcontinent.

C.S. Venkatachar, who belongs to the Indian Civil Service and was Secretary of the Ministry of States during the most critical post-independence period, also looks at the diversity of India, within the geographical framework of the country in chapter 5. He puts the perspective of the unity of India by viewing North from the tip of peninsular South. He recapitulates the physical geography of India and looks at the impact of demarcation of the North and the South by the Vindhyan range and how the two distinct political regions have emerged due to this physical division, and how it gave rise to the riverain civilization in the Gangetic plain with a continental outlook. Another riverain civilization came into existence in the deltaic parts of the eastern peninsula which became the focus of the ancient civilization of the South where flourished the well-known southern kingdoms. In contrast to the deltaic civilization the riverain civilization of the north rose to great heights of achievements and splendour and it is here that Aryavrata, the sacred land of Indo-Aryans, originated. Because of its fertile alluvial soils, availability of river and rain water, and tropical climate, it became an area of the most sedantary and static populations of the world. He also notes the movement pattern of population in the past governed by physical geography of India.

Tamaskar's pioneering work on the concept of boundaries in India still remains a major landmark in political geography of the country. He was a professor of political science who made a systematic study of boundaries, a subject which was among the core of political geography studies, almost upto the time of his writing the article. This was yet another failure of political geographers of India not to have built upon his work substantially. In chapter V he deals with evolution, delineation, manifestation and demarcation of boundaries since the ancient times in India when "large congeries of states ..... separated from each other by belts or zones forming frontiers between them." He finds that in the early stages of the evolution of boundaries there was a marked preference for natural barriers to delimit them, like stretches of mountains, courses of rivers and canals, outskirts of deserts, forests, swamps, etc. He investigates further how boundaries were marked between villages as well as between states, and the mechanisms of settling boundary disputes. He also deals with classification and functions of boundaries in India.

Tamaskar's several other contributions need particular attention from political geographers intending to study India. For example, his study of geopolitical thoughts of Kautilya (Tamaskar, 1962) can be a good starting point to study the evolution of indigenous concept of geopolitics in India.

The reorganization of internal administrative structure of India in 1956 was of profound importance for movement of goods and people and development within each state of the country. Deshpande analyses in chapter V geographical consequences of reorganization of India into 14 states and 6 union territories, and whether the new states were viable from economic, social and administrative point of views, and whether they had a balanced geographical structure. He points out the possible frictions and conflicts arising out of the border areas and bilingual states, and consequences of eccentrically placed capital cities, the particular orientation of each state towards development policies according to its altered geographical setting, and the inter-state relationships under the new administrative framework. This theme needs pursuing further under the newly developing field of administrative geography of India.

Section II of the volume takes into account the bases of extremely diverse and pluralist politics in contemporary India and the difficulties of nation building here. The understanding of geography of socio-cultural groups in India is necessary to explain the intensity of some of the recent conflicts and agitations based on the politics of caste, tribe and religion in large parts of the country. Niharranjan Ray traces the origins of the term *caste* as used by sociologists and anthropologists working on and concerned with India in chapter VII. He clarifies that the term *caste* is not of Indian origin. It was used by the Portuguese to mean a social organization based on hierarchical ranking of people determined by birth or lineage. Other Europeans who followed them in India accepted and adopted the term and the concept that went along with it. Later on some European ethnologists, especially British, got enmeshed in the cobwebs of textual interpretation of the traditional Hindu- Brahmnical social organization on the one hand and the functional or behavioural pattern of what they had come to know as *caste* as they saw it in operation before them. For a long time they tried to explain the latter in terms of the traditional concept of *varna* as recorded in the early and later vedic texts.

Ray then goes on to explain what tribe means, and the origins of the term. He notes that with the rise and growth of socialism in Europe, the term tribe came to be used in denotation of a particular stage of socio-political evolution of community of people within a given territory and language area. Clan, tribe, nation, etc. thus came to denote the various successive stages in the progressive march of people aspiring towards nationhood. But the so-called 'tribals' of India are the indigenous and autochthonous people of the land, in the sense that they had long settled in different parts of India before the Aryan-speaking people penetrated the country. In Assam and the north-east, people of Tibeto-Burman stock migrated into these areas. Due to these reasons the western term 'tribe' does not explain the social situation in India. He argues that the term *jana* derived from Indian languages is more appropriate for its biological meaning 'to be born' or 'to give birth to', or community of people, from which is derived the term 'janapad', the territorial extent of a community of people. Ray also gives a brief history of the geographical distribution of tribes in India.

As a case study of tribes in regional setting, their geopolitical importance, problems of development and their integration with the heartland of the country Roy Burman analyses the situation of hill tribes of north-east India and the approach to be adopted for their development in chapter VIII. He explains the geopolitical importance of north-east India, socio-cultural perspective of the region, its political and administrative structure and economic situation in the north-east. He argues that along with economic planning, there should be social and political planning in an integrated manner. The implementing agencies should also ensure that their focus in dealing with the tribals of the region should not take them as primitive people.

Haqqi finds that the political geography of India is 'neither neat nor tidy' in chapter IX. The political subdivisions designated as states and union territories of different shapes and sizes, inhabited by persons belonging to various religious and linguistic/cultural groups, have been constituted not only on linguistic lines but also on historical and political socio-cultural distinctiveness, ethnic and economic considerations, religious, script and communal sentiments, and integration of princely states and viable groupings. The constitution of India was so framed as to bring unity in such diversity. The politicians also have attempted a pluralistic approach in political appointments by which the highest offices of the President and Vice-President have rotated between the South and the North. The country also opted for a pluralistic model of development for the solution of the problems of a highly pluralistic community.

Such a high degree of pluralism in a country generates conflicts of various kinds and could lead to balkanization of a country. The possibility of fragmentation of the country always exists there. But Iqbal Narain argues in chapter X that balkanization of a country is not essentially a function of cultural pluralism as of failures at the managerial level, both economic and political. He studies in the specific Indian context the pattern of interaction between cultural pluralism on the one hand and national integration and democracy on the other. His central argument that cultural pluralism manifests itself, among others, in regionalism (inclusive of linguistic regionalism), religion and caste, need not necessarily have a negative correlation with national integration and democracy. From the point of view of political geography of India, Narain analyses the stages of growth, components and types of regionalism in India. He emphasizes that geo-cultural components is the heart of India's historic cultural pluralism, and the bedrock of the phenomenon of regionalism. It may be pointed out here that cultural geography of a country is essential for the study of its political geography, which needs special attention like historical geography of India. Among the different components of regionalism that Narain identifies the first is the factor of geographical boundaries, followed by the factors of history, caste, language, economic conditions (leading to regional imbalances and pulls and pressures of distributive justice) and the political component of regionalism, in which politicians sometimes exploit regional deprivations and unrest and convert them into movements to forge and strengthen their individual and factional support bases.

His typology of regionalism — supra-state, inter-state and intra-state — is a useful framework which has been employed to study the complex amalgam of geo-cultural, politico-economic and also psychic factors. It is after the discussion of the nature of regionalism in India that Narain turns to the vexing issue of its relationship with nation building.

The regional diversities of India within her vast spatial extent found expression in her adopting a federal constitution. Adhikari looks at the geographical bases of India's federal polity in chapter XI. He notes that because of its large geographical size, India is bestowed with physiographic and climatic diversities which have largely impeded the process of cultural and linguistic integration. Unequal distribution of resources and specialization in varied nature of economic activities have given rise to regionally grouped economic diversities and unequal economic growth. He identifies several core areas in the country which have been the foci for separate political-religious ideas operating in the opposite direction. But federalism, which is the centripetal force for India, is keeping the centrifugal forces under control.

The evolution of political party system in India, its impact on the political system, location of different parties within the ideological spectrum, and their future prospects have been studied by a number of political scientists. For instance, Rudolph and Rudolph examine the dominance of the Indian National Congress among India's several parties in chapter XIII and analyse why India's national parties are centrists and pluralist rather than right or left class parties, and why, in a country with prominent religious communities, confessional parties are like those of the Middle East or Western Europe that appeal to religious sentiment and identity ineffective and defensive. They advance several reasons for national centrist parties to constitute the main-stream or party politics over a rather long period of time. These include the tenuous quality of national class consciousness in the country where class has not been and may not become the principal medium for representation. Even organized labour's apex organizations have little control over subordinate units or influence on national policy or politics. Second, the fragmented nature of the support base for a national confessional party of the Hindus is more likely to support secularism. The Hindu majority is also competitive along class, caste and regional lines than are India's minorities. Thirdly, by contrast India's minorities have had a greater propensity to vote en bloc. Fourthly, with the first wave of land reforms the power of the neo-feudal landlord class was broken and a centrist political class of "bullock capitalists" and backward classes has emerged as the leading representative of agrarian interests. Fifthly, the goal of capturing power in Delhi or wielding political influence there compels different factions and personalities to come together. Sixth, the constraints imposed by cultural diversity and social pluralism, and finally, the formal rule governing electoral competition among parties have favoured centrism in Indian politics.

The peculiar characteristics of Indian society and polity are complemented by the mixed economy that India decided to adopt as a key policy of the Indian state. Seth analyses the nature of mixed economy and the strategy of development and regional planning that has emerged after Independence of India in chapter XIII. He explains political philosophy of the mixed economy under the Industrial Policy Resolution 1956, which has been termed by some as the economic constitution of India. Though there have been several changes in it since then, the basic industrial policy has remained the same even now. Seth traces the process of planning since 1951 and the thrust of different five year plans. Under the industrial policy and the five year plans spatial pattern of industrialization was due to such policy instruments as location of private-sector investments, and the use of input subsidies and other financial incentives to attract industrial investment in the backward regions.

Seth also recognizes the importance of the role of the federating units, i.e., the states, in industrializing of the country. But in the kind of administrative set up that has developed in the federal structure of India the central government can influence the spread of industrial sector through government regulations, incentives and policies. The decisions of the government, however, can be influenced through the political pulls and pressures under the federal set up, particularly since the reorganization of states with respect to education runs parallel to the distinction between the states with respect to gross indicators of development.

The internal organization of India— of its society, economy and polity, within its given geography, and their interplay with each other over a long period of time making it a unique historical experience — has given it a particular character which is essentially Indian. With so many diametrically opposite forces operating in the system, in different regions of a country of subcontinental extent, the task of nation building is not easy. These forces have their linkages outside the territorial extent of India which exert their influence within the country. These characteristics of India are the basis of its strength and weaknesses of the Indian state, with which it conducts its external relations and attains a particular ranking among other countries of the world. Section III of the volume deals with the external relations of India, its ranking as a regional power in Asia, and the role of geography in it.

Venkatachar begins with a historical perspective of India's relations with her neighbours in chapter XIV. He points out that it was because of the physical geography of India, particularly the Indian Ocean and the Himalayas, which barred easy movement from outside. A large population living away from the sea was responsible for India having a continental outlook. Also the vast spatial extent deterred mobility from region to region within the country, which had the effect of imprisoning the minds of her people. He notes further that if Indian attitude was static across the mountains, it was more dynamic only for a time across the seas. The natural exit out of India is by sea. On the western seaboard three points of contacts with the Fertile Crescent and the Mediterranean world manifested themselves from the first century B.C. On the eastern seaboard India developed more intimate contacts with South-East Asia through various islands of Andaman and Nicobar and the Indies, due to which voyages were easy and frequent. Later on new power pattern was developed by the Western countries who broke into the Indian Ocean.

Panikkar also takes up the role of sea in Indian history in chapter XV. He shows how the history of peninsular India has been intimately connected with the Indian Ocean from the earliest days. After describing the geography bays and islands, connectivity of the Indian Ocean with other seas and oceans, and the wind pattern in and around the Indian Ocean, Panikkar takes up the political history of southern kingdoms of India and their relationships with countries across the sea. He demonstrates the role of Indian Ocean in the ultimate colonization of India by European countries, and the present trade pattern which is largely and predominantly maritime.

Andersen looks at India's relations within a broader framework of Asia as they were at the end of the 1970s in chapter XVI. India had gone to war with Pakistan twice and with China once. Bangladesh had come into existence. The Indo-Chinese war had changed equations among the Asian countries and their relations with the superpowers on this issue. Insurgency against the Marxist regime in Afghanistan had generated tensions in Pakistan-Afghanistan relations. Domestic unrest continued in Iran after the collapse of Pahalvi dynasty. Conflicts in Indo-China had stalled the Sino-Indian normalization process. Andersen describes how the state of India conducted its foreign relations in Asia, apart from balancing China and the USSR, tackling the Indian Ocean concerns to achieve its two broad goals there: removing the superpower naval forces and bases and building up its navy to give India an independent capacity to defend her national interests. In the case of western Asia, from where two thirds of the oils consumed by India came, Pakistan and inter-Arab squabbles were complicating factors.

Misra looks at the regional security aspect of the South Asian region in chapter XVII. He observes that studies on security generally limit their scope to national security only. But he extends his study to include the entire South Asian region consisting of India, Pakistan, Bangladesh, Nepal, Bhutan, Sri Lanka and Maldives. This study also sets the background to the formation of South Asian Association for Regional Cooperation (SAARC). These seven states have nearly one fifth of the world's population, but have 3.3 percent of land, making the region as one of those having very dense population in the world. Among these India is the only country which has common border, either on land or in the sea, with all the countries of South Asia. Because of its large size, huge population and vast resources India stands as collosus among its neighbours in the region. Misra points important reasons where unity in the region lies: common historical experience, close cultural affinity, similar problems of poverty and social evils, some complementarity of resources, etc. He finds that regional cooperation can solve many problems of the region, with which threats to regional security can be met effectively.

Kapur also deals with subcontinental level in chapter XVIII to study the contemporary structure of power at the end of the eighties. He traces the power relations in the subcontinent from the late 1940s to 1980s. He finds that South Asian international relations have passed through several stages of development — from conditions of anarchy in 1947 to orderliness in regional diplomatic and political affairs. These relations are based on the attitudes and policies of the three part cluster of South Asian states: 1. India 2. Pakistan, and 3. Sri Lanka, Nepal, Bangladesh, Bhutan and Maldives. These relations are based on regional affairs, depending on the internal political developments in the states, their economic and military strength, their relations with India, and the cultural sensitivity to, and the political utility of Indian and Pakistani conceptions about regional security for the elites of the other South Asian states.

The importance of Indian Ocean for security and economic interests of India has been noted earlier by Panikkar and Venkatachar. The location of India astriding the 28 million square miles of Indian Ocean, through which pass some of the very important sea routes for world shipping, and which is the repository of some of the most valuable minerals, metals, energy and food, make India as one of the most interested nations in the events in the Indian Ocean region. The growth of superpower rivalry, naval bases of big powers, and presence of nuclear weapons are some of the concerns of the littoral states, particularly of India.

Bhattacharya analyses economic interests of big powers— Western maritime powers, the Soviet Union and China — in Indian Ocean region in chapter XIX. He makes an assessment of raw materials available in the region which are vital for the growth and development of the major manufacturing industries of the big powers. These include raw materials like rubber, tea, coffee, sugarcane, jute, cotton and wool, and mineral wealth like tin, gold, chromium, iron ore, lead ore, tungsten, nickel ore, zinc, antimony, thorium which is an essential radioactive mineral, and above all, oil which is the most vital mineral which plays a crucial role in the development of modern industries and which has been at the centre of global power politics and war. Due to industrial backwardness and lack of scientific knowhow the countries around the Indian Ocean import sophisticated machinery and manufactured goods from the developed countries. He further analyses the pattern of foreign investment in the region to understand the level of economic interest of western countries and Japan in it.

The rivalry of big powers in establishing and keeping their presence in the Indian Ocean and efforts of the nonaligned countries to get a declaration of this Ocean as a zone of peace created considerable tension among the countries concerned. Vivekanandan evaluates the problems and prospects of Indian Ocean as a zone of peace and international politics associated with it. He shows in chapter XX how for different reasons the Indian Ocean region is an area of vital interest for many countries, and examines the differing prospectives with which some of the littoral countries are approaching the concept of peace zone in the Indian Ocean.

The collapse of the Soviet Union and the end of Cold War have altered the role of countries of the Indian Ocean region.Bratersky and Lunyev examine in chapter XXI the prospects for the evolution of India's strategy in the subcontinent and the Indian Ocean. They remark that India became a centre of force in the South Asian region gradually. In the 1980s it became a regional naval power, while in the 1990s it will emerge as a major regional power in Asia. Their prediction is based on a number of premises and analysis of the strengths and weaknesses of India in emerging as a regional power. They analyse military expenditure of India in the eighties and the nature of weapons and their technological level, and India's position as a nuclear state. The end of the Cold War has brought about a significant reassessment of their roles in the region by the superpowers. The United States seems to have concluded that it is mainly interested in two strategic positions in the Indian Ocean — the Straits of Hormuz and Malacca, and has got accustomed to the fact that India is steadily becoming a regional power with which it can have some complementarity of security interests. They further analyse the political, cultural and psychological factors shaping India's strategy, India's ocean policy and economic measurements of India's foreign policy strategy.

Muni also takes up the opportunities and challenges faced by India in the post-Cold War world in chapter XXII. After giving an account of the changes that have come about after the dismantling of the Cold War and its concomitant processes of relaxation of tensions between the two superpowers, Muni draws attention to the foreign policy dilemma of India of the early 1950s re-emerging. The dilemma was choosing between joining the Pax-Americana and the charting of an independent Nehruvian course. The Indian choice is clearly for the former now. In the changed global context the quiet ejection of nonalignment has also changed the course of Indo-U.S. relationship. Muni reminds that the issues which nonalignment addressed most were international economic issues. These would now be articulated on the North-South polarization of countries where there would be competition among developed countries for resources and markets in countries of the Third World. He expects that some of the developing nations of Africa and southern Asia may experience yet another form of economic pressure resulting from their increasing marginalization in the post-Cold War order. The greater regionalization of the world economy with regional economic groupings like those of Europe, North America and the Asia Pacific region, in addition to the existing ASEAN, due to greater economic competition among the global economic giants, will change the political and economic context of the world. To exist and grow stronger in such a context India cannot follow the policies of the past decades.

The above contributions of different scholars are merely illustrative of the unity of several disciplines in the study of interlinkages of geography, society, economy and polity within territorial containers and their combined evolution over time. There are many gaps which have not been even touched upon in this volume. For instance, electoral geography and the state of the environment in India. Elections are occasions when the interplay of the above factors is reflected in perceptions and opinions expressed by the voting electorate. The state of the environment on the other hand reflects the concerted action and interaction of man operating through the combination of these factors at given points of time. These are therefore essential aspects of political geography of a country, which need equal attention. It is hoped that the above framework will trigger off interdisciplinary studies of political geography of India, which will help in understanding the complex reality of this part of the world.

REFERENCES

1. Ackerman, E. (1963), "Where is a Research Frontier?", *Annals,* Association of American Geographers, Vol. 53, No. 3, 429-440.
2. Basu, B. (1944), "Evolution of National Frontiers", *Calcutta Geographical Review*, Vol. 6, No. 3.
3. Bose, S.C. (1949), "Cease Fire Line in Kashmir", *Calcutta Geographical Review*, Vol. 11, No. 2.
4. Chatterjee, S.P. (1947), "The Partition of Bengal", *Calcutta Geographical Review,* monograph no. 8
5. — (1982), "Evolution of Political History of India As Influenced by Geographical Factors", *Geographical Review of India*, Vol. 44, No. 1, 20-37.
6. Chaudhuri, M. (1947), "The Evolution of Ideas in Political Geography", *Calcutta Geographical Review*, Vol. 9, No. 1-4.
7. — (1947), " Geopolitical Aspects of World Map", *Calcutta Geographical Review*, Vol. 11, No. 3-4.
8. Delby, S. (1992), "Ecopolitical Discourse: Environmental Security and Political Geography", *Progress in Human Geography,* Vol. 16, No. 4, 503-522.
9. Dikshit, R.D. (1992), "Another False Dawn?", *Political Geography*, Vol. 11, No. 6, 523-525.
10. Mohiuddin, Iqbal (1987), *Political Geography of India,* Delhi: Adam Publishers.
11. Mookerjee, S. (1945), "Who Advised Hitler?", *Eastern Express,* Calcutta, Vol. 1, No. 1, Annual Number.
12. — (1946), "A Case of Geopolitics", *Eastern Express,* Annual Number.
13. — (1947), "The Ideas of Aviation", *Eastern Express.*
14. Pithawala, M.B. (1944), "Geography and the World War", *Indian Geographical Journal*, Vol. 19, No. 4.
15. Reynolds, D.R. (1992), "Political Geography: Thinking Globally and Locally", *Progress in Human Geography,* Vol. 16, No. 3, 393-405.

16. Rudolph, S.H. and L.I. Rudoph (1972), "Regional Patterns of Education", in S.H. Rudolph and L.I. Rudolph (Eds.), *Education and Politics in India: Studies in Organization, Society and Policy,* Delhi : Oxford University Press, 51-67.
17. Schaefer, J.F.K. (1953), "Exceptionalism in Geography", *Annals,* Association of American Geographers.
18. Singh, C.P. (1981), "Geography and Electoral Studies", *Transactions,* Institute of Indian Geographers, Pune, Vol. 3, No. 1, 81-87.
19. — (1993), "Democracy and Privatization: A View From India", in P.J. Taylor (Ed.), *Political Geography of the Twentieth Century: A Global Analysis,* London: Belhaven Press, 257-261.
20. Singh, G.S. (1969), *A Political Geography of India,* Allahabad: Central Book Depot.
21. Spate, O.H.K. (1948), "The Pattern of India and Prospect of Pakistan", *Geographical Review,* New York, January.
22. Sukhwal, B.L. (1971), *India : A Political Geography,* New Delhi : Allied Publishers.
23. — (1986), *Modern Political Geography of India,* New Delhi: Sterling.
24. Tamaskar, B.G. (1962), "Some Geopolitical Concepts of Kautilya", *Quarterly Review of Historical Studies,* Calcutta, Vol. 3.

Section I | Geography and Evolution of the Indian State

Physical geography of India has played an important role in guiding the movement of people from Central Asia and what is now known as the Middle East. Their adoption of a sedentary life and developing a high level of culture in what came to be known as Aryavrata, pushing the original inhabitants to their periphery or subjugating and assimilating them in lower rungs of their society, is to a large extent responsible for the politics of forward, backward and scheduled castes and scheduled tribes. The later Muslim invaders and rulers, who mostly settled down in urban centres of north Indian plains found the east-west stretching Vindhyas an obstruction in the effective control of territory to their south. The deltaic areas along the eastern coast of India became centres of secondary culture and seat of several southern kingdoms.

The vast stretches of the northern plains bounded to their north by the Himalayas, which were far from the sea, were responsible for a continental outlook of the Indian state in the past. Fertile soils, variable climate with no sudden extremes, and perennial rivers reinforced a self-centred, less hostile, and largely defensive attitude due to which no serious attacks were mounted from here for colonisation of distant lands. The area was a fertile ground for the growth of such concepts as non-violence and peaceful coexistence. It is not surprising that Hinduism, Buddhism and Jainism were born in this area, where spread Islam and Christianity rather recently. This is still the largest area of vegetarianism in the world. The co-existence of all the major religions of the world in India was responsible for an independent India declaring itself as a secular state.

The shape and length of the coastline of southern India on both sides of the Deccan plateau, astriding the Indian Ocean, was responsible for development of overseas contacts of India with the Middle East, eastern Africa and South East Asia. In contrast with north India, physical geography of south India helped to develop a seafaring attitude among the people along the coast. But even they did not mount any offensive for colonisation of the countries across the seas. These contacts were responsible for trade, and in the case of some of the island states of South East Asia, for a peaceful spread of Indian religions and culture. But the protrusion of India into the Indian Ocean was responsible for the Europeans to be able to reach Indian shores and penetrate into the interiors from different points till they subjugated the entire Indian subcontinent. This particular location of India in the Indian Ocean is still important for her world trade, security and is a factor in her emergence as a regional power.

II | India's Geography

K.M. PANIKKAR

Geography has been at all times one of the great and most regrettable gaps in Indian knowledge. Worse than even our lack of interest in things historical has been our utter neglect of geography. This is something peculiar to India, for both the Chinese and the Arabs were expert geographers in their time. They wrote not only detailed descriptions of their own countries but of others they visited or knew about. But we may search the whole range of Sanskrit literature for even passably accurate descriptions of India's main geographical features. Of the mountain systems that Indian books speak, Sumeru would appear to be wholly mythological: Gandhamadan is equally a vague definition which may be presumed to be the Tibetan Plateau. Only of the Himalayas, the most dominating physical feature of the country, is any detailed knowledge available in Indian literature. This great mountain range we shall consider separately. Compared to the knowledge of the Himalayas, or at least its lower range, the knowledge available in Indian lore about the Vindhyas, the Aravalli range and the Sahyadri—all of them indeed familiar to our people— is scanty.

States and kingdoms are of course mentioned but even about them there are but few geographical details available. Sometimes the details given are contradictory. The only great writers who show adequate geographical knowledge are Kautilya, Vatsyayana and Kalidas. Kautilya in dealing with gems and other products mentions the area in which they are found. For example, he mentions the Tamraparni, Pandya Kavata and the Churni river in the Kerala country among places where gems are obtained. Most parts of India are alluded to but there are hardly any geographical data of value that could be extracted from them. Vatsyayana, again, in his *Kamasutra* while enumerating the special customs and practices prevalent in different parts of India shows considerable familiarity with the geography and political divisions of the country. His general division, of *pracya— the eastern parts, of madhya desa,* central India, or *dakshinapatha* or the south, is not very helpful. He however, mentions many states and shows a fairly accurate knowledge of their location. But his contribution to geography ends here, for though his references to

Reprinted with permission of the publisher from K.M. Panikkar, *Geographical Factors in Indian History*. Bharatiya Vidya Bhavan, Bombay, 1955, pp. 17-32.

customs and social behaviour provide invaluable data to the social historian, they are of little use to the geographer.

Kalidasa stands on a different footing. In his *Meghaduta* and *Raghuvansa*, he shows a detailed knowledge of geography, of hills, rivers, plains, mountains and cities and describes them accurately. From Ramagiri to the Himalayan gate through which he asks the cloud to ascend, his knowledge of India would seem to be that of an expert geographer. The cloud in his passage from Central India is directed with an exactitude of geographical knowledge, exceptional for a poet, to keep to a route and the rivers, hills, dales and cities which lie on the way are adequately described. When the messenger reaches Bidisa, Kalidasa does not fail to mention, not merely the river but the caves on its shores. As Ujjain does not lie on the direct route the messenger is asked to deflect his course in order that he may see that famous city. In *Raghuvansa* again the poet shows his knowledge of Indian geography. The *digvijaya* of Raghu takes him to different kingdoms, the characteristics of which are described with a degree of knowledge that makes one think that the poet was familiar even with the southern countries of Pandya and Kerala for he talks not merely of famous rivers like Tamrajarni but of lesser known rivers of the West Coast.

These are however works of literature and not of geography. Of books giving or professing to give geographical knowledge there is total absence in Sanskrit which is striking. Such geographical knowledge as can be gleaned from the *Ramayana, Mahabharata,* the *Puranas* and the *Katha Saritsagara* are vague and often inaccurate, though the last- named work strangely enough provides some data about countries outside India.

With geographical interest so wholly absent it is not surprising that India has up to date taken no interest in cartography as a science. While the Arabs and the Chinese produced most interesting maps, from very early times, India has been extremely backward in this respect.

Even today geography as a course of major study is neglected in India. Such geography as we are taught in schools is elementary and deals with physical conditions and political divisions. No attempt has so far been made to connect the facts of geography with India's historical evolution or to provide a basis for political thinking.

In terms of pressure on space India's geographical problem is different from that of European countries. She has no necessity to expand, from the point of view of security or defence. Her size, natural resources, her location in reference to oceans and land masses make a policy of territorial expansion of her boundaries unnatural. Kautilya defined *Chakravarti Patha* or the empire state as extending 2,000 *yojanas* from the Himalayas across the peninsula of India, and the idea of extending the territories beyond the natural frontiers of India. It may be assumed therefore that the question of space never worried Indian political thinkers. *Himachala Setu Paryantam*— from the Himalayas to

Rameswaram — was their conception of India. Every ruler with imperial pretensions tried to extend his authority over the whole area. The *Mahabharata* in its present recension representing the political ideas of the 3rd or 4th century A.D. takes the whole of India into its imperial design. The Mauryas, the Andhras, the Guptas and all other monarchies representing the imperial tradition attempted to extend their sway over the whole territory. But outside the natural frontiers of the Indian sub-continent, they never tried to build up effective rule. Chandragupta Maurya no doubt extended his empire to the Hindukush as a result of his treaty with Seleucus misnamed Nikator or the victorious, but we do not know whether it involved only a sphere of influence with autonomy for local rulers or was directly administered from Pataliputra. In any case, it seems to have been given up after Ashoka. No other Indian King seems to have held territory outside the Indian sub-continent, though many monarchs of Central Asian origin from Kanishka to Shah Jahan held the Hindukush valley with their capitals in India. That this was an unnatural geographical agglomeration was again and again demonstrated when after a generation or two the extra Indian territories had to be given up in favour of India. So one of the main preoccupations of European geopolitics, that is the desire for space, has not the same significance in India.

Next to space what concerned the Western geopoliticians was the question of resources. The experiences of blockade and submarine warfare had taught the continental powers and Britain the necessity of having under their control the natural resources which will ensure not only food and other essential commodities, but raw material for industrial production and strategic material necessary for defence.The very conception of Heartland which Mackinder developed was related to this. His Heartland was an area which had not only ample space but adequate natural resources and therefore was not dependent on sea-trade and was immune from the effects of blockade.

It is a striking fact that there is no such thing as equality of opportunity among nations, because fertility, mineral resources, strategic location, climatic conditions, etc., are not evenly distributed on earth.A nation's greatness depends to a large extent on these factors. The continuance of its greatness in history depends equally on the way these advantages are conserved and utilized. Fertility of land, for example, may deteriorate in time, leading to disastrous consequences. A distinguished historian, Prof. Sinkovitch, has produced impressive evidence to prove that the gradual degradation in the productiveness of Mediterranean lands during the later Roman Republic and the Empire was one of the major factors which led to the downfall of Rome. Erosion of soil, deforestation, encroachment of the desert, uneconomic exploitation and other factors may easily reduce the fertility of the soil. Wasteful exploitation may cut short mineral resources. Generally speaking the endowments of nature are of as great importance to nations as to individuals.

India is fortunate in this respect also. Her soil is fertile. Though neglected for centuries, it has shown evidence of quick recovery under treatment. She is well-provided with coal, iron and other minerals, necessary for industrial development and has ample sources of electric power. Even in the matter of nuclear minerals she is not, I am told, badly placed. Her major deficiency is in fuel oil and here also, there is reason to believe that her resources have not been fully studied. In any case, with ample deposits of lignite, the problem of fuel oil is not so serious as it was once considered. India's main problem is her food supply dependent as it is on the monsoon whose vagaries are notorious: but the experience of the last seven years has proved that with a considerable extension of irrigation, reclamation of land, protective measures against floods, and above all scientific utilization of land, the problem of supplying food to the people of India from her own land, can be solved at least for the next few decades. But the problem does not end there. The population of India increases at the rate of four millions a year. Is it possible for India not merely to continue to feed the millions annually added to her population, but to feed them at more than a starvation level? Geographical factors themselves have no value except in relation to men. The problem of population, their standards of living, their mental and physical efficiency and their productive capacity are the bases on which nation's life has to be ultimately built up. While therefore India may be said to be well-endowed by nature, her deficiencies in some essential respects are important and require serious study.

If unlike European countries India is not concerned about space and at least in the immediate future about resources, the geographical factors about India's structure have clearly to be studied from other angles, especially her internal stability, the integration of her large and ever-increasing population, which shows wide variations due to differences in climate, food habits, etc., the closer knitting together of her natural regions, apart from a study of the influence of the two dominant features of her geography, the immense mountain wall of the north, the Himalayas, and the sea that surrounds her on three sides.

Now the land mass of India lying to the south of the Himalayas falls into five well-defined areas, the Gangetic valley including the Deltaic region of Bengal, the vast uplands of central India with its Vindhyan backbone, the Deccan plateau reaching down to Mysore, a desert area from the south of the Punjab to the plains of Gujarat and a coastal tract in the south expanding into a plain with its central portion drained by the Cauvery. For our present purpose it is not necessary to consider the island of Cutch, an extension in fact of the desert area, and the western littoral between the ghats and the sea which will be considered separately. The main area of India is included in these five regions.

A look at the population map of India will show that the concentration of population is in the two valleys watered by the Ganges group and the Cauvery. The Gangetic valley accounts for over 110 millions. The littorals and the deltas

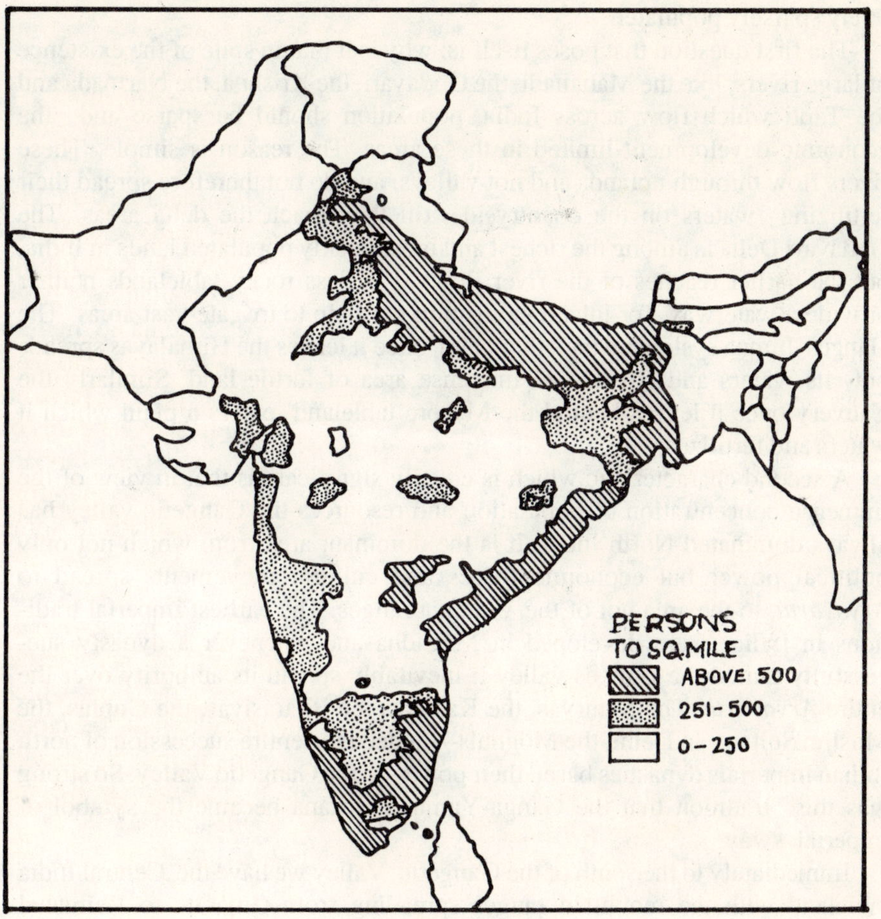

Fig. 2.1. India-Density of population

of the other rivers, the Mahanadi, the Godavari, the Krishna group, the Narmada and the Tapti account for another 50 millions while the Cauvery valley though extremely small accounts for 25 millions. The vast area which forms the interior from the Gangetic plains to the Mysore tableland is comparatively sparsely populated.

The first question that poses itself is: why is it that in spite of the existence of large rivers, like the Mahanadi, the Godavari, the Krishna, the Narmada and the Tapti which flow across India, population should be sparse and the economic development limited in these areas. The reason is simple. These rivers flow through uplands and not valleys, and do not therefore spread their fertilizing waters on the countryside, till they reach the delta areas. The Godavari Delta is among the richest and most thickly populated lands in India, but the earlier reaches of the river flowing across rocky tablelands neither provide a waterway for inland navigation nor help to irrigate vast areas. The Ganges-Jumna system on the other hand, once it leaves the Himalayas spreads out its waters and covers an immense area of fertile land. Similarly the Cauvery once it leaves behind the Mysore tableland enters a plain which it waters and fertilizes.

A second characteristic which is equally significant is that in view of the immense concentration of population and resources the Gangetic valley has always dominated North India. It is the dominant area from which not only political power but economic forces and cultural movements spread to *Aryavarta*, to the area not of the Vindhya ranges. The earliest imperial traditions in India were developed in Magadha and whenever a dynasty successfully united the Ganges valley it inevitably spread its authority over the entire *Aryavarta*. The Mauryas, the Kanwas, the Bharasivas, the Guptas, the Muslim Sultans in Delhi, the Moghuls— indeed the entire succession of north Indian imperials dynasties based their power on the Gangetic Valley. So strong was this tradition that the Ganga-Yamuna torana became the symbol of imperial sway.

Immediately to the south of the Gangetic Valley we have the Central India tableland with the mountain ranges spreading from Gujarat to Rajmahal constituting its backbone. This tableland slopes gently towards the Gangetic Valley. Till recent times it was widely and densely forested for we have it on record that the Moghul armies found the forests of Orcha so thick that they had to undertake a special operation to clear the way for them. With the upper Vindhyas and the Aravallis bisecting the area, the pressure from the Gangetic Valley is dispersed and finds its line of expansion normally into Malwa or Gujarat, and only to a less degree into Bundlekhand and other areas which have thus remained outside the main currents of Indian development.

The dominant position of the Gangetic Valley which makes it the core of India from every point of view is so obvious as not to require any proof to establish it. That Hindu opinion always viewed it as such is seen from the fact that not only is Ganges the sacred river *par excellence* of India but the area from Gaya to Mathura, from Sangam to Hardwar, is recognized by every one to be the holy land of Hinduism. Every dynasty, even in the middle India, once it established its authority in its homeland dreamed of consolidating its rule by establishing itself in this area. The Satavahanas of Pratisthan, the Kings of Avanti and even the Peshwas of Poona realized that their imperial sway could be maintained only if they controlled the Gangetic Valley. The British advancing from the sea came to the same conclusion.

If the Gangetic Valley gave control of *Aryavarta* and led its master to attempt to bring the rest of India under its control, the geographical features of the Deccan Plateau always baffled them, even if through overwhelming military strength, they were able to conquer it for a time. The unity of the Deccan tableland is as much an obvious geographic fact, as the unity of the Gangetic Valley.

Deccan is geographically a plateau commencing with the Ajanta Range "where the ancient undisturbed rock begins to extend over the centre of the Peninsula right up to the Nilgiris". This plateau, writes a recent historian, Harron Khan Sherwani, "is partly covered by a huge semi-circle of ancient lava flow, extending like a ten days moon with one horn at Nagpur and the other at Goa....... The struggle for Berar between those who hold Malwa and those who hold the Deccan Plateau is explained by the fact that Berar forms almost the first of the lavaic crescent and is always the butt of the greedy eyes of their northern neighbours."While the plateau ends abruptly at the Western Ghats, towards the east it slopes gently to descend to sea level.

The Deccan has always formed the great middle rampart of India and the Gangetic Valley was not able at any time to establish its authority over it on a secure footing. At the beginning of the Christian era we discover it as a flourishing empire under the great Satavahana dynasty of Pratisthana. How this dynasty which patronized the *paisachi* language became in course of time the great carriers of Sanskrit civilization is one of the strange facts of history: but one thing is clear that though the Satavahanas assimilated in time the culture of north and became its champions, the Deccan area was not brought under the political control of a northern power. Even the imperial Guptas did not exercise effective authority there. Samudra Gupta in his expedition to the South avoided the Deccan tableland and his Allahabad Prasasti by Harishena makes it clear that his route lay via Sambalpur, Orissa and the coast. The Vakatakas held sway over the uplands and they shared with the Guptas the Empire of India at the time. While the imperial authority declined in the Gangetic Valley, the Deccan maintained its territorial integrity under a succession of great dynasties.

The establishment of the Delhi Sultanate did not affect the independence of the Deccan. Alluddin Khilji indeed sent expeditions to the South destroying the power of the Yadavas of Deogiri and humbling the pride of the kings of

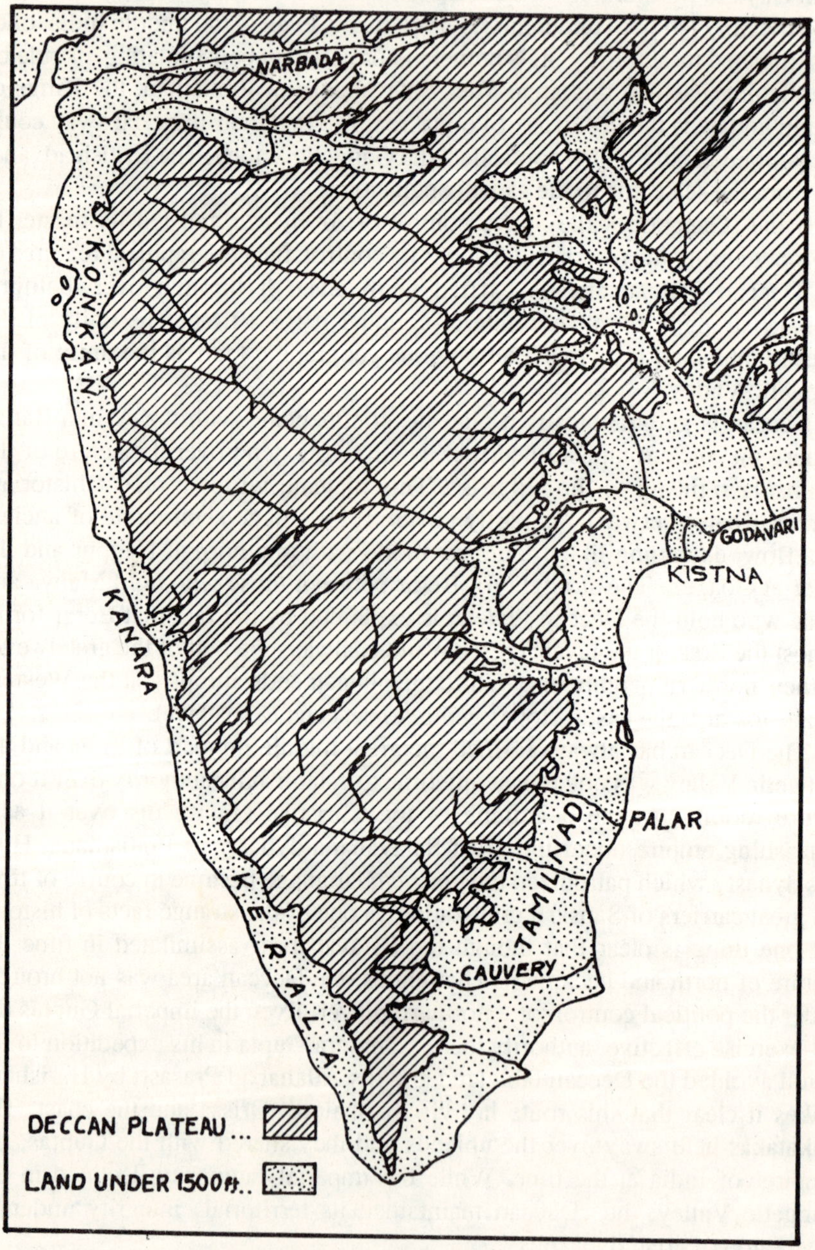

Fig. 2.2. The Deccan and South India

Warrangal. But soon the facts of geography asserted themselves, the upper tablelands of the Deccan being organized east to west under the Bhamini Kings, and the lower tableland being organized north to south under the Vijayanagar Emperors. No power in North India from the middle of the 14th century was able to challenge the independence of those kingdoms for over 250 years. It was only when the Gangetic Valley was again organized as a great military power under Akbar that the iron topic of history drove him against the ramparts of the Deccan. In the last decade of the 16th century Akbar opened his campaign against the Deccan Sultanates and over a century later when Aurangzeb died, the empire was still fighting in the Deccan. Technically the plateau had been brought under control and annexed to the Empire but 32 years after the death of Aurangazeb geography asserted itself once again and the tableland declared its independence of Delhi.

The Cauvery Delta, the core of Dravidian life and civilization remained equally outside the political control of the North. Its history indeed does not follow the course of events in North India, for, from the beginning, its activities were controlled by its peninsular character and were more orientated towards the sea. Hornell in his paper on the origin and technological significance of Indian boat designs has established the Polynesian origin of some of the backward races of South India and has also demonstrated that the outrigged canoe and the coconut were introduced into India from Polynesia. The excavations at Arikamedu have clearly proved, what was long known from Tamil literary records, that the people of the South were in close commercial relations with the Mediterranean peoples, that long before the time of the Romans their ships had sailed up to the Red Sea and the Persian Gulf, that intimate commercial and cultural contacts existed between them and the peoples of Western Asia and Egypt. From the fifth century (A.D.) the culture of the North spread, all over the region through the influence of the Pallavas, another dynasty like the Satavahans, who were the carriers of Aryan civilization. Kanchi becomes a great centre of learning, but there is no evidence of any substantial political relations with the North. The land policies of the southern Kings are only in relation to the States of the Deccan, the Chalukyas, Kadambas and others and not with the Empires of North India.

The main political interests in historic times, apart from their local rivalries, were in further India, in Malaya, Siam, Indo-China and Indonesia. The influence of the Pallavas in these areas is very marked and is now fully acknowledged. Again during the period of Chola greatness South India looked across the sea for its political activities. A hundred years of overseas expansion and naval warfare by the Cholas are striking features of South Indian history which has no parallel even in the imperial traditions of North India.

Apart from the policies of Kings and dynasties the people of peninsular India were as we know from epigraphic records more concerned with maritime activity than with political expansion. Of the Nanadesis, the Valangai and the Elangai, an inscription in Baligami in Mysore (*Epigraphica Carnatica* Vol. VII P. 118) speaks thus: "Brave men born to wander over many countries since the beginning of the Krita age penetrating regions of the six continents by land and water routes and dealing in various articles such as horses and elephants, precious stones, perfumes, and drugs, either wholesale or retail". The Manigramam Chettis were equally well-known. Their operations extended all over South-East Asia.

The vast desert area lying to the south of the Punjab requires no description. It is in fact an extension of desert girdle, a geographical intrusion into India of the Middle East. This area is bisected by the Aravallis. To the west of the range lie the low lands of Marwar and Gujarat and to the east it is a highland consisting of Hardoli, Kotah, Bundi, Udaipur and Malwa. This was at all times an area of refuge, the geographical conditions of which enabled communities displaced by invasions to take shelter and carry on their life unmolested by the people of the Ganges Valley. The Minas, the Bhils and other indigenous tribes first took shelter there. In more recent times when the Muslims occupied the Gangetic Valley it is to this inhospitable area that the Rajputs withdrew and carved out new states.

These are the five basic areas of India, separated by geographical factors but united by a common culture. Apart from these clearly marked regions there are the marginal areas which come into Indian history only in a limited sense. There are the tableland of Chotanagpur and the Chattisgarh States, still occupied predominantly by pre-Aryan and in many cases Dravidian tribes—an area which never came under the effective authority of the Rulers of the Gangetic plain; the peninsula of Kathiawad, which was effectively occupied from Gujarat, but geographically jutting out into the sea, and having dominantly maritime traditions and the Kerala littoral of the Western Ghats, a monsoon area of great fertility cut off by fairly high mountains with impenetrable forest and therefore remaining on the outer fringes of Indian history.

An important feature of this north to south division into three regions and of intervening tableland is the marked absence of direct routes of communication between the North and the South. Samudra Gupta, as we saw, went on his *Digvijaya* along the eastern coast. The description of the *digvijaya* in *Raghuvamsa* emphasizes the same fact. The pilgrimage routes also pass along the marginal areas. Nor are there in India any rivers flowing from North to the South or *Vice versa.* The normal flow of rivers in the Gangetic Valley is from West to East as the land slopes gradually towards the Bay of Bengal. The West to East flow of rivers, in a country which lies north to south creates what may be called horizontal areas, and the geographical regions of the country have therefore less connection among themselves than if the rivers had flowed from north to south. In the Gangetic Valley, especially, this eastward direction leads to the integration of population in a close area, bound together by the river and its main tributaries, thus converting the area into dominant force in *Aryavarta.* From the time of the Kuru-Panchalas to today this has been the capital and governing fact of Indian history.

It is an axiomatic fact, but seldom recognized, that an extension of land towards the south, because of the change in climate following latitudinal differences, always creates problems of integration. Speaking in general terms, this means that an east-west extension within similar climatic conditions, leads to greater integration than states formed in depth from north to south. Climates vary, generally, according to the distance from the equator, with the result that racial characteristics, economic conditions, etc., lead to differences in social organization which become persistent factors in history. The history of every country viewed from north to south, shows this characteristic. Even in so small a country as Great Britain, the union between the north and the south, realized as a result of many centuries of effort, has not been fully achieved. Scotland remains persistently Scottish. In the U.S.A. with its great depth, the differences between the semi-tropical south with its different economy and the north led to a bitter civil war and even after the defeat of the secessionists, though the political union was re-established, the differences between the north and the south have not been lessened. In Italy, the political and social differences between the Lombardy plains and the south have long been recognized to be of a major significance. This is equally true in China. Though racially, culturally and politically united for centuries the differences between the people of south China and of the dry plains of the north are most marked.

These facts are to some extent exaggerated in India by the geographical factors discussed earlier in this section. While from first century A.D. at least the cultural unity of India was being worked out by the penetration of Sanskrit and of the Hindu social system, ethnically, politically and economically the three tiers continued to be separate. Only with the British conquest was the political, administrative and economic structure of the three areas unified. The Deccan and the South still continue to be racially Dravidian with the least admixture of the blood of Aryans and other races which penetrated into *Aryavarta.*

III | Routes of Invasion

K.M. PANIKKAR

Like every other country in the world, India has been subjected to many invasions. Some of these were the results of large-scale population movements, while others were the military action of ambitious generals. Of the latter, the most notable examples are the raid of the Macedonian Alexander into the Persian satrapy in the Punjab, of Timurlane and of Nadir Shah, all isolated but spectacular incidents, leaving no doubt, confusion and chaos behind them, but without permanent effect on the country. The invasions which were the results of population movements stand in a different category. The change in the nature of the population, introduce new factors into the civilization of the country, displace existing social groups— in short they are revolutionary in their character. Some of these movements which have a direct bearing on the growth of Indian life have now to be considered in terms of India's geography. The first movement of populations of which we have definite and sufficient evidence is that of the Aryans. The earlier movement of Alpine race into India across Baluchistan which is postulated by many scholars is too vague for us to draw any conclusions from it. Nor is it possible to speak with authority about the Polynesian movement which is said to have peopled South India. All that we can say is that there might have been such movements but we know very little about them. About the Aryans the position is different, because this movement produced an outstanding literary monument in the *Rig Veda*. The *Rig Veda Samhita* is looked upon by the Hindus mainly as a religious book to which they fondly trace the origins of their metaphysics and their philosophy. But to the historian and to the student of geography it is a unique collection of documents, covering a period of many centuries, dealing with the geography, the history and the economics of the penetration of the Aryans into the Punjab from the Kabul Valley. When it ends they are in the Ganga-Yamuna Doab. The celebrated hymn to the Rivers (X.6.7) in the *Rig Veda* sings thus—

"Accept this my praise Ganga, Yamuna, Saraswati, Shutudri, Parushni, Maruvriddha with Asikni and Vitasta, Esten Ayikiya with Sushoma."

Of the rivers to the East of the Doab, that is the Kuru Panchala country, the Aryans have knowledge only of Sarayu and it is doubtful whether the Sarayu

Reprinted with permission of the publisher from K.M. Panikkar, Geographical Factors in Indian History, Bhartiya Vidya Bhavan, Bombay, 1955, pp. 72-88.

alluded to is the river which flows near Ayodhya. What is more interesting is that long before the end of the Vedic period the momentum of the population movement had already weakened. The Aryans in the Doab and those in the Punjab, perhaps stiffened by newcomers, were at loggerheads. In the battle of Parushni celebrated in the 7th Mandala of the *Rig Veda* Anubaka II (VII 2.1) the dwellers on the Yamuna and Tritsus, under the leadership of Sudas, are said to have destroyed over sixty thousand of their enemies among whom were the followers of Turvas, *son of Yadu,* and Bhed the turbulent chief. Also we come across allusions to Aryan enemies, for example in *Rig Veda,* VI-3. 10, where it is said—"Thou, hero Indra, destroyed both classes of enemies, both Dasa and Arya adversaries."

With the battle of the Ten Kings, the Aryan invasion may be said to have come to an end. The earlier colonists in the Kuru Panchala territory with the help of indigenous local rulers had fought and destroyed the attacking Aryans and finally put a stop to the invasions. It is notable that in this hymn, Indra, the great god, the Wielder of the Thunderbolt, the destroyer of the walled towns, is not on the side of the invaders but on the side of Sudas and his allies.

The battle of the Ten Kings is one of the greatest events in history, an event which indelibly stamped its influence on India and shaped the future course of Indian development. It may well be claimed to have laid the foundation of Indian and Hindu civilization for here it was that the synthesis of neo-Aryanism, which we describe as Hinduism was laid. The Aryan racial strain became thinner and thinner while the neo-Aryan civilization spread to the Indo-Gangetic Valley. By the time it reached Bihar, it had practically ceased to exist. Even in the Indo-Gangetic Valley where neo-Aryanism or Hinduism became predominant, the population consisted mainly of indigenous tribes as is well established by the fact that even a thousand years afterwards, it continued to be peopled by tribes which were only imperfectly Aryanized.

That neo-Aryanism or Hinduism, as we may now call it, was the outcome of synthesis between the pastoral Aryans and the indigenous people may be shown by instancing two well-known facts which form the basis of the new religion. The *Aitreya Brahmana* which is the most significant supplement of the *Rig Veda* was, it is recognized, the work of a man who was the son of an Aryan Rishi by an indigenous woman— an offspring whom he refused to instruct in sacred knowledge. Even more significant is the fact that the arrangement of the Vedas into Samhitas was the work of Badarayana known as the black whose mother equally was an aboriginal. The inter-mixture of races at the level of the priests themselves had become a major fact, before the Vedic period had ended.

The Neo-Aryans spread not only in the Gangetic Valley but also made their way slowly to the plains of Malwa and Gujarat, leaving the interior to the aboriginal races. But in Gujarat though they were able to impose their culture, the Aryana racial strain was almost negligible as the recent head measurements by anthropologists and examination of kinship organizations by Mrs. Irawati Karve have clearly demonstrated. For centuries neo- Aryanism did not reach the Deccan Plateau at all.

The distribution of the indigenous races even today in the uplands of South Bihar and in the Eastern areas of Madhya Pradesh, and the persistence of the Bhils in the Aravalli and Vindhya ranges show that as a population movement the Aryan invasion ceased to have any momentum after it reached the Gangetic Valley. The gradual spread of Hinduism all over India and with it the Aryan speech should not blind us to the fact that even in North India, outside the Punjab, the Aryans contributed only a racial strain. In Gujarat and in Maharashtra the neo-Aryans were able to impose their language but in the Deccan and in the south, the Dravidian speech not only held its own but was able to drive out the Austric and other linguistic elements. The spread of Aryanism and Sanskrit to the South, originally associated with Agastya's crossing of the Vindhays, became an accomplished fact only in the first centuries of the Christian era, as may be seen from the earlier Paishachi tradition of the Satavahana Emperors of Pratisthan.

The Aryan invasion was the first known and the greatest of all invasions of India and we have seen that its racial effects were not very considerable outside the Punjab. The next invasion of which we have some knowledge was of the Achamedian kings of Persia. It merely annexed to the Persian Empire the trans-Beas areas of the Punjab and had little influence on even the Indo-Gangetic Plain, except in the realm of art and architecture. Alexander's raid was into the India satrapies of the Persian King and was insignificant in every way, except in the eyes of the Europeans who pretended to see in it an early version of an alleged European conquest of India

The really significant movement of which we have any information after the eruption of the Aryans is that of the Yue-chi. This was a barbarian tribe who had their homes in the vacant belt of Koarsu and Sinking in Chinese Turkestan and were driven from their homes by the Hun conqueror. They in turn drove the Sakas before them and in the first century B.C. the Sakas descended on India, through the Bolan Pass and occupied Sindh and from there it would seem threatened Ujjain itself, as the Arabs tried to do at a later time. Their defeat and expulsion in 57 B.C. was according to Hindu tradition the achievement of Vikramaditya of Ujjain, who to commemorate this fact established the Vikram Samvat era. But though the Sakas were driven off, the Yie-chi who had settled down in the valleys of the Hindu Kush began to make their power felt on India. Under Kanishka, their empire extended to Mathura.

The Yue-chi had arrived in the Hindu Kush valley as a result of the pressure from behind. But during their settlement in that area, the pressure relaxed and the Kushans as they now came to be called became a settled people. The empire of Kanishka was therefore a political phenomenon of an ephemeral character and not a population movement involving great and

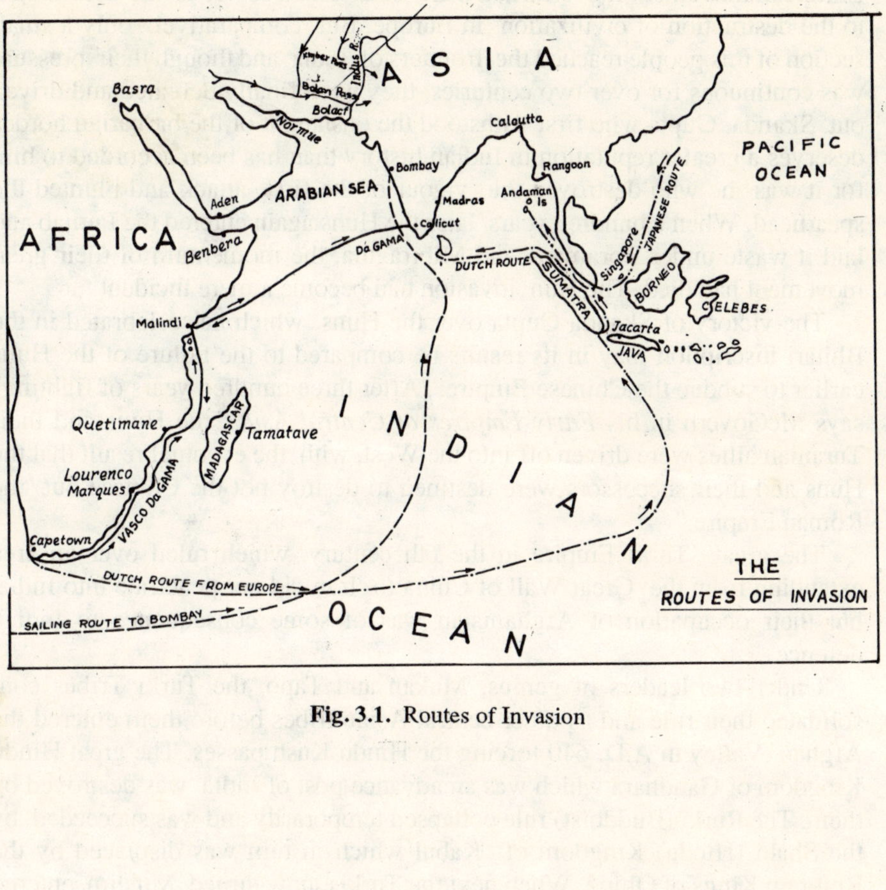

Fig. 3.1. Routes of Invasion

revolutionary changes.

The Hun invasion was of a different character. It was in fact an immense movement of population across the vacant spaces of the Eurasian steppe. In its European movement it overturned many established societies and contributed to the destruction of civilization in Europe. But comparatively only a small section of this people reached the frontiers of India and though their pressure was continuous for over two centuries, they were finally defeated and driven out. Skanda Gupta who first withstood the onslaught of the barbarian hordes deserves a greater reputation in Indian history than has been accorded to him for it was he who destroyed the vigour of the first attack and blunted the spearhead. When a hundred years later the Huns again entered the Punjab and laid it waste under Toramana and Mihiragula, the momentum of their great movement had died. The Hun invasion had become a mere incident.

The victory of Skanda Gupta over the Huns which he celebrated in the Bhitari inscription may in its results be compared to the failure of the Huns earlier to subdue the Chinese Empire. "After three hundred years of fighting" says McGovern in his *Early Empires of Central Asia,* "the Huns and their Turanian allies were driven off into the West, with the eventual result that the Huns and their successors were destined to destroy not the Chinese but the Roman Empire."

The great Turki Empire in the 6th century, which ruled over an area extending from the Great Wall of China to Iran did not penetrate into India. But their occupation of Afghanistan was of some consequence to Indian defence.

Under two leaders of genius, Mukan and Tapo, the Turki Tribes consolidated their rule and as other central Asian tribes before them entered the Afghan Valley in A.D. 640 forcing the Hindu Kush passes. The great Hindu Kingdom of Gandhara which was an advance post of India was destroyed by them. The Rurki (Buddhist) rule collapsed temporarily and was succeeded by the Shahi (Hindu) Kingdom of Kabul which in turn was displaced by the Brahmin Kings of Ohind. When next the Turkis, now turned Muslim, entered the Afghan Stage the Hindu monarchies of the area had become weak and disunited.

After the Huns the next serious invasion, potentially the most dangerous from every point of view, was that of Mahommed Bin Kassim on Sindh under the orders of the Khalif of Baghdad in A.D. 712. This was the great period of Arab expansion when from the borders of India to the border of France the Arabs had overturned kingdoms, established dynasties and conquered and governed the most powerful empire of the time. This invasion by the greatest organized military power of the time has not received sufficient attention from historians. The general view, sedulously promulgated by Muslim chroniclers and following them the European historians, is that the Arab invasion of India through Sindh fizzled out for lack of support from Baghdad and the failure of initiative in Sindh. But what are the facts? After the battle of Raos in 712 when Ibn Kassim defeated Dehir, the Hindu King of Sindh, the whole of the lower Indus Valley from Multan to the sea fell under the Arabs. With their base firmly established on the Indian soil, they attempted to penetrate into Rajasthan from the land side and from the sea side to occupy Gujarat, Kutch, Saurashtra and Broach. If Junyad's expeditions had been successful Western India would have become a part of the Arab Empire, but the Chalukya Rulers and the Pratihara Emperors of Gujarat and Rajasthan were able in continuous fight to destroy Arab power and to contain its remnants to the Sindh deserts.

Leaving aside Mahmud Ghazni's plundering raids, the conquest of the Gangetic Valley following the invasion of Mohammed Ghori, by Kutubuddin starts a new chapter in Indian history. But the general idea that these were Islamic invasions, part of the great upsurge of Islam to conquer the world of the faith, is historically untrue.

With the decline of the Khalif of Baghdad the expanding force of Islam from the religious point of view, ceased. The Khiljis, the Tughlaks and the Lodis who conquered North India and established their rule in the Gangetic Valley were Central Asian Turks, who in the historic process of their movement had become Muslims, very much in the same way as the Yue-chis or the Kushans, in their movement across the same area, had become Buddhists. The affiliations of the Delhi Sultans were not towards Arabia, or to the West but to Central Asia. Infact after the first generation, they settled down in India as exclusively Indian rulers with no interest outside.

The fact that they were Muslims had however one outstanding significance. While their predecessors from Turkistan, whether Kushans, Sakas or Huns were assimilated into Indian society and accepted Hindu culture, the new people from Turkistan because of their exclusive religion and the rigidity of their beliefs maintained their communal separateness. It was impossible to assimilate them, as time could not eliminate the differences, as in the case of the earlier peoples.

It is however interesting to note and necessary to emphasize that the geography of their dispersal followed the same pattern as the geography of Aryan dispersal. Their racial and cultural influence was greatest in the Punjab. By the time they settled in the Gangetic Valley, the dynamism of invasion was exhausted and the movement became almost exclusively political, the Sultans of Delhi, claiming to be Indian Rulers of Hindustan. They extended their authority to Bengal leaving the tablelands well alone as the Aryans had done before them. Another branch spread out to the South-West, holding Malwa and occupying Gujarat. But these were political developments and not population movements for, after a short interval we see the Rulers of Malwa and the Sultans of Gujarat allying themselves with Hindu kings. Finally the power based on the Gangetic Valley, while able to Carry to raid into the

Deccan, was unable to conquer or hold peninsular India. The dispersal from the Gangetic Valley finds its term on the Vindhyas and the tide which had already lost its force breaks into billows against the mountain ranges of the Deccan.

The history of the Moghuls tells the same tale.This small backwash of the great population movement of the Mongols finally reaches India not as a world-shaking event as the conquests of Genghiz and his immediate successors were, but as a military adventure. But its geography is the same: from Ferganah to Kabul, from Kabul to Agra and with Agra the Gangetic Valley. Under Akbar the empire is consolidated, Bengal is firmly held: a lighting campaign annexes Gujarat. The old areas of dispersal are in Moghul hands. Then begins the struggle. The armies of North India, with all the latest equipment available to a great power, open their campaigns against the Deccan. For over a hundred and ten years the might of the Moghul empire is pitted against the geography of India and when Aurangzeb dies and the Empire is practically ruined. The Mohguls had not penetrated beyond Tungabhadra. Within twenty-five years, the Marathas are on the move towards Malwa. The greatest and most deliberately organized campaign to bring the Deccan under the control of Delhi fails after an expenditure of men, treasure and effort to which Indian history shows no parallel.

Why is it that these great foreign invasions from the Aryans to the Moghuls spent themselves in the Gangetic Valley and were unable to touch the heart of India in the Deccan ? This is the major secret of Indian history. The great population movements which overran Europe as in the case of the Aryans and the Huns, destroyed and laid waste everything up to the Carpathians as in the case of the Mongols, established new homelands in Western Asia and held half Europe under fee for 300 years as in the case of the Turks, exhausted themselves in the Punjab, spread out as a backwash in the Gangetic Valley, but failed to have any influence on the Deccan tablelands or on the South. The reason is to be found in the basic geographical factors. As Spykman, the noted American geographer, points out: "From Central Asia came the early invasions into the border zone, the eruptions of nomadic barbarians into Europe, Persia, India and China. Most persistent has been the pressure on Europe and China because *geography forces the flow of power into the East-West channel.* To the north lie Tundra waste and the polar sea. To the south stretches a zone of barren desert and the highest mountain barrier in the world (the Himalayas), in between an easy grass-land belt provides an area of mobility. Along this broad way the ancient horsemen of the plains exerted pressure on Vienna and Peking". This barbarian horde can find an outlet only across the Khyber Pass into the valley of the Indus and even to reach the pass they have to leave the broad highway, cross the Hindu Kush and establish themselves in the valleys of Afghanistan, by which time, deprived of the continuous flow of population, the movement loses a great deal of its momentum. India gets only a thin stream, whether it be of Aryans, of Huns, the Turks and the Mongols, for the main stream passes along the broad grass-lands.

The Hindu Kush is the outer area for the defence of the Indian sub-continent. But it is not an impassable range. There are three high-altitude passes from Central Asia into the Kunduz and Shorbard Valleys of Afghanistan which invaders, throughout history, used to cross into Afghanistan. The Indus Valley comes under threat once a military power is established in this area for there are routes from the Kabul Valley into the Punjab which have been known from the beginning of history. The two most famous are of course the Khyber Pass and the Passes into the lower Swat Valley. The British have given a mystic significance to the Khyber Pass, possibly because of their own experiences in the unfortunate Afghan campaigns.But Khyber was not the route favoured by the invaders of the past. Baber for example moved up the Kumar and debouched in the Lower Swat Valley. Alexander also seems to have preferred this line of advance.

It will be seen that the Afghan quadrilateral which looks a massive defence fortress became more or less a half-way house for invaders. The Aryans, the Kushans, the Huns, the Turks and the Moghuls used it in turn for this purpose. An anonymous geographer writing in the 10th century whose work has been edited by Prof. Minorsky states that Afghan area had four approaches. The same geographer affirmed that the gates of India were Bust, a site on the junction of the rivers Helmand and Arghandeb, a little to the west of Kandahar.

The other route into the Indian sub-continent, the Bolan Pass, though it seemed to have been used only once in history, by the Sakas, is not without importance. From Seistan it is not difficult to approach. The reason why in the past this route was neglected was perhaps entry through it led an invading force into the deserts of Sindh and Rajasthan. The Sakas who under pressure from the Yue-chie entered India via the Bolan Pass, were unable to establish a firm foothold in India.

What has preserved Indian civilization in the past and saved it from being uprooted as the civilizations of the Nile and Tigris Valleys were, was her geographical position which lay out of the way of the main routes, protected by mountain barriers which only permitted a trickle of the great population currents to reach her; once having reached the densely populated Gangetic Valley, these invaders, without constant reinforcements from their homelands, got submerged in the local population and further, even when their momentum carried them forward, the Deccan masif operated as barrier strong enough to prevent their penetration to the South. Thus even when the Gangetic Valley fell under the domination of foreign culture as it often did during the last 2,000 years, Hindu civilization was firmly entrenched in the South to maintain the continuity and provide the inspiration for a general revival.

So far we have dealt with invasions across the land frontier. The invasion by the sea nomads to which India was subjected during the last 450 years of her history presents different features. In 1948 Vasco da Gama's armed fleet arrived in the Indian seas round the Cape of Good Hope and sailing across to Calicut from the African coast. Within a short period of 15 years his successors were able to establish a maritime empire which exercised steady and continuous pressure on the economy of India for centuries afterwards. The Portuguese closed all the sea routes to India and exercised a monopoly of Indian trade by the simple system of building bases round the Indian coast, from where they patrolled the Indian seas and drove out all others who claimed the right to sail or trade. With naval stations in Malacca, Ceylon, Ormuz, and the African coast, and with land base in Goa they exercised effective sovereignty over the Indian seas. But the Indian States of the time were too well organized and too powerful for the Portuguese to extend their influence on the land.

But pressure from the sea has a relentlessness which the sporadic threats from the side of the land do not possess. Once the coasts of a country become open to pressure from the sea, they remain perpetually under threat, for though the particular power which exercised the pressure might itself be destroyed, the indivisibility of the sea enables any other power, however distant, to step into the vacant place. Thus it was that England which had eliminated her rivals and established the mastery of the seas, stepped into the seats of power which Albuquerque had erected. From the middle of the 18th century, Britain was able to exercise continuous pressure on India's coast line. She was able to build up bases of power at Surat, Bombay, Madras and Calcutta when the land-based power of the broke down at the centre and as a result the new maritime economy, financial power shifted to the coast, sea power was able to undertake a systematic conquest of India such as no land power had ever done.

The routes of this invasion were also very different. Their power being based on the sea, the English first expanded into the two great river valleys, the two focii of Indian civilization, the Cauvery Valley in the South and the Gangetic Valley in the North. From Madras they extended their authority first in the Tamil areas, and along the deltaic portions of the Andhra Coast. From Calcutta, they moved up slowly bringing under control the Indo- Gangetic Valley without any major military campaign. In a great pincer movement from the South across the Mysore tablelands and from Bengal across Central India, the landbased power of the Marathas was crushed in campaign which culminated in Assaye. Slowly, for the first time, the whole of India passed under foreign rule. The relentless pressure of the sea, capable of bringing continuous reinforcement over thousands of miles, and moving steadily forward from its bases on the coast brought within a period of one hundred years the entire territory of India under occupation.

There are certain aspects of this conquest which require to be emphasized. In the first place, the most notable fact is that during the entire period of conquest and occupation (from A.D. 1857 to 1947) the invader remained a foreign power, the source of his authority and power remaining in a far off land. This is very different from the position of land powers, who whether Khiljis, Tughlaks or Moghuls became Indians in the first generation with the sources of their authority and power exclusively in India.

Secondly, the reinforcements from Europe were continuous and unending. From the time of the first fleet of Vasco da Gama there was an uninterrupted flow of power from the European ports to India, which continued even during the period of British occupation. As we noticed the position was totally different in the case of land powers which by this very fact were forced to become national.

Thirdly, all parts of India were equally under foreign rule though the form varied in the case of territories which were under the suzerainty of the Crown. While the bases of British strength continued to be the sea ports and the coastal areas, the weight of Britain's military strength was concentrated in the Punjab and the North-West Frontier so that both continental and Peninsular India were equally under control. The land powers based as they were on Delhi and the Gangetic valley were never able to do this, with the result that except when the Centre was exceptionally strong under an able ruler, the periphery was but weakly held. The British had in fact no periphery, as the coastal and the Deccan areas which are the normal periphery of a land power, constituted their bases and their military strength was concentrated on the North-West Frontier.

Fourthly, their administration also continued to be mainly foreign with the superior personnel recruited from abroad and maintained, at least after the great Rebellion of 1857, in a constant state of preparedness.

Finally, with the government directed from a distant land and carried on by a predominantly foreign administrative machinery which had its origin in trade and economic exploitation, India became a colony, that is a territory governed in the interests of the metropolitan power. Her economic life, her trade and the exportation of there natural resources were controlled by the conquering power. Invasion from land never had these results and could not in the nature of things lead to such developments.

In short, the British conquest of India was the first real foreign conquest of the country. It was the first time in India's long history that she was ruled from outside her own territory. Small portions of India in the North-West had in the past been sometimes ruled as portions of a large empire. The Indus Valley was thus a part of the Empire of Persia. Portions of the Punjab had been annexed by the Rulers of Kabul at different times, for example, in the time of Ahmed Shah Durrani and Zaman Shah. But neither the Gangetic Valley nor the rest of India had ever been ruled from an outside capital by monarchs whose interests were external to India. Only the sea power was able to do this, for the indivisibility of sea enables any power which can establish undisputed mastery of the oceans to exert her power anywhere in the world and by long distance control rule over areas where land armies cannot easily reach.

The short-lived Japanese invasion of the Indian water came from the eastern route, through the Straits of Malacca. Japan is situated in the Pacific many thousand miles away but once the control of the sea passed to her through the conquest of Singapore, Indian Ocean lay at her mercy. That control which enabled her to send her ships to the Bay of Bengal was challenged not in the Indian Ocean but in far off Guadalcanal and the possibility of landing in the South India was thus averted. Japan's control of the sea was only a limited one, but in view of India's geographical position it was sufficient to enable her to land troops anywhere and create a base, if her power had not been challenged elsewhere.

We may briefly consider in this connection how air power has modified the situation as we have analysed it: the practical invulnerability of the Indian sub-continent from the land side, and its dangerous vulnerability from the side of the sea. To an invader from across the land frontier a superior air force will give additional strength, ground support, destruction of communications in the interior, dislocation of economic life through bombing, etc. But a conquest of India would be possible only through sustained military effort across long distances through difficult passes and under trying climatic conditions. The Himalayas with their dreary uplands, the Hindu Kush and its ranges will still remain major problems. If India has also an effective air force capable of continued defensive operations, the danger from the land side is greatly minimised as the routes of invasion are limited and can be effectively defended. The position is altogether different in the case of pressure from the side of the sea. With adequate air cover, an invading force could select its own place of landing, can be continuously reinforced and hold and enlarge its base. The physical features of the Indian Ocean do not, as we have seen, give island cover to the coast to enable serious interception operations at sea. Even if it did, the American campaign in the Pacific from the Guadalcanal to the very shores of Japan, to Okhinawa, would demonstrate that overwhelming sea power has an island or a peninsula at its mercy, if not for conquest at least for effective attack. The coastal waters of every country belong in effect to the state that has control of the seas, which involves also air power.

IV The Unity of India

C.S. VENKATACHAR

Geographically, India is one. No other area of continental dimension has such strategic significance as geographical India. The outer geographical framework of India was permanently fixed by Nature. This has given unity to India but within it there have been free play of internal forces giving rise to linguistic and ethnic diversities.

What promotes or hinders India's unity is an internal and not an external problem. The historical angle on this matter has been most obtuse. For this western Imperialism is the real culprit and it must bear a large share of blame. In its understandable anxiety to perpetuate Western dominance, it presented a picture of the terrifying invasions along the traditional routes to India through the Khyber Pass and the unending prostrations of a helpless India before the invaders. It became an obsession to view India and the problems of her unity from the Khyber. To obtain a balanced perspective of India's unity, the most intelligent way would be to view the North from the tip of the peninsular South.

Peninsular India below the Narbada consists of a central tableland from which is isolated the western coastal littoral area. The eastern coast-land is deltaic in character. The rivers which have their sources on the Western Ghats flow through the tableland, descend to the low lying coastal regions and discharge into the sea. An important feature of Indian physical geography is the Vindhyan Range in the central regions which divides the North and the South and from the remote past till modern times the Vindhyan system proved an effective barrier of separation between the two areas. The North, that is the historic plains of Hindustan, is a vast stretch of unrelieved monotonous alluvium, completely encircled, in its northern boundaries for over 1500 miles, by the majestic Himalayas and watered by the rivers of the Indus and the Gangetic systems.

These characteristics of Indian physical geography have their homologue in the Indo-Chinese Peninsula. The north Indian plain with its river systems has its counterpart in the Yellow River basin of North China which is the cradle of Chinese civilization. The deltaic regions of the eastern parts of India may be

Reprinted from C. S. Venkatachar, "Geographical Realities in India", 1955, *The Eastern Economist*, Pamphlet No. 34, pp. 54-76.

compared with corresponding regions of South-East Asia. The latter have rice growing basins, deltas and irrigated dry zones. The lower reaches of the Irrawady, the Menam and the Mekong have been the bases of organized kingdoms and higher up in the wilder mountains and dense forests live hill men pursuing their primitive life. Similar is the case in the Assam hills, the upper reaches of the Mahanadi and the hilly portions of the east coast of India.

The lie of the axis of the higher range of the mountains and the direction of the flow of the rivers have determined the regions for the growth and development of civilization in the Indo- Chinese Peninsula. Very early in the history of the Indian civilization three great political divisions were mapped out. The Indus Valley lies close to the mountain orifices through which passage was easy for the movement of small streams of nomads and military adventurers. This valley lay open to the political and cultural influences from Central and South-West Asia. If the gate-way of India was successfully pierced, then the way was open to the plains of Hindustan which has always been the largest and the most populous of the political divisions of India. The third was peninsular India south of the Narbada.

The North-Indian rivers are snow-fed from the Himalayas and have undergone immense changes in historical times. The Indus system has undergone great transformations in the upper Himalayan reaches and in the plains where the changes in the soft alluvium have been most frequent. Some rivers like Hakru have ceased to exist; others like Kurram and Saraswati are now feeble and insignificant streams. Where the rivers of the Indus debauch into the plains, there some kind of precarious agricultural life came into existence along the river banks by utilizing the inundations; otherwise the Central and the Southern parts of the Punjab remained largely pastoral. The problem of the integration of the pastoral life with the advanced agriculture society of the Gangetic Sown remained unsolved and uncared for. The Indus system collected the human sediments of the waves of nomadic flood spilling over the Hindukush. The Indus basin together with the Punjab to the west of the Sutlej was just an antechamber for passage to the Gangetic Valley. Strangely, this area came to be regarded by the orthodox Hinduism of the midlands as an alien or impure cultural area. The Indus Valley and the adjacent tracts of the Punjab remained arid areas, sparsely populated by semi-nomadic pastorals until the face of it was changed by modern irrigation towards the end of the nineteenth century.

The Vedic Rishis may never have tired of singing the praises of the mighty India but Indo-Aryan civilization decreed otherwise by drawing the line of civilized life to the east of the Saraswati and Sutlej, closer to the upper waters of the Jamna. Thus India's riverain civilization developed in the Gangetic plain and a second independent civilization in the deltaic parts of the eastern peninsula. The latter became the focus of the ancient civilization of the South where flourished the well-known Southern Kingdoms. Generally, the population of a small deltic area shows no tendency to spread. It lacks massiveness and strength to thrust itself upwards. Also its gaze is sea-ward and not land-ward.

In contrast to the deltaic civilization the riverain civilization of the North rose to great heights of achievements and splendour. Successive attempts were made to build around it a universal State embracing the whole of India. This attempted unification from the North met only with a partial success even though the balance of forces favouring unification lay in the North. The North had an inadequate appreciation of the geographical character of peninsular India and it had no sea sense. At no time did a northern land power solve the problem of maintaining effective lines of land communications and military posts between the North and the South. The sisyphean attempts of the North in the pursuit of its ambitious task of political unification of India failed largely on this single issue of inability to develop successful lines of communication by sea, water transport or by land.

Peninsular India brings home that India faces the Indian Ocean and has an extensive coastline. The sea, however, exercised a negligible influence in welding the South and the North of India. Transportation by sea, except in the coastal areas round the southernmost tip of India, was practically negligible. Further, with the exception of Ceylon there are no islands dotted along the coastline. The peninsula is not deeply indented and there are no inland seas. From the sea there are no routes of inland navigation. Hence the absence of water-way communication from the peninsular area piercing the northern plains.

Irrigation system was developed very early in the North as well as in deltaic areas but its continuous progress appears to have been arrested. There is little evidence to show that advancements in methods of flood control and hydraulic engineering were progressively achieved. There has always been a limited extent of transportation on the great northern Indian rivers but in the absence of network of canals, transportation and communication were insufficient to weld together large areas under a central administration.

A brief delineation of the contour lines of land communication will indicate the nuclear areas of political power which have persisted for long in Indian history. The Mauryan Empire understood the importance of communications under a centralized empire. The imperial highway from Pataliputra to Taxila connected the Central Asian routes through the Hindukush. A concomitant to the rise of a centralised empire in the north was the revival of the trunk communications from east to west. This happened under the rule of Sher Shah Suri and of the Moghuls and in the nineteenth century under the British. In the opposite direction, that is from Central Asia to India, the traditional route crossed the Indus at Attock, hugged on to the submontaneous regions to the north, and, after intersecting the five rivers of the Punjab, the route converged to Delhi. The Central and Southern regions of the

Punjab were arid areas and attracted fewer adventurers, and if any one entered through the Gomal Pass, then the Indus was crossed at Dera Ghazi Khan and through Harappa the road to Delhi was joined at Kurukshetra.

There was another route from the midlands of the North to the western sea-board whose ports connected with the outside world. This route manifested itself from the Buddhist times. Originating from Bihar it passed through Allahabad and then to Bhilsa and Ujjain, both now in Madhya Pradesh. After crossing the Vindhyas, it followed closely the bank of the Narbada to its mouth at Broach, the port of which was known on the ancient maps as Barygaza.

Yet, a third and an important route was from the Malwa Plateau to the Deccan. This was an easy line of communication for migratory movements from the midlands to peninsular India. It is not without significance that the Deccan lava is spread over the Malwa Plateau. The traditional route from the North to the Deccan followed the lava belt of the Southern tableland and very early it was used by the cultural missionary from the North, to be subsequently followed by military conquerors. The Aryans, the Moslems and the Moghuls found lodgements in the Deccan lava area where they reproduced the conflicting patterns of the North. One consequence of pursuing the easy line of communication to the South was that the area known as Gondwana remained isolated till recent times.

These three trends persisted in the Moslem and Moghul periods. When power was organized on Delhi-Agra axis, the movement of Islam to Deccan was through the lava area. Gujarat was reached from Delhi through the Aravalli gap; the same route is closely followed by the railway line from Ajmer to Ahmedabad. From the west coast of India the main line of communication lay through Khandesh territory from whence the Narbada was crossed and the Malwa region was entered near Mandu. The Moghul highway then lay through Ujjain, Sarangpur, Narwar, Dholpur and Agra. This was the route taken by the first Jesuit Mission under Father Monserrate to Akbar's court at Fatehpur Sikri in 1580 A.D.

Such was the partial development of lines of communication having regard to the basic geographical needs of the country. History had to wait till the eighteenth century when Britain successfully demonstrated that it was possible to unite India starting from the peninsular South. British power, using the Deccan tableland as a base, penetrated into the Gangetic plain from the coastal areas thus reversing the long established tradition of the North where power had first to be organized in the North-West and then advance eastwards. Power organized on the Delhi-Agra axis fanned outwards whereas power organized on sea communications fanned inland. Later on when India was closely knit by modern communications, the age-long difficulty experienced in advancing to peninsular South by land communications was finally overcome.

II

Political and military organizations require communications but not so the drift of population. It would be reasonable to expect that the densely populated areas of the Gangetic basin would have played a preponderating role in the physical unification of the South. The northern regions had space, population and elements of higher civilizations, and these in combination should have united India into one organic pattern. Why did not this happen ? A categorical answer may not be easy as no set explanation can wholly explain the complex forces of Indian history. There are however two persistent trends in the political geography of India which have an interpretative bearing on this question. Firstly, the axis of the movement of North-Indian civilization and the population spread lay from West to East; it is difficult to establish the contrary principle that it lay from North to South. The movement of population was lateral and not vertical. Such a movement exerted little pressure on even the midriff of the continent, let alone the more distant South.

As civilization developed in the North, it fell into three separate segments. In the Punjab and the Indus Valley, Indian civilization had resigned itself, which it should never have done, to the fate of yielding substantial ground, culturally, and politically, to the nomads. This area became a very important marchland between Central Asia and the seat of the hard core of Indian civilization which was located in the mid-land portions of the North. On its progress towards the east, Indian civilization lost considerable momentum due to the emergence in the very early times of Buddhism and Jainism — both dissentients from the Hinduism of the midland. The eastern parts of India were in cultural content incompletely flooded. Hindu culture had not sufficient strength to absorb the tribal elements which it encountered in the higher reaches of the Bengal plains. More significantly, at the peripheral regions of the east, Hinduism failed to confer a protective cultural armour on the local population. In consequence, Islam, long after, was able to penetrate in depth in this outlying area, whereas in the vast intervening regions Islam was unable to make any impression on the Hindu masses in spite of the advantage Islam derived from its political ascendancy for many centuries. The loss of vigour in the eastward march of cultural Hinduism towards the geographical frontiers of India may be safely presumed.

The early manifestations of these three divisions are today vividly illustrated on a map of India showing communal concentrations. The northwest region is Islamic, the midland and the eastern portion are predominantly Hindu, and at the eastern extremity Islam, tribalism and Hinduism jostle in confused medley.

The central or the midland of the north, known as the Aryavarta— the sacred land of the Indo-Aryan — was traditionally deserved as lying between the Himalayas and the Vindhyas. By destiny it was the key area for political centralization and for economic power; by fulfilment, its history is a tragedy of partial successes and of partial failures. There was no doubt that from it emanated a great cultural radiation, the rays of which impinged in different directions on all parts of India, but in relation to the unification of India, this alone was not sufficient. The midland weakened its western and eastern flanks. As a compensation it should have utilized the resources of the Narbada Valley and the Deccan tableland, if not the southernmost part of peninsular India, Before disturbances set in the northern plains, Indian civilization had a thousand years of peaceful evolution and had built up a stable secular society. It had dynamism, strength and viability to leap across the puny Vindhyas and spread over in strength into the Central Indian hills and jungles, subduing the land for agriculture and assimilating culturally in to the fold of Hindu civilization the peoples of the forests. Instead, for some reasons not yet apparent, this great key area of the North chose to be the central but the fragmented part of the vast northern space setting a limit to its expansion up to the Vindhyan barrier. In the journey to the Deccan tableland the movement from it followed not the path of the adventurous pioneer but of an easygoing wanderer. The weight of political authority was permanently placed in the North, and the partial utilization of this advantageous position gave a wrong turn to the working of the demographic forces from the earliest times. In consequence we witness even today the existence of tribal people in the confused jungles and hills of Central India and of the still continuing process of the displacement of the tribal languages by the dominant Aryan languages.

Several reasons may be attributed to account for the ineffective functioning of the midland region of the North. There was India's disinclination to adopt a pragmatic political philosophy to deal with the recurrent invasions of the Central Asian nomads and adventurous intruders. India has no internal barriers to impede the movement of men, thought and ideas; the real barriers are her own outer geographical framework of the Sea and the mountains. Inside India there was no possibility of erecting any real barrier between the nomads and the Indian society. If there was no active policy to prevent the nomads from entering India, then it was necessary to adopt a vigorous political attitude towards the intrusive presence of the nomads within India. Isolationism and exclusiveness were hardly the answer; in the long run they are dangerous weapons as they injure the body- politic of the assaulted higher society. Eschewing the political solution of the problems of the nomads, India turned the nomads over to a slow moving and slow grinding machinery under the care of her socio-religious system. The working of this absorption-sponge of Hinduism must have expanded a great amount of internal energy and since the process of absorbing the foreigners was spread over for more than a thousand years, we must expect the weakening rather than strengthening of the process of internal political consolidation.

The enervating and the debilitating effects of climate as a factor in weakening the energy of a dominant minority may easily be over-exaggerated.

Theories based on climatic factors may have some application to the people of the cold climate, more particularly in the industrial phase of the European civilization. But their application to the ancient agricultural empires are wide off the mark, for there is sufficient justification to assume that the energy-content of the people who developed riverain civilizations was sufficient to achieve great many things within the competence of the then state of their civilization. Their technological equipment may be poor, judged by the present-day requirements, but the mind and spirit of man had soared great heights assaulting even the ramparts of Infinity. It is not to the climate factor but to the somewhat peculiar conditions of the Indian society and its deep internal stratification that we may look for our next reason.

Hindu society was securely based on caste stratification but it purchased security at the great cost of depriving itself of initiative and restlessness. It became one of the most sedentary and static populations of the world. This was an undesirable development for a very highly civilized society of Antiquity. Small colonies of sedentary people became virtually the slaves of the energetic nomads, but it was unusual to have a high concentration of sedentary population in one area. So sedentary was the North Indian population that it had no traditions of any internal migration. Even the pounding attacks delivered by Islam resulted in some scattering of the Hindu castes from the Gangetic Doaba to the eastern and central regions of India. Otherwise the Aryavarta of the past was a big inland lake of static humanity; nothing moved it, neither the barbarians who came from the northwest regions nor the dissolution of the centralized empires which were located within the heart of the region. The Indian world outside Aryavarta excited no interest, not even its contiguous parts; the people of the peninsular South were as distant and strangers as the foreigners outside India.

Then again Indian temperament is not favourable to the emergence of an Indian Caesar. No great organizer arose in the North who set out to unify the whole country by drastic means. Unification by crude use of force was not an easy matter. The South was not inhabited by uncivilized tribes. Peninsular India had already reached a high degree of civilization ante-dating the rise of the North Indian Aryan civilization. Any attempted unification by strong and vigorous measures should have been with circumspection and with political insight. No North Indian organizer could have afforded to imitate, even if he wished, Shih-huang-ti who in 214 B.C. sent an expeditionary force to "Sinicize" the southern part of China up to Canton. This Emperor rounded up the vagrants from the North and Central China to populate the territories from the Long estuary to Canton. In this method of colonization the Chinese had long before anticipated the Europeans who also began to populate vacant spaces, which they had discovered, by convicts.

The Indo-Aryan civilization of the Aryavarta would have been cut off from the outside world but for the balancing role played by peninsular South which for long stretches of periods was not even in political union with the North. The people of peninsular India promoted oceanic voyages and established contacts far beyond the confines of Aryavarta. This balancing factor of peninsular India is a permanent feature of Indian geo-politics.

III

It may appear unfair to lay all blame on a geographical region of India and charge it with failure for not fulfilling that which we have assumed to be its manifest destiny. As a corrective to such an assumption we may examine certain factors which have dominantly persisted towards unification. In selecting a few of them, some bias may have to be exercised, for they must have affiliation to physical geography. These indices may be: (a) the conflict between the Steppe and the Sown and its effect on internal conditions of India and China, (b) the establishment of some form of institutionalized bureaucratism, (c) transportation and communication including the development of hydraulic engineering, and (d) common language and culture.

Earlier we have examined the conflict between the Steppe and the Sown. How it affected the internal conditions of India and China remains to be assessed. Broadly this question has two counterparts though they shade into divisions or fragmentation of territorial units. The other is the location of a key economic area, a term used in a broad sense to include an area of agriculture and manpower resources which give enough power to dominate and control the rest of the country.

To withstand the weight of attack of the nomads the political structure must have had cushioning to take off the weight as well as some underpinning. Once the nomads entered into China or India they could not be held back by any artificial barrier, but the nomadic movement did set in motion other movements territorially to the geographical peripheral regions: in China, from North to South, in India mostly from the Punjab to Bengal. In Chinese history we see further trends developed which we de not see in Indian history. Three such are of significance in Chinese history. Chinese political experience realized that the southern regions of China, though not the seat of the original Chinese culture, could be turned into a natural fortress impregnable to attack by the advancing horse-riding hordes from the Steppe. China realized defence in depth and utilized her space in times of adversity in her fight against the invaders. Secondly, the Yangtze Valley was developed into a key economic area to maintain independent centralization when the Yellow River basin was temporarily lost to an alien authority. Thirdly, China has often shifted her capital from North to South. As far back as the fourth century A.D. Nanking in the lower Yangtze basin became the capital when the northern areas were under the barbarians. Since then the capital has been shifted on several occasions, the last being in 1950 A.D.

On the Indo-Gangetic plain on a west to east axis, the shock of the nomadic attacks was something like lightning forks. There was no hard media to penetrate but it did not mean that the Gangetic plain should not have provided itself with some underpining. Nature itself had provided the Gangetic plain with large contiguous territories in the central parts and in the peninsular South as natural fortress. An extreme instance of the failure to make use of a rallying area is that incident in the thirteenth century when Mohammed Bhaktiyar Khilji rode in advance of a small army with only eighteen horses and subverted the Hindu Kingdom of Bengal which from that time till the Battle of Plassey was ruled as province of some ruler in Delhi. If there was a second key area from which forces could have rallied, Bengal or any other part might have had another tale to tell.

The vast spaces of India and China presented from the beginning of their civilized existence a challenge for political unification and the integration of their respective societies. In a preindustrial and commercial age of their agricultural civilization, regionalism was an inevitable feature of the agricultural economy. In thousands upon thousands of villages were embosomed the minuscule of national life. Urban centers were comparatively few and production and distribution were highly decentralized. A certain degree of fragmentation of territorial units and wide range of diversities in national life and society were very natural. A high degree of centralization was impossible to attain without aids which simply were not there in the earlier phases of every civilization the world over. A device like the key economic area, in the particular context and course of the national history, was a convenient aid for the political unification of the country. When the unity of the northern area of the Yellow River basin crumbled after the Han rulers, the Central regions of the Yangtze Valley attained enough economic power and viability to assume an independent centralization. For nearly seven or eight centuries the centripetal and centrifugal forces operated as a see-saw and in this process North and South China were forced to find balance and equilibrium. China underwent internal partition at least on four different occasions till all splintering within it came to an end, and that was about the twelfth century. Such has been the persistent tradition to unity in spite of vastness of territory and regional diversities.

The fixation and stabilization of territorial frame-work within vast spaces and the control of tendencies to excessive fragmentation are still passive factors. The active form is that which assumes the shape of a powerful institutional organization capable of surviving through periods of turmoil and carrying over from dynasty to dynasty, from one epoch to another. This is the running thread of unity, whatever may be the catastrophical changes in the fortunes of the nation. It is here that China made a great discovery and assured to herself a continuity of cultural life and tradition for nearly two thousand years. Chinese genius understood and discovered that centralized state power with potentialities to break up any possible coalition of feudal lords wand with capacity to guard the frontiers and carry on large public work projects had to depend on a competent and learned civil service the recruitment to which should have nothing to do with birth. To its conception was wedded the Confucian classical thought and this alliance worked a miracle for China. "The **carriere ouverte aux talent** was a Han discovery." The emergence of a Central bureaucracy, learned and loyal to the State, prevented the establishment of feudal families and effectively frustrated the ambitions of revolting servants of State by converting their offices into hereditary rulerships, such as happened with the Carolingain Counts in medieval Europe or the Muslim and Mahratta 'war lords' following the collapse of the Moghuls in the eighteenth century India. The feudal bureaucratic system continued its course unaffected by the disappearance, at the top, of authority by frequent dissolutions of dynastic rule. A stable bureaucratic structure made it feasible and practicable to administer territorial units under the direct authority of the central government. After making allowance for the working of the centrifugal forces till the twelfth century, the Chinese pattern of administration settled down in favour of directly administered units from the centre, and thus there was no room for territorial fragmentation by subordinates exercising varying degrees of sovereign authority.

India had no such continuity. This institutional idea did not develop in India beyond initial stages. The administrative machinery under the Mauryan rule appears to have broken up and not reassembled again in any recognizable form. We are unable to trace at least in outline a going concern which could have been taken over by a succeeding authority either local or foreign. Hindu India did certainly contribute ideas and doctrines, forms and symbols, but not any continuous frame-work of organization. This is clear from the difficulties which the early Muslim rulers in Northern India encountered. They had to hold down a rebellious and refractory agricultural population over whom they ruled as a dominant minority, but they had no large-scale apparatus of administration at their disposal as a going concern. In developing their feudal system the Muslim rulers imported into India certain ideas of primitive Islam from Arabia and before the Moghul power was established Islam had not found an answer to centralization over the northern plains. In fact fragmentation increased over the northern and central parts of India.

The Moghuls did better. Akbar's genius had perceived that a stable administration by a foreign agency rested on a sound treasury and a contented peasantry. His plan was to govern through what may be described as a system of district collectors. His successors failed to develop his great idea. Foolishly they handed over the peasantry to the rapacious revenue farmers and intermediaries at whose hands they suffered unheard of miseries. His successors fell back upon the feudal system of jagirs. The situation demanded not the increase of the intermediaries but the exercise of power directly under the authority of the centralized empire. Islamic tradition in India created an aristocracy, not of wealth but of office. This left no room for carry-over of any tradition in administration. Akbar alone had seen as did the French Kings of the Grand Monarch that every absolutist state needs the services o the holder of an office as its skilled and subtle instrument for administration. "Know," said the Scotsman Law in the beginning of the eighteenth century, "that this kingdom of France is governed by thirty **Intendants.**" The British who followed the Moghuls had a clearer conception of the groundwork of organization for a country of India's size and they governed two-thirds of India through two- hundred and fifty district officers, and in the remaining one-third through hereditary executives represented by Princely India. Institutional organization of government as a means of promoting unity was very much a late-comer to India.

Chinese feudal bureaucratic society came to rest on intensive agriculture requiring irrigation. Hydraulic engineering and water control became a function of State authority and this augmented the centralized State power. Political power acquired a close connection with regional geography. Unlike the collector under the British Indian administration who had to look after the fiscal interest of the State and maintain law and order, agriculture and bureaucratic administration attained close co-ordination from the Han period onwards. Whoever might rule, the scholar gentry had to administer the affairs of the State in his region. The administrator possessed mastery over written language and he was the repository of all technical knowledge relating to hydraulic engineering. This promoted continuity and progress of agriculture and irrigation even during the periods of fragmentation of political power and unrest.

The magnitude of the water-way communication under such persistent efforts will be appreciated when told that China had 200,000 miles of canal system, the greater number of which was located in the Yangtse Valley. This gives a measure of the transportation and irrigation system and its bearing on the agricultural life of the community. By way of illustration a quotation may be given from a Chinese work of the sixth century A.D. giving a glimpse of early irrigation and transportation system.

> "From Shanyang to the Yangtse River, the water surface of the canal was forty paces wide. Roads were constructed along both banks and planted with elms and willows. For every 2,000 **li** from eastern capital (Loyang) to Chiangtu (modern Yangehow), shadows of trees overlapped each other. An imperial resting place was built between every two post stations, and from the capital of Chhang-an to Chiangtu there were more than 40 such pavilions."

As culture and writing developed in China, the written characters became universal; the existence of dialects and mutually incomprehensible forms of speech ceased to be an impediment. In such a case the tendency to unity must dominate. Moreover paper was invented as early as the first century A.D. and China led the way for the civilized world by the development of printing. Confucian classics appeared in wood blocks from about 930 A.D. leading to wider diffusion of books and consequent spread of learning throughout the country— a fact of tremendous significance in the cultural unification of China.

IV

Northern Indian civilization did not spread as a result of political unification by the North; or by population movement and migration or by conquerors imposing their language and culture. Southwards, it spread by cultural drift. The phenomenon of the cultural assimilation of the South by the North is illustrated by an excellent dictum of the American anthropologist Franz Boas:

> "A people may remain constant in type and language and change in culture; that it may remain constant in type but change in language; or that it may remain constant in language and change in type and culture."

The physical type of the Aryan speaking people was submerged in the somatic character of the indigene. The lasting gift bequeathed by the Aryan speaking people of the North was neither a higher material nor a superior physical type by a more excellent language and the mentality which generated it.

It is not therefore the demographic factors or the political history but the linguistic history of India to which we have to turn our attention to understand how cultural unity was established between the North and the South. There is a complete linguistic break between what is termed the Aryan languages of the North and of the Deccan lava and the agglutinative languages of the Dravida group. This major linguistic division should have proved a formidable barrier hindering the unity of civilization of the entire sub-continent and should have set up wide cultural differences such as the Slavic, Teutonic and the Romance languages have done in Europe. Fortunately for India it has not happened and is a fascinating chapter in the linguistic history of India.*

The Dravida languages represented a very high stage of development in the evolution of human speech but their role was confined to familial and social purposes. The pride of place and primacy was given to the more

*Toynbee in his 'A Study of History'—Vol. IX, pages 79-80 and footnote 3 to page 80 makes some very interesting observations on what he calls the 'irrepressible vitality' of 'this ever-green Sanskrit language' and literature and the allegiance it captured of the people speaking its derivative languages.

dominant Sanskrit. The general tendency amongst languages appears to be for the easiest and the socially more useful to supplant the more difficult or the less useful . The Dravida speaking people of the South met the most challenging of all cultural problems by choosing to be bi-lingual.

The spread of Sanskrit as the vehicle of superior culture is primary contributory factor towards the unity of India. This aspect of the unification of India by a dominant language is by no means a novel or a fanciful view; it is rooted in the peculiarities of Indian history. Under the Moslem rule the North was overlaid with a thick paste of Persian culture; Sanskrit ceased to be its living language. When political India was divided into the Moslem dominated North and the Hindu South, Sanskrit remained the cultural link between them. The emergence of English under British rule as the common **ligua franca** re-emphasised the previously well-established principle that unity between the North and the bi-lingual South could only be maintained by a language culturally so dominant that it is the highest vehicle for thought and culture.

One has to take a text book of Indian History to see how the political dichotomy between the North and the South is laid bare. The history of the North and the South is unintelligently divided into two separate worlds of kings, dynasties and chronicles. A point which has received little attention is the divergent historical experiences of the two regions and their imprint on the minds of the people.

Nature designed Peninsular South— the bastion of the North. Unaided by South, the North was unable to accomplish its historic role of meeting the challenge of the open spaces and of the barbarians. The situation worsened when a weak and politically divided North faced Islam in the twelfth century. The North alone had not the strength nor the defence in depth to deal with the Islamic onslaught. Only with the moral resources of the country as a whole could India have confronted the Turkish marauders from the twelfth to the sixteenth century. The impact of Islam shattered the entire face of North India.

That India had to confront the world-wide movement of primitive Islam was not something peculiar to her. She occupied an intermediate position between China and the Mediterranean world. The Mediterranean became the moat between the Saracens and the Christians when Islam in less than half a century overran the Roman world in Asia and North Africa. It is not the primitive Arabian Islam but the Central Asian Islamized Turks who confronted India in the twelfth century. Thus the Islamic challenge came later to India. Islam entered China in the Tang period and the Moslem population in China was replenished at intervals by small streams of migration from Turkistan. "Islam in China had to conform to Chinese notions of the place of religion in the State." The Chinese Moslem communities were politically assimilated without losing their cultural identity. The Christian peoples of the West fought for centuries the politics of Islam and, under the adversity of Islamic attack,

Western Christendom attained a certain common European outlook and later emerged, in Western Europe, into national societies.

It was a politically feeble and disunited northern India which confronted the politics of Central Asian Islam. Northern Hinduism had to fall back, for overthrowing the political dominance of Islam, on a long process of attrition in time and in numbers. The politics of Indian Islam responded to this challenge by fortifying the Moslem religion with the culture of Persian language and equipping itself with a strong protective cultural armour against the insidious inroads of dominant Hinduism. It is true a kind of synthesis of culture took place; the so-called composite culture—an imperfect blend of Hinduism and Islam—somewhat narrowly restricted to political and social purposes. Islam stubbornly remained an unassimilable force.

There ensued for long an incessant duel between Hindu and Islamic minds. Loss of political freedom apart, the Hindu soul was badly seared; there was obvious cultural degradation. The Hindu in mind and spirit had to renew himself if he had to survive. Hindu revivalism in the North expressed itself largely in the Bhakti cult and religious emotionalism to give strength and sustenance to a withering soul. It is very significant that all movements of Revivalism and Renaissance in the Islamic period and even in the 19th century were confined to the North with little application to or influence on the South. We have yet no clear idea of the profound effect of the alien impacts on the structure of Indian society. Certain it is that the glorious mansion of Hinduism in the days of its pristine glory was forced to admit many socially undesirable tenants. The North has no respite from the process of absorption and the consequential social adjustments.

During these periods of trials and tribulations in the North, the South remained culturally most stable. There was no inner conflict in the mind of the Southerner for the northern conflict had little repercussion on the South. The South settled down to a long period of the Hindu way of life, the traditional and ancient forms of which were enforced rigorously by the Brahminical hierarchy with all the authority and prestige of the ancient lawgivers of the Aryan North. The South was able to retain a mature and poised mind, able to assess the fundamentals of spiritual and human problems with rationality and without any frustration. Elsewhere, wherever the conflict between Hinduism and the alien forces was intense, the resulting clash of culture and the contact of races developed a split personality.

V

This then is the final summing up of India's long and determined quest for unity. The physical and population differences between the North and the South did not lead to a major break in the social and cultural continuity of the two regions. Within the framework of a universal civilization, shaped and conditioned by history and geography, there came into existence from very early times two foci of culture in India. The North attempted but failed to unite the South; the South by itself had not the resources to unite with the North. North and South were never divided on a mutually antagonistic basis. On the other hand, the cultural life of North and South developed on a co-ordinated existence, and jointly—never separately but always in unison—they advanced the fundamental unity of India.

The unity of India is a passionate theme. The disunity of India was the standard imperialist argument against the nationalist claim for a united India. It suited the imperialists to deny the unity of a country over which they held dominance. The theorists of the Austrian empire were wont to prove that Italy was a geographical expression; the Englishman, Sir John Seeley, asserted "that India is not a political name but only a geographical expression like Europe or Africa." Such a theme was naturally repudiated by an exposition of the fundamental unity of India's culture and civilization amidst its prolific diversities. The working of the historical forces in the last one hundred years has demonstrated that the political and economic unity of India is natural because it is 'in response to its geography.' Aided by Western Science, these forces in the course of transformation of Indian society have applied correctives, removed obstacles and facilitated unification. Of these five stand out in prominence. Firstly, distance has been conquered by the network of railways, telecommunication system and aviation. Physically, India has been knit so closely that no soldier or politician or administrator would say that the physical character of the country makes it hard to hold or govern it. With the withdrawal of Britain's imperial rule over India, the political rule of the subordinate sovereign authorities came to an end. The people of these areas joining hands with the people of the rest of India made a silent revolution by wiping out in an incredibly short time the age-long curse of territorial fragmentation. Thirdly, India possesses one of the finest administrative machineries in the world, and its defence services have a record of magnificent services of which any country may be proud. Fourthly, those areas in the North, long lost to Central Asian Islam, seceded from India by agreement when it was realized that it was not possible to maintain national unity over them. Fifthly, Indian society is no longer continental. It is rapidly being fused into a cohesive national body. Nationalism as a vital force is the basis of Indian unity. Secular nationalism based on democracy is the surest guarantee for holding together the elements which go to make India a great nation.

"India," wrote Mr. Nehru, "was like some ancient palimpsest on which layer upon layer of thought and reverie had been inscribed and yet no succeeding layer had completely hidden or erased what had been written previously." There are **three** such writings of the past which continue to have a bearing on India's unity. The first of these is the national language. India's past has shown that what promotes powerful unity is the acceptance of a language of such a dominant character as to meet her highest cultural requirements. The position in this respect has not changed during the last two thousand five hundred years. The regional languages have not hitherto promoted groups rivalries. In Europe language in association with ethnical and racial groups has led to explosive nationalism. India fortunately has no racial problem. The national language and the cultural problems associated with it will have to be handled with understanding, sanity and maturity.

The partial functioning of the demographic factors has left large islands of primitive people in the north-eastern, central and southern parts of India. The assimilation of these elements into the present-day democratic society is a problem of importance, and this legacy of the past must be handled with care and understanding.

Thirdly, the role of peninsular India must be understood; more so with the amputation of the Islamic limbs in the North. The peninsular part in the past orientated India's gaze to the world beyond the shores of her seas. The period has been one of the greatest epochs in the history of India. To quote a recent geographical authority:

> "How false is Lyde's still current concept of an almost solely inward looking India, with its sea contacts with those of alien traders, may be seen from the fact that the decline of Sailendra was due in large part to an attack by Rajendra Chola II (1012-44); the Cholas were driven out after a century of intermittent war but in the 13th century a disastrous expedition against Ceylon fatally weakened the power of Shri Vijaya. These armadas presupposed high standards not only in navigation and seamanship, but in naval organization on both sides of the 1,200-mile wide waters between Coromandel and Sumatra. Certainly no European power of the day would have dreamt of such oceanic adventure; only the Viking voyages are as impressive, while the Crusading fleets were in comparison mere coastal forays. On the terraces of Borobudur, the carved ships of Shri Vijaya still sisal, mobile and endlessly, over their seas of stone."

The peninsular regions remind India that her interests are oceanic. The sea lanes are linked with the maze of internal transportation system of modern India which handles the vast bulk of sea-borne trade vital to her economic life. India's vision should not stop at the Himalayas and the Indian Ocean. India has to be intensely curious of the happenings beyond her own world. An exterior conception is always essential for the maintenance of peaceful international relations.

V The Concept of Boundaries in India Through Ages

B.G. TAMASKAR

Many and varied visible and latent traits and dimensions of boundary concept in India are still not fully understood, particularly during the ancient and medieval periods of the country's history. In this context, it may be borne in mind that there is "underlying unity and continuity in Indian history given by its geography and culture, by the spirits and achievements of its civilization"[1]. This particularly holds good in the case study of the concept of boundaries since its inception in the dim past of the ancient Indian traditions. This academic exercise on spatial analysis of boundaries attempts a preliminary and tentative exposition of their little explored character so far.

The objective of this paper is to present a succinct delineation of the concept of boundaries, its practical manifestations, the classification of boundaries, the solution of boundary problems, demarcation of boundaries on land and sea, their implications in relation to territorial sovereignty and allied attributes and features upto the pre-Independence period in India.

The boundary concept in India through its various phases of evolution has shaped itself into the existing pattern and development by slow degree, with visible traces of concomitant features, attributes and implications from the ancient past of the country. Some of these characteristics may sound surprisingly too modern for the contemporary times. A penetrating and deep insight into the evolutionary processes of boundaries in India will always remain elusive without taking into consideration interacting forces of their indivisible component, namely, territorial sovereignty, one of the inseparable, integral and interlinked constituents of the State. No international boundary can remain fixed, stable and persistent whitout effective backing, constant and vigilant watch of the state concerned and firm, spontaneous and sincere respect for territorial sovereignty and its inviolability by neighboring states.

The concept of manifold attributes of boundaries, their modifications,

Reprinted from *India political Science Review*, 1978, Vol. 12, pp. 59-72.

transformations and changes have been investigated and highlighted on the basis of material and substantial empirical experiences and theoretical constructs and evidences, embedded in ancient Vedic literature, epics, Puranas, treatises on *dharma-sastras* and politics, and recently published relevant historical works and papers in a bare outline, ending upto the pre-Independence period in India.

It will do well, at the outset, to spotlight the Indian idea of spatial sovereignty, before the exposition of the boundary concept can be taken up for in-depth discussion.

Possibly, the first clear and well-defined notion about the spatial *all-embracing* sovereignty can be traced back to *Aitaraya Brahman* (an ancient Indian treatise)[2] in the Vedic times, from which Kautilya seems to have drawn his guidance in the following statement in his famous *Arthsastra:* "The land (space) which extends north to south from the Himalayas to the sea and measures east to west a thousands yojanas (7,250 Km.) is the kshetra (that is, the sphere of influence of a *Chakravartin* or *all-embracing* sovereign)"[3]. However, D.C. Sircar points out cogently that the conception of this ideal ambitious sovereign territorial extent, encompassing the whole of India, distinctly separated by physical barriers and sea-margins, had yielded place to two sovereign dominions in India, north and south of the Narmada river.[4]

There are earlier and contemporary references to the spatial extent of sovereign State in India in the religious works *Baudhyans's Dharmasutra,*[5] (500-200 B.C.), *Vasistha Dharmasutra* (300-100 B.C.)[6] and *Sankha-Likhita Dharmasutra* (300 B.C. to 100 A.D.)[7], which need a close investigation in their original texts. The conception of horse-sacrifice (*Asvamedha yagna*) often translated into practice in the past history of India by ambitious aspirants to the goal of ideal territorial sovereignty in India cannot stand in isolation from the inviolability of boundary concept, as the former breathed out constant threat and menace to the latter. The number of feuds, battles, wars, clandestine or open encroachments beyond recognised or agreed boundaries, tacitly or by mutual bilateral agreement, treaties and foreign invasions and incursions is a legion to be even mentioned here. Evidently, intranational boundaries within the country had been always in a state of changes and shifts, till the promulgation of the British power in India.

The international boundary in the north along the Himalayan mountain barrier had remarkably remained stable till the beginning of the 19th century and its problems became actively alive in the recent Sino-Indian conflicts of 1962, though in the beginning of the 20th century, the British power in India and the Chinese government were drawn closer together than in the past for recognition of the Chinese suzerainty over Tibet for the first time."Even precious little is known about sea frontiers and boundaries, though the ancient Indians had indulged freely in sea-faring activities, till the advent of European nations by sea-routes to India towards the end of the 15th century A.D.,

However, *Kautilya's Arthsastra* contains the germs of the concept of territorial waters in the statement: the controller of shipping "should demand duty from ships sailing on sea when they come within the *domain*". (Kangle: 2.28.11). It is pleasant to hit upon the reference which states,"Power was then vested in him as the lord of every seaport and custom house between the first and third century A.D."[9] The Moghul rulers had neglected the sea frontiers, as they were land-bound. Their rule was weak at the boundary regions or outskirts of their territory,[10] particularly, their hold on sea-boards was loose and shaky.

The explicit mention of boundaries rarely appears in the Vedas, Upanishads and Smritis and even in the Puranas, and Buddhist literature, though names of a large congeries of states in India occur in them, most probably separated from each other by belts or zones forming frontiers between them, because of the amorphous character of existing boundaries. Large areal .extents of landscape remained unhumanized on account of the process of colonization in progress and the occurrence of vast, dense and impenetrable forested tracts, interposed between colonized areas.[11] So long as tribes or groups of people did not lose their nomadic habits,[12] the question of any crystallization of boundary concept remained a remote possibility. Extensive sprawls of interposed forests persisted even in the 16th century, particularly in the Deccan. Hence, *areal* frontiers separating political dominions and realms in India[13] persisted even after the appearance of the Moghuls on the political scene, though *linear* boundaries had already been stamped on some parts of the political landscape of the country, since the dawn of Indian history. We get the first glimpses of perceptual constructs of *linear* boundaries in the 4th century B.C. and their actual manifestations on the landscape for the first time probably on Alexander's retreat from India. Alexander had erected twelve altars to mark the farthest point of his advance.[14] Evidently, these altars could not demarcate the entire farthest eastward limit of his conquests. Likewise, there is another instance of erecting pillar inscription for demarcating boundaries between two neighboring states. The demarcation of boundaries between two feudatary rulers is recorded in Maharaja Hatin and Maharaja Sarwanatha, who erected a pillar at Ambloda to mark out the boundary between their territories about 510 A.D.[15] These appear to be an incipient beginnings of concept of boundary-markings on land in India, between two neighboring states.

As the concept of boundaries is bound up with sovereignty of the State, let us turn to the examination of the prevalent notion of sovereignty in the ancient past of the country. According to the *Matsya Purana,* the sovereignty comprised of seven organs of the State, *viz,* "the lord (the ruler), ministers, republics or confederations of peoples (*Janapadas*) forts, political power (*Danda*) treasury and friends (friendly neighbours)."[16] While the Puranas lay emphasis on friendly neighbours of a state, Kautilya could not envisage the peaceful existence of vicinal states, implying thereby implicit ever-changing boundaries of intra-national states. Kautilya advocated the creation of a geopolitical sphere around the sovereign state (would-be conqueror) and expediency of pursuit of his six-fold external policy. As the details of his international geo-politics have been highlighted elsewhere, they need not detain us any longer.[17] Manifestly, the prevalent notion of boundaries conforms to boundaries of confrontation,[18] though peaceful infiltration by creation of pilgrim centres and purchase of territory of the peripheral state have been not lost sight of by Kautilya.[19]

With the recognition of and focus on the economic utility of the sea in the 17th century Deccan, the notion of territorial waters added a new dimension to territorial sovereignty into the sea,[20] which is expressed in the accretion of new organ to the notion of sovereignty, *namely,* navy with the rise of Shivaji to political power.[21]

The new-born facet of territorial sovereignty led to the impressive spatial organization of the littoral, manifested in the erection of ports, creation of naval forces and merchant fleets, resulting in increased commercial and economic activities.

(*a*) Natural Boundaries

We shall, now, turn to the fact that in the early stages of the evolution of boundaries, it is not surprising that there had been a marked preference for natural barriers to delimit them. The boundary bases, aligned with continuous features of natural landscape, were selected to follow stretches of mountains, courses of rivers and canals, outskirts of deserts, forests, swamps, etc.[22] Generally, mountain ranges and stream courses served to separate cultural groups by virtue of their comparative well-defined character.

As so-called natural boundaries could not and cannot be assured of their protection by themselves, Kautilya, as a consequence, had urged strongly the erection of defensive fortifications on all four quarters of boundaries.[23]

(*b*) Boundary-Markers

In the ultimate analysis, all boundaries are man-made geographical occurrences.[24] As such, they ought to distinctly appear on the landscape by such devices as marks, cutlines, fences, notices, etc.[25] Kautilya enjoined the headman of a village to cause "a fence to be made with pillars all round the village at distance of one hundred *dhanuse.*" The practice of concealed boundary marks is quite interesting whereby pots filled with chatt, coal, bricks, bones and similar other, not easily corrodible, things are buried inside the earth.' A row of large-sized stones should be employed to indicate the boundary. In the 4th century B.C., perhaps, even earlier, artificial boundary marks had come into vogue with checkposts and custom barriers under the supervision of boundary gate keepers.[26]

The practice of demarcation of boundaries of spatial units in India, obviously, must have persisted throughout subsequent ages. The land-boundary between the erstwhile Sawantwadi princely state and the Portuguese territory of Goa has been delimited on ground by a narrow clearing in the adjacent forests.[27] We have come across various land-measures in time and space throughout the country, particularly in specific reference to revenue surveys. The present author has incorporated the regional land-measures in one of his papers, existing in one of the Deccan states of the first quarter of the 17th century.[28] Land-measures occur in Kautilya's *Arthsastra* also.[29]

It is interesting to note that land-measures were used for surveying land[30] and perhaps, cadastral maps were prepared therewith.[31] In this connection, the attention of the reader may be drawn specifically to the professional *Jaribdars*, engaged in the measurement of land on payment of specified rates or fee.[32] Far more astounding may sound the existence of Indian cartographers. The following reference to the 17th century cartography has been culled out as the convincing evidence in its support: "The map of Shivaji's kingdom, prepared by Parvat Rao, was shown to Emperor Aurangzeb."[33] This map is referred to in the Persian manuscripts, kept in the Royal Asiatic Society, London. The Marathi translation of this interesting reference occurs in *Farasi Riyasat* of Sardesai.[34]

One is inclined to believe that some of the contemporary general, and cadastral maps and even sea-charts of the past may be lying buried in foreign manuscript libraries and museums, but also in similar learned institutions and societies in India. Some of these have been unearthed recently. The search for indigenous cartography awaits the selfless and painstaking labour of some patient and tenacious scholar, or scholars, with the financial aid of such institutions as are able to render it, before they are lost to us for all time to come.

One patent fact emerges from the extract cited above that the cartographer belonged to the Deccan and it is likely that he and others of this profession might have prepared other maps. These maps may prove highly useful for the indepth analyses of the contemporary boundary concept and its application in the real world.

Notice of village boundaries and reference to boundary disputes occur from ancient times. At least from the 4th century B.C. onwards, meticulous care has been bestowed to spell out the allocation and delimitation of village boundaries, followed by their actual administration.[35]

A categorical mention of boundary-markers has been made in the period preceding the 1st century A.D. In addition to the man- made boundary-markers of villages, trees, large stones or rocks, etc. permanent natural features were also utilized for delimitation and demarcation of boundaries on the landscape. It is illuminating to note that village boundaries were demarcated on the land in the presence of witnesses thereto.[36] This healthy and useful practice of the demarcation handed down to us must be reckoned among one of the valuable and venerable legacies of our ancient forefathers.

In village and land-grants, occurring in recorded inscriptions also, the allocation and delimitation of boundaries have been explicitly mentioned in the order to avoid possible future disputes.[37]

(c) Classification of Boundaries

We have already made a reference to implicit and explicit notices of natural boundaries. We may, now, turn to the administrative units, which had their own boundaries. It may be noted that large spatial units in ancient India were known by several appellations such as *Rashtra, Janapada,* and *Desh.* Their divisions were known by such names as *Visaya, Mandal, Bhukti, Prants,* etc. with their varied sizes and shapes. These were sometimes used interchangeably. We come across with an unfamiliar term, *Khanpanak* in the 13th century in the Deccan,[38] in connection with the *Khanpanak* of Nidhivas (Modern Nevase). Another similar rare administrative unit, referred to in ancient India,is *Pauras*, which in the modern terminology may be identified with the township.[39]

In the Moghul times, the administrative units, namely, *Subha, Sarkar* (District), and *Pargana* (also styled as *Mahal*) appear in the descending order of administrative hierarchy.

In the Deccan of the sixteenth century a varied number of administrative terms was in vogue which included *Tape, Mamle, Pargana, Tarf, Mahal Karyat,* etc. Shivaji introduced in his dominion only three categories, *viz. Prants, Taraf* and *Mauja* (village) in the second half of the 17th century. The old administrative divisions did not easily become obsolescent for a long time causing undue confusion in identifying territories. In the far south, the Tamil-speaking area, prior to the 9th century seems to have been dotted over with administrative units known as Nadu. Nothing has been said about the size, compositions and functional jurisdiction of this spatial unit.[40]

The following table of the administrative units in Karnatak and Tamil Nadu gives their corresponding names during the regime of the Vijayanagar Kingdom for a comparative study:[41]

Karnatak	*Tamil Nadu*
1. Stahl (Village)	1. Nādū or Simai.
2. Sīma or Nadu	2. Paru or Manlandu
3. Valita or Binte	3. Koltam.
4. Rājya or Cavadi	4. Mandalam or Rājyam.

The names of administrative divisions of pre and post-Independence period in India are quite familiar to bear any repetition.

In a nutshell, the evolution and significance of administrative units varied in space and time in India, which in themselves constitute an independent theme for intensive and fruitful research project.

In the end, we may refer to minuscule boundaries, setting apart forts (*Machis* in Maratha history), royal places, custom houses, military establishments, etc., from general encroachments.

(*d*) Boundary Functions

The early concept of boundary was one of barrier character. Deep afforestation, elephant forests and wild tracts along boundaries are suggestive of expediency of separating factors to fix them permanently and to impede integration across the border line.[42]

It may be recognized that boundaries were and are jural laws. They were and are one of spatial expression of the given legal system. As such, we find that the flow and circulation of goods and persons, of indigenous and foreign origin, was regulated by strict rules. The provision of passports was scrupulously enforced, particularly, in case of foreign visitors. The unfettered flow of commodities was forbidden by law and a higher range of toll than that on indigenous goods was levied on imported goods as a protection against unhealthy competition to establish commercial relationship and home-made goods were exported to foreign countries. Thus, measures were adopted for imposing effective checks on smuggling,[43] during the 4th century B.C.

At a later stage in the Indian history, it appears that persons were not only required to obtain passports for crossing established boundary but also to pay fee. We may cite the recorded instance of the Dutch traveller, Van Ravestyn, who was allowed to cross the farthest limit of the Nizamshahi boundary on payment of a formal tax in June, 1613.[44] Another instance of the similar category is recorded towards the end of the 18th and the beginning of the 19h century in which there is also a definite and categorical mention of payment of fee on reaching a certain place.[45] The regulation of entry and exit of individuals and goods seems to have been in force both in normal and war periods.Outlaws or runaways from the law were forbidden to remain in the territory of any state, after crossing its boundary at the point of entry. There is an explicit reference to the search efforts made by the Nizamshahi Premier in 1621 A.D. of a debtor, who had been trying to escape from his obligations.[46] Early in the 19th century or even earlier the question of extradition of insurgents and criminals seems to have been raised.[47]

The above discussion points to the irresistible conclusion that the evolution of boundary functions witnessed several stages to attain their present pattern and functioning.

(*e*) Boundary Disputes

The existence of boundaries implies that disputes could arise at any time and place, but their nature could differ from place to place and area to area. International boundary disputes can be classified into territorial, positional and functional. For the sake of economy of space, we may only call out one example of each category. In the first category, we may cite the boundary dispute in Afghanistan's frontiers in which four parties were involved in 1872-73, namely, Persia, Russia, Britain and Afghanistan. The British administration in India directed its efforts for the precise definition of Afghanistan boundaries and earnestly helped to acquire natural and scientific boundaries with her neighbours in west, north and north-west.[48]

The above dispute arose from the previous superimposition of boundaries upon the physical landscape. We may, now, give an instance of partial evolution of boundaries, resulting in positional disputes. It is highly interesting to note the preposterous and extravagant claims advanced by Feringi pirates of Chitgaon (in Arakan) in 1665 A.D. upon the Moghul territory which ultimately ended in the conquest and capture of Chitgaon in the following year,[49] shifting the eastern boundary of the empire across the Bay of Bengal. In the last category of disputes, we may cite the example of the International boundary on the Terekhol River between the former Portuguese territory of Goa and the erstwhile Sawantwadi princely state, which had been a perpetual source of disputes, due to large-scale smuggling resorted to by the peoples on either bank of the river, because of the improper functioning of the boundary.[50]

The impressively large number of rural settlements in India has witnessed frequent and numerous village boundary disputes, since the ancient times. We come across the earliest examples of these disputes in the 4th century B.C.[51] They are also repeated in the period, preceding the 1st century A.D.[52] Village boundary disputes are not traceable till the beginning of the 10th century A.D. from the historical sources available to the author, except the reference embodied in the *Yagnavalkya Smriti.*[53] Such recurrent village disputes are noticeable in the first half of the 17th century in the Nizamshahi kingdom of the Deccan,[54] and also in Shivaji's times later on.

(*f*) Evolution of Boundary Disputes

For the sake of convenience and clarity, we may divide this topic into two broad categories: (1) International Disputes and (2) Village-boundary disputes.

The common, crude and ready method of solution for international or intra-national boundary disputes is the resort to the use of arms since times immemorial.But peaceful methods have also been utilized for the solution of certain boundary problems with extraordinary degree of success at times. Examples of peaceful solutions of boundary problems are cited below from studies of the Nizamshahi kingdom between 1600-1626 A.D.:

(1) the bilateral negotiations and stipulated retrocession of disputed territories or places; retrocession of the fort of Kandahar (in the

Deccan).[55] (2) the establishment of condominium into the border of Shirwal Pargana.[56] (3) sharing of land-revenue in the proximity of Chaul between the Nizamshah and the Portuguese.[57]

The modes of settlement of Village-boudnary disputes may be summarized below.

(1) *4th Century B.C.:* "In all disputes regarding the boundary between any two villages, neighbours or elders of five or ten villages shall investigate the case to be furnished from natural or artificial boundary marks."[58]

(2) *1st Century A.D.:* An elaborate and systematic procedure seems to have been adopted and may be mentioned as follows: (a) examination of traces of the former boundary marks, when and if necessary, (b) respondents and dependents to produce witnesses, (c) the deed of the settlement with particulars of the boundary fixation, duly signed by witnesses thereof and (d) the decision to be made in the presence of the ruler.[59]

(3) *17th Century A.D.:* (a) Boundary dispute between the village of Chatan, Tape Karkamb and the village of Bhosar, Tape Udargaon (23.11.1608), decided by *Gotsabha* by the use of an ordeal (Sanads and Letters, II, Article 6.):(b) boundary dispute between Kedgaon and Pargau of Pranta Punas, decided by *Rajsabha,* first decided by ordeal of fire, followed by taking ball out of boiling fat and finally marking it on the ground. (Shiva-Charitra-Sahitya, Part -II, Article 93 dated 1610 A.D.); (c) boundary dispute between the township of Supe and village of Vadane, decided by *Gotsabha* (3.12-1619) on the basis of previous papers and ordeal, followed by final demarcation of the boundary by members of the Sabha by traversing the boundary (V.K Rajwade: *Maratha Itihaschi-Sadhane,* (in Marahti) XVIII, Article 4); (d) boundary dispute among Malsiras (Pargana Supe) and Tekwadi and Pisrava (Traf Karhepathar) (27-9-1621), decided by *Rajsabha* on the basis of an ordeal by taking out ball from boiling fats. (Sanads and Letters, II, Article 39). The dispute started in the Nizamshahi regime and was finally closed in the times of Baji Rao Thorale.[60]

The transportation of soil by a river, separating two villages, had given rise to an interesting solution of the boundary problem. The accretion of transported soil belonged to the village to which it became attached but this was only when the soil carried away had no crops growing in it, but if the soil with growing crops was separated from a village and joined to another by river in flood, the former owners could reap only the crops and land would belong to the village or field to which it had become attached.[61]

(*g*) New Trends

The new trends express themselves in (i) the orientation and extension of political and economic activities towards littoral regions, and (2) the perception and realization of cultural diversity.

The appearance of the European traders on the Indian sea-coasts markedly invited the attention of some Indian rulers, particularly of Shivaji, to the gains and advantages of marine resources, martitime trade and commerce and resultant economic activities and prosperity. As a result, navy came to be reckoned as the separate and independent organ of Maratha Kingdom with the powerful realization of the fact that the mastery of the sea could be in hands of any political power which possessed the strongest navy[62], though its humble beginning had been already anticipated and made in the preceding Nizamshahi kingdom (1600- 1626 A.D.). The remarkable upshot of the institution of navy was the equal and successful competition of the Maratha rulers in naval and maritime activities in the Arabian Sea against their determined and avowed European rivals, particularly, the Portuguese strongly entrenched on the western coast of India, claiming to issue navigation permits, known as *Cartaz* to all Indian navigators. "Right from the beginning of their coming to India the Portuguese regarded themselves lords of the seas.[63] The mastery of the seas was challenged by the Maratha rulers with the result that the concept of pushing coastal boundary seawards into territorial waters could be sensed and began to crystallize into existing pattern unconsciously and consciously to lay claim on all wrecks in their ports.[64] The increased ship-building and commercial activities, fishing, salt-making and collection of custom duties improved the state finance to a considerable degree on this account. We have not come across any empirical evidences recorded in historical materials so far as to the theoretical constructs, nature, scope and function of territorial waters.

Another notable but unfortunate trend, that is, dormant, latent and lurking undercurrent of cultural diversity in the past gave vent to its ugly and unhealthy pattern and burst into open activity, even before the firm imposition of the British rule, which had been fanned into fire by it to prolong its political strangle-hold on the country. The work, entitled *Satpāncāsidde's Vbhaga,* assigned to a period not earlier than the first quarter of the 18th century, incorporating a list of fifty-six *deshas* in India, gives a vivid depiction of cultural diversity of India.[65] "The earliest work containing a list of fifty-six *deshas* seems to the *Candragarbha-sutra* or *Candragavaipulya* (translated into Chinese by Narendrayasas in 566 A.D.), mentioning them in connection with Buddha's manifestations in Jambudvipa."[66]

It may be pointed out that India took an unfortunate decision to adopt ethnic basis for state reorganization into the present linguistic spatial units, with resultant pockets of minority groups to resort to peaceful and vociferous agitations for minor adjustment of intra-national boundaries, administrative hindrances, obstructions and inconveniences on borders with little consideration of economic and industrial economy of the country. In sum, the ancient physical base of boundary delimitation and demarcation has yielded place to the recent cultural basis.

In the end,it may be concluded that the observations made above verify substantially the hypothesis posed by us in the beginning, *namely*, the slow and imperceptible evolution of boundary concept in India, with some of its astoundingly modern traits, subsisting in the ancient past. It has not been possible to do full justice to the chosen captivating and interesting theme under study in this sketch in bare outlines.

REFERENCES

1. S.V. Puntambekar, *Presidential Address in the Proceedings of Indian History Congress (1947),* 396.
2. B.G. Tamaskar: 'Some Geo-political Concepts of Kautilya' *Quarterly Review of Historical Studies,* Calcutta, Vol. V, No. 3, 142.
3. R.P. Kangle, : *Kautilya Arthsastra,* English Translation, (1972) 9.1.17-18.
4. D.C. Sircar, *Studies in Geography of Ancient and Medieval India,* (Delhi 1960), 1-15.
5. P.V. Kane, *History of Dharmasastra* Vol. I, Pune *(1930), 25.*
6.
7. *Op.cit.,* 79. H.N. Sinha's work entitled, *Sovereignty in Ancient India* calls for a probe from geographic perspective to note whether there is any notion about spatial co-ordinates components of sovereignty.
8. V.B. Karnika, *Chinese Invasion: Background and Sequel,* (Bombay, 1966), 95.
9. Baij Nath Puri, 'Administrative Data in the Baddhist Literature, I-III Century A.D.' Proceedings of Indian History Congress, (1947), 146.
10. S.V. Puntambekar, "Presidential Address," Indian History Congress, (1947), 422.
11. B.G. Tamaskar, "Geographical Knowledge in the Upanishads." *National Geographical Journal of India,* Vol. II-2. (1966), 111.
12. V.A. Smith, *The Early History of India* (Oxford, 1914), 250. The author refers to the Yuch-chi, losing their nomadic habits and "became a settled nation" about the year, 10 B.C.
13. B.G. Tamaskar, "Historical Geography of Malik Ambar's Territory", *Journal of Indian History,* Vol. LIII, Pt. III, 256; Kanchanmoy Mojamdar, "Nepal's Relations with Indian States," *Journal of Indian History,* XLIII, Serial No. 128. The author makes a very pointed reference to the barrier character of Nepal's boundary.
14. V.A. Smith, *The Early History of India* (Oxford, 1914), 76.
15. J.F. Fleet, *Corpus Incriptionum Indicarum,* Vol. III, No. 24, p. 110.
16. Shriram Sharma Acharya's Hindi translation of the Matsya Purana, (Bareilly, 1970). Part II, Ch. 98, Verse 19, 299. Kautilya's ideas of elements of sovereignty in the English translation of the original *Arthasastra in Sanskrit* by R.P. Kangle (1972) (6.1.1). *Cf Manusmriti* with Shri Kulluka Bhat's Sanskrit Commentary, (Bombay, 1959) Ch. IX, Verse 294.
17. B.G. Tamaskar, "Kautilya's Theory of Inter-State and International Geopolitics". Symposium on Political Geography. International Geographical Congress, Nagpur. (1968). 6-15.
18. J.W. House, *Frontiers and Boundaries. op. cit.* 19-21.
19. B.G. Tamaskar, *op. cit.* 11.
20. Balkrishna, *Shivaji the Great,* Part IV (Kolhapur, 140), 102.
21. Ramchandra Pant Amatya, *Agnapatra,* 36. Quoted by Bhaskar Waman Pandit: *Shivajichi Rajniti* (Political Policy of Shivaji in Marathi). (Dhule., 1941), 236.

22. R.P. Kangle, *Kautilya Arthasastra, 5 Manusmriti,* VIII. 250-1, *Manusmriti* (composed between 2nd century B.C. and 2nd century A.D.) with Shri Kulluka Bhat's Sanskrit commentary, (Bombay, 1959) Ch. VIII, Verses: 245-266: S.M.Ali, *The Geography of the Puranas,* (New Delhi 1965) mentions *Maryad Parvat* (limiting mountain) 110, and rivers "formed the boundaries of Janapadas," 170.
23. *Kautilya's Arthasastra op. cit.* 2.3.1. It is interesting to note that in the Matsya Purana there is a distinct reference to water forts among different categories' of forts in Ch. 95, Verse 7. *Cf. Manusmriti, op. cit.,* Ch. VII, Verse 70, has another reference to water for tand perhaps, sea-fighting in Verse 192 of the same chapter.
24. Landis D. Kristof, "The Nature of Frontiers and Boundaries" *Annals of the Association of American Geographers.* Vol. XLIX-3. (1959)
25. J.R.V. Prescott, *The geography of Frontiers and Boundaries* (London, 1965), 92.
26. *Manusmriti* VIII, 250-1.
27. *Smricandrika,* II, 228.
28. R.P. Kangle, *The Kautliya's Arthasastra,* 2.21.1
29. B.G. Tamaskar, *The International Boundary on the Terekhol River* (Unpublished)
30. G.G. Tamaskar, 'Mark Ambar's Financial Adminsitration', *Quarterly Journal of Mythic Society,* Bangalore, vols. XLIV-2 & 3, and XLVI-2, 1954 and 1955.
31. R.P. Kangle, *Kautilya's Arthasastra,* 2.20, 2-25.
32. "Shiva-Charitra-Sahitya". (vol. X) in the Quarterly of the *Bharat Itihas Samshodhak Mandal,* Pune, Vol. XXXX-3 & 4.
33. *Shiva-Kalin-Patra-Sar-Sangrah.* (Shri Shiva - Charitra, Krayalaya, Pune. 1930) Article No. 1301.
34. Ibid., 336.
35. R.P. Kangle, *The Kautilya Arthsastra,* 2.1.3.
36. *Manusmriti* with Kullaka Bhat's Commentary (1959), Ch. VIII, Verses : 249-251.
37. M.G. Dixit, *Selected Inscriptions from Maharashtra* (Pune, 1974) 12: and 90: V. Srinivasan, "Srivilliputtur: History and Epigraphy". *The Quarterly Journal of Mythic Society,* (Bangalore,1953), Vol. XLIV -2, 66.
38. M.G. Dixit, *Op. Cit.* 90.
39. Radha Krishna Choudhary, "Judicial Importance of Representative Institutions in Ancient India" Proceedings of Indian History Congress (1947), 166.
40. K.S. Vaidyanatha, "Muttaraiyar-Nadu," *The Quarterly Journal of Mythic Society.* vol. XLIV -2 (1953).
41. Shridhar Shankar Patake, "Administrative Structure of Vijayanagar Kingdom" (In Marathi). *The Vijayanagar Souvenir volume.* Marathi Edition, Pune, (1936-37), 352.
42. R.P. Kangle, *The Kautilya's Arthasastra,* 2.31.1.
43. Ibid., 2.21, 22-23 and also 2.12. 29-30.
44. "Malik Ambar from Dutch Sources." Quarterly of Bharat Itihas Smashodhak Mandal, XI-I.
45. M.H. Apte: Kashi to Pune (Travel Account), (Rajwade Samshodhan Granthamala, vol. VII. (Dhule 1939). 41-54.
46. *Abstract of Letters from Robert Jefferies to the Surat Factors,* Oct. and Nov., 1621 A.D. *during his mission to Malik Ambar,* (Factory Records Miscellaneous, vol. I, 91).
47. Virendra Kumar, "East India Company's Negotiations with Goa (1844-47)"; Transactions of the Indian History Congress, 1941.
48. R.L. Malhotra, British Policy Towards Afghanistan's Frontier Demarcation, 1870-84." *Journal of Indian History,* vol. L, Serial No. 148, 61.

49. Jadunath Sarkar, *Studies in Mughal India* (Calcutta, 1919), 118-162.
50. G.G. Tamaskar, *the International Boundary on the Terekhol River,* (Unpublished). There is similarly recorded example of dispute in ancient India along a river in northern India, arising out of utilizing a major share of its waters for irrigational purposes by inhabitants on one bank, depriving their counterpart on the other bank, along the river Rohinia which formed the boundary between the territories of the Sakya and Kowyas.... There was a quarrel between the Sakya and Koliya cultivators who lived on opposite sides of the river with regard to the right of its water for use in irrigation *(Jataka* Ed. by Cowell, V., p. 219; B.C. Low, *Ksatriya Tribes of Ancient India,* pp. 208-10.)
51. K.P. Kangle, *Kautilya Arthasastra,* 3.9.11-25
52. *Manusmriti* with Shri Kulluka Bhat's commentary. (Bombay, 1959) Ch. VII, Verse 245.
53. P.V. Kane, *History of Dharmasastra* Vol. I, (Pune 1930) 178.
54. B.G. Tamaskar, *Malik Ambar's Life and Work* (under print), Chapter on Judicial System.
55. *Futuhat-i-Adilshahi.* (Shiva-Charitra-Sahitya, Part - III) Quarterly of Bharat Itihas Samshodhak Mandal, Pune, vol. XXIV-4.
56. *Shiva-Charitra-Sahitya,* Part - I, Article 15.
57. *Shiva-Charitra-Sahitya,* Part - IX, Article 18, Quarterly of Bharat Itihas Samshodhak Mandal, Pune, vol. XXIV-4.
58. R.P. Kangle, *Kautilya's Arthasastra, 3.9.11.*
59. *Manusmriti* with Shri Kulluka Bhat's commentary, (Bombay, 1959), Ch. VIII, Verses 253-260.
60. B.G. Tamaskar, *Malik Ambar's Life* (Under Print) Chapter on Judicial Administration.
61. *Smriti of Brhaspati* (Sacred Books of East Series, Vol. XXXIII, Verse 14, p. 352), P.V. Kane, *History of Dharmashstra,* Vol. III, Ch. XXII, pp. 502-510 incorporates elaborate procedural details for the settlement of village boundary disputes.
62. Balkrishna, *Shivaji, the Great,* Part IV, (Kolhapur, 1940), 49.
63. B.S. Shastry, "The System of Cartz,"*Journal of Indian History,* Vol. LI, Part -II, Sr. No. 152.
64. Balkrishna, *Shivaji, the Great,* Part IV, (Kolhapur, 1940), 102
65. D.C. Sircar, *Studies in the Geography of Ancient and Medieaval India,* (Delhi. 1960), Ch. IV, 68-82
66. *Op. Cit.*

VI India Reorganized: A Geographical Evaluation

C.D. DESHPANDE

With the reorganization of the States, India for the first time in her history wears a rationalized and streamlined appearance. In place of the 26 States of unequal status and size, now there are only 14 States and 6 Union territories. The 6 Union territories consist of small and detached portions of Delhi, Himachal Pradesh, Manipur, Tripura, Andaman and Nicorbar Islands and Laccadive Group ; though geographically separated from each other,they have together an area of 27,740 sq. miles and a population of 5,122,000.Thus the centrally administered areas represent only a small portion of the territory and population of the Indian Union; and the significance of the States reorganization can be judged from the fact that nearly 80% of the territory and 98% of the population are now being governed under 14 major administrative units.

To commemorate this historic event, several newspapers and journals issued special supplements full of valuable information and maps, but the attempt to evaluate the component states in terms of their size and population, and 'rank' them as first or second, needs to be deprecated. It gives a wrong perspective of the whole arrangement and sets a premium on false values. The best way to judge the new arrangement would be to recollect the words of the Prime Minister on the eve of the appointment of the S.R. Commission, and see how far the new administrative structure would be capable of promoting "the welfare of the people of each component unit as well as the nation as a whole".

From the SRC, Report to the S.R. Act

The Commission was appointed in December 1953. It submitted its report in October 1955. The draft of the Bill was placed before both the Houses of Parliament on March 16,1956. The report of the joint Select Committee was

Reprinted from *Indian Geographer*, 1957, Vol I & II, pp. 164-169.

presented to the *Lok Sabha* on July 16, and it was given a third reading on August 10. It was finally passed by the *Rajya Sabha* on September 11. The new States came into being on November 1. During the difficult days from the publication of the Report to the final passing of the Bill, regional agitations and public passion influenced the sponsors of the Bill and the Bill was modified to some extent. The prevailing trend, however, was in favour of larger states and determination to 'see it through'. Inspite of the many obstacles, it does credit to the Govt. that the new arrangement came into being only a month later than the date previously fixed, and it speaks well of the increasing democratic spirit and discipline of the country as a whole that the change affecting an area as large as Western Europe, was taken by and large, calmly and dispassionately.

The New Administrative Framework in its Geographical Setting

The basic problem of the States organization stems out of the geographical fact that the country has a large size and a varied pattern of regional culture. So far the constitutional and administrative structure of India was determined by the nature of political conquests and the administrative ability of the conquerors. With the gaining of self-rule it is but natural that people should desire a new orientation of the administrative structure and the Government should respond to the wishes of the people.

The framework as recommended by the S.R. Commission was a nicely balanced structure arising out of various considerations — geographical, economic, social and administrative. The legislators had to contend with public opinion and agitation, and certain bold measures had to be taken. Examining the framework, as it has emerged from the final anvil, against the geographical background of the country, certain major gains are obvious. There are now only 14 States as against six hundred and odd units 12 years ago; the new States enjoy equality in constitutional status. From the point of view of their area and population, there is, barring Kerala and Assam, not that disparity among them, that was so characteristic of the India of the British Raj and Indian Princes. They are geographically compact. Excepting the territories under Portuguese rule there are no 'pockets' (barring Pondicherry). They are culturally more homogeneous, and even in the Bilingual Punjab and Bombay, the two language groups have been brought under one administrative unit, with no major 'irredentist' element outside. Lastly, all of them are, to use the administrator's jargon, 'viable' units.

Landscapes at Cross Roads

It is well known, however, that the solution has not pleased one and all, nor did any one expect to discover a solution that would please all. At least there are two major issues over which there has been a good deal of friction: the border areas and the bilingual States. The border friction arose from the fact that boundaries of the new States were largely demarcated on district bases, and that, in any case, administratively and practically it would have been difficult to draw a line of micro-precision that would please all the people and would satisfy the administrative and economic needs of the locality. Thus, to take one concrete example, the Belgaum City and some northern parts of the Belgaum district agitated for the inclusion in the Maharashtra, while their economic ties and regional associations more strongly leaned towards the rest of the Belgaum district which is Kannada. Here is a good example of a landscape, due to its position and space relations, standing at the cross roads. Cultural affinity makes Belgaum seek its inspiration from the north, while its economic ties are closely linked up with the Kannada south and east ! The friction between Madras and Kerala over parts of the Western Ghats backbone: Andhra's all round displeasure that not all the Telugu-speaking areas were included in Andhra, especially the Bellary district with its Tungabhadra irrigation; between Bengal and Bihar over a rich mineral area in the south and a 'corrider' in the north: these are some of the more important instances where regions have developed ambivalent attitude. Janus like, they look to the past as well as to the future. They were subjected to 'pulls' from opposite directions: one from the historic past which tries to hold them back within their old boundaries and the other from the future which holds out brighter hopes in a new framework ! What else is the explanation of the dispute between Bihar and Bengal, which at one time almost brought forth a merger of the two States under the Roy-Sinha plan as the only sensible solution.

The bilingual States of Bombay and the Punjab offer another set of examples where landscapes are at cross roads, and yet they are held together by superior economic or political forces. The case of the Bombay States illustrates this point with clarity. Realizing the economic status and strength of the Bombay Metropolis, the S.R. Commission recommended a bilingual formula where the Gujarat and Marathi elements would just balance each other, with the Greater Bombay as rallying point for all for administration and economic development. Subsequently, the three-state formula was proposed, but it was soon abandoned in favour of the larger bilingual formula. It would not be too much to say that in spite of linguistic ideologies and forces, the economic strength of Bombay that proved too powerful a factor to ignore! The Punjab was wiser by Bombay's experience and accepted the bilingual solution as the only practicable one in the prevailing widely diffused distribution of the Sikh and Hindu elements in that area. The desire to avoid another harrowing experience of partitioning the Punjab was the prevailing force, though keeping out the Himachal Pradesh made matters more easy.

Regionalism in the New Set-up

That geography still influences a great deal the new set-up can be seen in the problems that have emerged with the creation of these States that are at once larger in area and population, and predominantly unilingual. On the one hand there is the problem of the State Government's control over a larger area with varying geographical conditions; on the other hand there is also the problem of the effective control of the Union Government over the State. To take an extreme example, the Bombay State must look after regions far and wide from dry Palanpur in the north to the monsoonal Ratnagiri in the south, from the semi-arid environs of Okha to the Cotton tracts of Nagpur and Chanda! The State accordingly must revert to the old administrative link between the District, and the Provincial Secretariat : Regional commissionerships, of which there are six approximately coinciding with the geographical subregions of the Bombay State. Similarly, Zonal councils have been interposed between the Centre and the new States to ensure national unity and inter-state co-operation. Although at present they are largely advisory bodies, their composition is entirely official and much will depend on the political experience in the various states in the near future. The zonal Councils would be undoubtedly strengthened if the State show a tendency to 'run away' from the Centre. Thus the new administrative alignment of the Indian Union has now a four tier arrangement : the Centre, the zonal councils, the State and the Regional Commissionerships. On these is based the administrative responsibility of the nation.

How persistent is the influence of the geographical factors, past and present, in the making of the new States can be seen through two leading aspects: the eccentrically placed capital cities and the particular orientation each State will have to give to its social and economic policies. Taking the States capitals first, it is interesting to see that many of them are now situated at one extremity of the State or another ! Bangalore is far in the south ; Madras at the northern extremity; Trivandram in the extreme south; Calcutta due to partition, in the extreme east. Nagpur, an ex-capital, is in a corner, and even Bombay would have suffered the same fate under any other arrangement. Thanks to the foresight of the Punjab leaders, Chandigarh becomes central. And similarly Bhopal resolves the rivalry between Indore and Gwalior. Most of these capital cities are situated in a corner (and are likely to be administratively inconvenient) is not due to any chance or accident; they are the vestiges of the past administrative set-up and economic centralization. Cities do not quickly respond to the fancies of administrators or the passing moods of people .They are slow and sustained growth, and they either impose their will on the new administrative structure or fade out into becoming small straggling towns. In the context of the situation of these capital cities, some administrative devolution of State Governments and migration of some of its offices to other towns may be expected.

Although the physical targets as envisaged in the Second Five Year Plan concern all the States and include development of agriculture, industry, transport and the like, it is to be noted that each State in its altered geographical setting will have to evolve particular policies to suit their areas; Thus, Madras must now depend on the industry and fisheries; Kerala will have good forests but that alone will not relieve the very heavy pressure of population: Mysore has a greater geographical variation, but it must be prepared for regional rivalries between the agriculturally poor but industrially developed south and the agriculturally rich and industrially backward north, between the heavily populated Mangalore area which sends a stream of migrants to Bombay City and the scarcity areas of Bellary anxiously awaiting the Tungabhadra irrigation! Andhra's strength lies in irrigation; Bombay not only faces the bilingual problem but a good deal of regional disparities which need to be eliminated; Madhya Pradesh will be concerned not only with administrative integration between the two former states, but will need giving a high priority to its backward areas in the east and south-east; West-Bengal comes out stronger in its new form, while other states of Assam, Bihar and U.P. and Jammu and Kashmir have not been materially affected by the reorganization ; with the merging of the Pepsu the Punjab promises to have a sounder administrative structure for developing its irrigation and industrial potential. Rajasthan with the addition of Ajmer becomes a homogeneous unit and seeks a psychological satisfaction in regaining Abu.

Interdependence and National Unity

It would be a fair judgment to say that the new States are on the whole more stable and are better fitted to carry out their responsibilities in building up a new nation. It is equally important to realize that the creation of these larger States implies a greater interdependence between them. The Punjab and Rajasthan are tied up with the irrigation waters of their rivers; Bihar and Bengal with their industrial belt; Orissa and Madhya Pradesh on the need to develop their economically backward borderland; Mysore, Andhra and Madras have common interests in irrigation projects ; between Madras and Kerala, and between Bombay and Mysore, irrigation and forest products. In the past, railways and roads were the forces which contributed to the unity of India. River valley projects, irrigation schemes and electric grids are their more modern counterparts. They promise to forge a more living unity in India through regional developments that pay no regard to linguistic or administrative borders. Economic progress, accordingly, is a unifying factor in the making of new India, but it hardly needs an emphasis to say that such forces must be adequately supported by, again in the words of the Prime Minister, the "psychological integration" of the people of India. Until a definite policy on the national language is implemented, linguism would remain a potential danger to the nation's unit. Until the political leadership in the most of the States rises to a full stature, there is the risk of the States degenerating into stagnant backwaters of petty strife and jealousies.

Section II | Pluralist Bases of Indian Politics and Nation Building

Politics is the product of a society, its economy and their geography. The combined interaction of the three creates peculiarities of politics of a place, region or nation. The peculiarities of Indian politics, which differentiates it from politics of other parts of the world, originate in its pluralist society based on religion, castes and tribes; economy based on the varied resources with which the country is endowed and which are utilized under the policy of mixed economy of the state; and their mosaic varying from place to place and region to region within the country. The two types of politics in India which have been dominant recently—politics of caste and politics of religion—are the product of social geography of the country as evolved over time. Agitations related with Mandal Commission Report, Ayodhya, and demands for greater autonomy or outright secession from the Indian Union have brought the centrifugal forces in sharp focus. These events have created serious misgivings regarding the unity of India and nation building in it.

The policy of mixed economy has produced mixed results. Economic development fettered with a systematic regime of controls and permits over private enterprise and the placement of the public (state) sector at commanding heights of the economy has allowed economic growth in India better than several Third World countries, but way behind the rich and powerful countries of the North and some of the countries of the Pacific rim in Asia. With the collapse of communism India finds itself in a better position with regard to her experience of limited privatization as compared with the former communist countries. Yet with her very low share in global trade she stands marginalized in the global economy, largely because of her internal economic policies of the past several decades. Internally, the regional differentiation of economy has also produced mixed results. Whereas a new entrepreneurial class has emerged in the urban and industrial centres, the number of those below the poverty line has increased further. Regional pattern of industrialization and development has been such that regional imbalances have accentuated.

Politics in such a fragmented society and mixed economy, varying from one part of the country to another, has its regional variations and peculiarities of its own. Added to these is the federal structure of government which sets the parameters within which politicians and bureaucrats operate, and determines whether Indian politics will remain centrists or move to the left or the right.

VII Tribal Situation in India

NIHARRANJAN RAY

Ever since the Chinese aggression in 1962, the tribal disturbances along our north-eastern frontiers and in Bastar in Madhya Pradesh, and rumblings of growing discontent and unrest in a number of other tribal areas of this vast country, Indian intellectuals in general, and anthropologists and sociologists, administrators and leaders of public opinion and makers of policy in particular, have been alerted to the vital significance of the extensive northern and north-eastern tribal border regions as much as of the 'tribal' peoples of other areas of the country. The cause of the security of the country as much as that of the unity and integration of these peoples with those of other regions and communities of the larger Indian society, it was increasingly being felt, called for a closer look at the total 'tribal' situation in the country in general and in the frontier areas in particular. It is now generally recognized that academic expertise on the human, social, economic and other factors in the life of these communities of peoples is not yet adequate, and that attempts should be made towards a systematic stock-talking of our existing state of knowledge so that significant problem- areas could be defined for further study and research, and certain basic policy guidelines could be evolved with regard to the problem of integration of these relatively isolated areas and communities and the specific problems of their all-round development, in other words, in regard to the problem of drawing them into the mainstream of Indian life and society. Here then is the justification, if justification was called for, for our convening this seminar.

With this end in view we have been privileged to bring together here outstanding academic experts in the fields of anthropology and sociology, economics and political science, history, human geography and linguistics, fields of knowledge and expertise that are necessary for the understanding of the very complex 'tribal' situation in our country. We have also been privileged to have in our midst a few leaders of our 'tribal' people and social

Delivered as Introductory Address at the seminar "The Tribal Situation in India" held at the Indian Institute of Advansed Study, Shimla.

Reprinted from K. Suresh Singh (Ed.), *The Tribal Situation in India,* Shimla: Indian Institute of Advanced Study, in collaboration with Motilal Banarasidsas, Delhi, pp. 3-24 with the permission of the Director, IIAS.

workers who have dedicated themselves to the work among the 'tribes', besides a few practicing politicians, administrators and publicists who are deeply and actively concerned with this particular problem. We are avowedly seeking a multi-disciplinary approach to the study of our 'tribal' situation.

On behalf of the Governing Body of this Institute and as initiator and organizer of this seminar, I have great pleasure in welcoming you all in our midst. We appreciate your generosity in responding to our invitation, and we are intently looking forward to a purposeful dialogue on a subject that is not only important academically but a very significant one from the national point of view. The Centre of Advanced Study in Sociology of the University of Delhi led by Professor M.N. Srinivas, is collaborating in part with this Institute in the organization of this seminar. I take it as an index of his and their intellectual involvement in the 'tribal' situation in the country, and we feel very grateful and indebted to Professor Srinivas for inviting this collaboration.

Since I began with an attempt at justification of the seminar I should, I believe, also try to say a few words in regard to what, in our view, the seminar should aim at. Frankly, the seminar, to my mind should aim at

(a) Exchange of detailed information on the basis of first-hand studies on problems of the 'tribes', region by region, with a view to locating the major gaps in our information and for cross-fertilization of ideas through discussion;
(b) Detailed suggestion of problems which need immediate attention, both from the purely academic as well as the national points of view. This should also include suggestion of approach and methodology; and
(c) Evolving certain policy guidelines regarding the general problems of integration of these relatively isolated areas and communities and on specific problems of their economic and cultural development, their health, education and administration, for instance. In other words, as I said a while ago, we should be able to formulate certain policy guidelines as to how these areas and communities could be drawn into the mainstream of Indian life and society.

II

What is this 'tribal' situation that we are speaking of? Who are these communities of people called 'tribes' that we are referring to? How did this situation come about? How long has this situation been with us? Is it one which is very modern in the sense that it has been brought about to relatively recent decades by social, economic and political factors lying outside of what we call the 'tribal' communities and emerging from the cumulative effects of forces at play in and the vicissitudes of history of the mainstream of our contemporary life of say, the last one hundred and fifty years?

We cannot, I am afraid, hope to find even inadequate answers to these questions without reference to history, that is, to the story of the near and distant past of our land. Indeed, Indian sociology and social and cultural anthropology cannot be pursued meaningfully, to my mind, without recourse to history. Non-recognition or inadequate recognition of the historical perspective has been, I believe, more often than not, responsible for the uncritical acceptance of ideas and concepts, methods and models that had a colonial origin in climes altogether different from our own.

Let me try to illustrate what I mean by referring briefly to one term, that is, one concept alone at the moment, that of what is known as *caste* by sociologists and anthropologists working on and concerned with India. *Caste* in the Indian context is, philologically speaking, a term of Portuguese orign, and its use does not seem to pre-date the seventeenth century, 1613 to be more precise, though there is a doubtful reference as early as 1563. Derived from the Latin *castus* meaning pure, unpolluted (*comp. chaste*), the Spanish *casta* or Portuguese *cast* came to mean just the 'breed' or 'lineage' that seems to have been socially considered as pure, unpolluted. In South America curiously enough, the Spaniards used the terms *casta* to mean mixed breeds of the Europeans, American Indians and Negroes. Historically, therefore, it is permissible to argue that the seventeenth and eighteenth century European writers on India saw in this land of ours a kind of social organization which must have been somewhat akin to what they presumably knew as *cast* (the modern spelling *caste* does not pre-date 1800), may be in their own society, or in that or those of their new lands which they had come to know by that time. We do not know what this society or societies were like, and hence we do not have any idea of the connotation of the term *caste*, either theoretical or functional, from the Portuguese point of view. The most that may perhaps be argued is that this social organisation known to them as *cast*, was based on a hierarchical ranking of people determined by birth or lineage. The Dutch, the French, and the British, who followed the Portuguese, went no further than to accept and adopt an already accepted term and the concept that went along with it, without asking themselves if these could be applied to the Indian situation. Till as late as 1766 one finds the terms *cast* being used synonymously with 'tribe' and as late as 1813 it meant to the English writers on India nothing more than a kind of ranking based on birth.

But then came the tribe of European, mainly British, 'Orientalists' versed in Indian lore or languages, especially Sanskrit, followed many decades later by that of ethnologists who were primarily administrators or missionaries trained in the school of ethnography current at that time. For a long time the 'Orientalists' lay imprisoned within the walls of an idealistic, priestly and essentially schematic interpretation of the traditional Hindu-Brahmanical social organization as they found it recorded in a set of hieratic and sacerdotal texts spread over a period of more than a millennium and a half. The ethnologists who were supposed to know better, got enmeshed, in their turn, in the cobwebs of textual interpretation on the one hand and the function or behavioural pattern of what they had come to know as *caste* as they saw it in operation before them. For a long time they tried to explain the latter in terms of the traditional concept of *varna* as recorded in the early and latter Vedic texts, but mainly in the *Dharmasastras* and *Dharmasutras*. At a later stage, they were confronted with another term, *jati*, but neither *varna* nor *jati*, two terms and hence two concepts which were by no means synonymous or interchangeable, could explain in terms of the text alone the very complex social phenomenon which lay actively open before their eyes. Here was an organization of society that had a strongly interwoven religious undertone, more or less well-defined society relationships and at the same time, also equally well-defined economic functions in a well-thought out production system. Obviously neither *caste* as understood in the nineteenth century ethnographic climate, nor *varna* as interpreted in the hieratic and sacerdotal texts, could explain such a complex system, increasingly becoming more and more complex and proliferating itself into a countless number of ramifications, and this under the pressure of an ever-expanding and ever-assmilating society and religion and of an ever proliferating production system.

If *caste* and *varna* could not explain the system, could *jati* do so? Perhaps it could, if we followed the logic inherent in the relevant chapters and *slokas* of the great epic, the *Mahabharata*, the Puranas, especially the later ones, in secular Sanskrit, Prakrit, Apabhramsa and regional literatures and in the events and situations in history. Etymologically, the term *jati* has a biological meaning; it has something to do with birth and biological heredity and hence with familial and social relationships. But behaviourally, as seen through history, it has also a social and economic purpose and function, regulated by birth and heredity. Here was thus a system which was directed not towards mobility but towards stability and security, and who would deny that it served its purpose efficiently and well until an altogether new production system started to strike at its very roots now for the last well-nigh two hundred years.

But be that as it may, what I have been trying to hint at is this: every social organization tends to create and envolve terms and concepts in its own language to explain and interpret itself in given times and situations, to suit its own aims and purposes, functions and exercises, ideas and aspirations. A time was thus reached in history, that is, in the evolution of Indian society, and this happened already before the beginning of the Christian era, when *varna* which seems to have been used and understood, generally speaking, in an ethnic sense, did not any longer suffice to explain the society to which the people belonged; it was simply inadequate and misleading. That it was inadequate, if not misleading, became evident atleast from about the beginning of the Christian era, when our *Smriti* writers had to explain the system by the invention of the concept of *misra* or *sankara varnas*. With the passage of time, through a process of permutation and combination, the number became simply unmanageable and the theory all but broke down, it seems. There was just no rational basis left of the system. Moreover, the way of life which was sought to be explained was not just *varnadharma* but *varnasramadharma* which connoted that the system of *varna* was not meant to stand by itself but on *varna* or *asrama*; in other words the four traditional *varnas* were to be understood in terms and in the context of the four traditional *asramas*. Hence was the term *jati* evolved. It did not replace *varna*; that indeed remained, but its concept underwent a sea-change and came to connote *jati* for all practical purposes. I would therefore make a plea that *jati* is a better term of concept than either *varna* or *caste*, and it can explain better the Indian situation. Not that I have any objection to the use of a foreign term. No, I do not believe in *svadesi* in matters intellectual. But the term is inadequate, if not misleading, in the Indian context. It has never been well-defined; its connotation never made clear and its conceptual frame was left in a haze of dust. Besides, looking at the countless numbers of academic and intellectual descriptions, analyses and interpretations of our social organization in terms of *caste*, I feel confused, more often than not, if what we today call the *caste* system in India is not a creation of the anthropologists of the late nineteenth and early twentieth centuries. May be it is my ignorance, but as a student of Indian history and culture I cannot but confess my bewilderness.

If I have dwelt on the term *caste* at some length I have done it on purpose. I could as well have done the same thing with other terms and concepts used by sociologists and anthropologists working on India. I could have asked myself: is 'family' the same as *parivara*, 'clan' the same as *kutumba* or *gotra* or *gosthi*? But I must not do so, especially in a gathering of experts. Yet I had to spend some time on *caste* for more reasons than one. First in eighteenth century writings on India the term *caste* has often been used synonymously with 'tribe' and later for a long time, not synonymously but in a cognate manner as in the phrase *castes* and *tribes*, as if they were cognate social groups. Secondly, with the rise and growth of nationalism in Europe, the term 'tribe' came to be used in denotation of a particular stage of socio-political evolution of a community of people within a given territory and language area. Clan, tribe, nation, etc. thus came to denote the various successive stages in the progressive march of a people aspiring towards nationhood. In the Indian context, *caste* was another intermediate stage thrown into this imagined story of progression, a stage that was presented to us as one of the greatest hurdles in the march towards Indian nationhood. It was in this context that the study of Indian anthropology and sociology began. Risley's *People of India* and his speculative comments on such terms as 'race', 'clan', 'caste', tribe", 'nation', 'people' etc. would illustrate what I mean. The situation, to my mind, has not changed very much since. Risley, who, we may not forget, was an important official high up in the British Indian bureaucracy and who was one of the first to plead for the first partitioning of Bengal in his days. The term 'tribe' in Indian anthropology and sociology seems to carry this background music even today, somewhat unconsciously perhaps. Phrases like 'criminal tribes', 'scheduled castes', 'scheduled tribes' are but echoes of this music.

But let me turn to the history of the term and concept 'tribe'. Derived from a Latin root, the Middle English term *tribuz* meaning the three divisions into which the early Romans were grouped, came to evolve into the modern English 'tribe'. With the Romans, the 'tribe' was a political division while the Greeks seem to have equated it somewhat with their 'fraternities' at times, with geographical divisions at others. In Irish history however the term meant families or communities of persons having the same surname. In certain other areas of the western world and certain periods of history, it stood for a division of territory allotted to a family or community. Today with the anthropologists and sociologists of western origin the term means, according to latest edition of the *Oxford Dictionary*, "a race of people; now applied especially to a primary aggregate of people in a primitive or barbarous condition, under a headman or chief". It is in this meaning, roughly speaking, that most of the western scholars working on India, have been using this term, with but slight change of emphasis here and there. I would humbly enquire: *does this definition meet the relevant social situation in India*?

I have also seen some of the definitions of the term given by our own experts including the one given by our esteemed friend who is giving the Inaugural Address this evening. I do not propose to go into an analysis of these definitions, but would like to approach the question from altogether different point of view. I would suggest to myself that if along the arrow-line of history there was something in our society analogous to what is called 'tribe', and if the communities of people in India recognized it as such, we should have in our Indian languages including Sanskrit and Prakrit, a term for it. Was this term known as *jana?* The term *jana* and *jati* are both derived from the root *jan* which means 'to be born', 'to give birth to', and hence has a biological meaning, we know for certain. We know also that there were *janas* or communities of people like the Savaras, the Kullutas, the Kollas, the Bhillas, the Khasas, the Kinnaras and a countless number of many others whom today we know as 'tribes', bearing almost the same recognizable names. By what term and concept were they known to the multitudes of people of this land of Bharatavarsha ? Not by what anthropologists and sociologists of today call 'tribe' certainly according to the definition I have quoted. What then was the term and concept by which these communities of people were known, communities whose list is indeed a very long one if we are to go by the evidence of the epics and *Puranas* and our secular literature? This term, to my mind, could be no other than *jana,* meaning 'communities of people'.

It is significant that in this huge body of literature whenever such and other communities of people find mention, they are always referred to in the plural number, collectively as a people or *janah, Andhra, Savarah, Kinnarah,* etc. An analysis of such names and the contexts in which they are referred to, shows very clearly and unmistakably that all such communities had each a territorial habitat of their own, which can still be identified. With the passage of time each of these communities lent their names to the territory inhabited by them, and the territories came to be known as *janapadas.* Most of the local names of districts, divisions and States that we know of today, have come down to us from the *janapadas* of old; but quite a number of them have also lost their name and identity in their larger and more powerful neighbors. *Jana* therefore, seems to have been, to my mind, the term for that we have been taught to know as 'tribe', and *jati,* the socio-religious *cum* economic organization that was supposed to sustain the *jana* and keep the given community of people together.

A careful analysis of the long list of *janas* in epic Buddhist, *puranic,* and secular literature of early and medieval times and the context in which they are mentioned, makes it very clear that hardly any distinction was made, until very late in history, between what we know today as 'tribes' and such communities of people who were known as the Gandharas and Kambojas, Kasis and Kosalas, Angas and Magadhas, Kurus and Panchalas, for instance. At any rate, in the whole body of historical data at our disposal; there is hardly anything to suggest that these communities of people belonged to two different social and ethnic categories altogether. In fact, in the literary sources I have referred to, the communities of people whom today we refer to as 'tribes' and those that we know from history as belonging to more advanced stages of socio-economic and cultural growth there is hardly any evidence to show that in the collective mind of India's communities of people there was any consciousness of a difference between the two sets of *janas* except in the matter of *jati,* that is, in the matter of socio-religious and economic organization alone. These *janas* whom we have been taught to call 'tribes', were indeed different from the other communities of people only in the sense that they continued to remain, for reasons which I shall soon come to, outside the control of the *jati* system of social organization.

The story of how and why they succeeded in doing so through the centuries belong to the domain of the history and sociology of Indian culture to which I now propose to turn briefly for a while.

III

The so-called 'tribals' of India, it is well-known, are the indigenous, autocthonous people of the land, in the sense that they had been long settled in different parts of the country before the Aryan-speaking peoples penetrated India to settle down first in the Kabul and Indus Valleys and then within a millennium and a half, to spread out over large parts of the country along the plains and river valleys. At a much later stage of history there came to this land civilized foreigners in small numbers like the Achaemenians, the Hellenistic Greeks, the Parthians and the Romans, nomadic-pastoral tribes in very large numbers and in wave after wave from the steppes of Central Asia and this over a period of more than a millennium and a half, beginning with the Sakas and closing with the Turks and Mongols, and finally, again in several waves, rural and semi-agricultural Tibeto-Burmans from across the north and north-eastern borders of the country. In between and later, there were small influxes from time to time of the Arabs, Iranians and Abysinians from East Africa and Tibeto- Burmans from Burma and Arakan regions.

By the time the Aryan-speaking peoples were settling down in the Indus and upper Ganga-Yamuna valleys, they were fast becoming rural and agricultural and hence a settled people, with a highly developed language capable of articulating very subtle and abstract ideas and thoughts. Secondly, they had by that time mastered the technology of agriculture, acquired the experience of producing a large variety of cereals, fruits and vegetables, and of animal husbandry; they had also evolved as many as eighteen identifiable crafts. Thirdly, they had also evolved by that time a relatively developed system of local administration and certain ideas and institutions of government like kingship, *sabha, samiti* and *parishad.* A few centuries later they were able to give them all a highly organized and sophisticated form. Fourthly, already by about the sixth and fifth centuries B.C. they had evolved, and in another five hundred years they had consolidated well throught-out and organized social and religious system that one finds recorded in *smriti* literature, the epics and the *Puranas* as what is generally called Brahmanical Hinduism. Fifthly, by about the beginning of the Christian era they were able to build into this socio-religious system of Brahmanical Hinduism, an economic structure based essentially on a production system managed and controlled on hereditary principles of group formation which has come to be known as the *caste* system.

There is linguistic and archaeological evidence to suggest that the pre-Aryan indigenous *janas* were settled originally on the plains and river-valleys of the land. But they were not all in same stage of development. Many of these were still in a food-gathering economy, not knowing, the use of metal; a few seem to have known the use of metal, a sort of shifting, *jhum* (slash and burn cultivation) and hoe cultivation, and were perhaps on the threshhold of a real food-producing economy, for instance. They seem to have lived in isolated settlements, in shelters at various levels of constructional activity, some even in natural or dug-out shelters. They spoke a variety of languages and belonged ethnically to a variety of physical types. Their religion consisted in the belief in the practices of what is called by anthropologists 'primitive' religion, and they seem to have lived in closely well-knit social units each presided over by a headman or chief and controlled by a group of elders.

What the pressure of a superior social organization with a superior techno-economy as that of the Indo-Aryan speaking peoples, will be on such under-developed peoples as I have just briefly described can well be imagined. In a word, these indigenous so-called 'tribes' were just slowly but surely obliged to move bit by bit, to farther and farther areas until they came to find their refuge in relatively more inaccessible regions of forests and hills and large mountain slopes, that is, in what the records call, *atavika rajyas, mahakantars* or great forest regions and *pratyanta desas* or frontier regions, on the fringes of agriculturally settled, organized and more developed areas. The process went on for centuries and millennia, in a very slow and steady but very relentless manner until very recent times since when state legislation has been trying to put a brake ot it because of rising consciousness amongst these peoples themselves.

But let us have a look at the rationale of the process, that is, of the logic of sociology behind the Indo-Aryan, Hindu- Brahmanical thrust forward and the slow retreat of the *janas* or *'tribes'* of under-developed economy. First, by about the beginning of the first millennium before the Christian era or shortly thereafter the Indo-Aryan speaking peoples seem to have come to the knowledge and experience of an advanced process of smelting and forging iron and making weapons and agricultural implements out of iron. This iron technology enabled them not only to enlarge their political control and establish kingdoms and empires but also extend their agriculture and build towns and cities. Secondly, agriculture and animal husbandry being the principal source of their economy, their hunger was increasingly for more and more land for pasture and cultivation, the more with irrigation facilities the better. The relentless nature and character of this land hunger can be seen clearly from an analysis of the land grants of ancient and medieval India. These documents also make it clear that there was an insistent and ever-increasing demand to bring more and more land under cultivation, especially in the unsettled, hitherto relatively non-agricultural areas, pushing the pre-food producing and low level food-producing communities farther and farther. The process was still at work as late as the late nineteenth century, in the Santhal Parganas, for instance, where taking advantage of settlement operations and other opportunities the Santhals were almost overnight dispossessed of their community lands. Thirdly, the inexorable laws of Indo-Aryan techno-economics and polity enabled these peoples to spread out and extend the bounds of their social, political and economic organization all over the country, slowly and stage by stage, but surely and inevitably. The carrier of this organization along with its ideas and institutions was the Sanskrit and Sanskritic languages which also carried with it the Hindu-Brahmanical socio-religious norms, ideas, institutions, pantheons of gods and goddesses etc., including the socio-economic organization of *jati*. The *janas* like the Andhras, Cheras, Dravidas etc., of India south of the Vindhyas, were successful in resisting or at least containing the pressure of Sanskrit and Sanskritic languages, by and large, but they could resist neither the political ideas and institutions nor the Hindu-Brahmanical socio-religious system, nor the socio-economic organization of which *jati* was pivot.

The cumulative and ever-increasing pressure of these relatively superior technological, economic, political, social and cultural forces, was not easy to resist or contain or even keep away from. For more or equally superior peoples like the Achaemenians, the Greeks, the Romans and few others who found themselves in India at later periods of our history, the question did not perhaps arise at all. They came in very small numbers and even those who chose to stay on, seem to have found it convenient, for some reason or other, to merge themselves altogether in the vast ocean of Indian humanity. By far the largest number of foreigners that entered India, in wave after wave and over a period of about 1500 years, was that of the Central Asian nomadic and pastoral-nomadic peoples whom I have already referred to. But the tribal economy of these peoples in their original habitats, was such as not to be able to sustain its people, which, along with others, was one of the main reasons for their periodic migrations to other warmer and more fertile lands. Once they were in India for two or three generations, they tended to give up their nomadic-pastoral way of life and settled down either as pastoral people or as agriculturists, and once they did so they were slowly but inevitably swallowed up by the Hindu social and economic organization and given a place in it, to the extent that they came to adopt the languages, religions and cultures of the Indo-Aryan speaking peoples and of Brahmanical Hinduism. One has only to remember what happened to the Sakas, the Kushanas, the Abhiras, the Jnatrikas or Jats, the Gurjaras or Gujars, the Huns or Hunas and other allied peoples, many of whom are mentioned as *janas* in the list of people as given in the epics and the *Puranas,* in Buddhist and secular literature and in land grants and other historical documents. The Islamized Turks, Afghans, Iranians and Mongols were certainly able to maintain their religious and cultural identity, even to extend the frontiers of Muslim society within India, but let it be noted that they too had to succumb to the socio-economic organization of Brahmanical Hinduism, to the same production system as that of the *jati.* Indeed, once they had fallen into the production system of the *jati* it was no longer possible for them to resist its social implications. The same thing happened to the communities of people of the Tibeto-Burman stock who trudged into Assam and our north-eastern regions as late as thirteenth century. Those of the *janas* of foreign origin that came to exercise political authority as kings and as members of royalty, nobility and the court, were given the *jati* status of Kshatriyas; those that eventually took to agriculture came to be self-styled as Vaisyas; but the large majority had to be content with very low *jati* status in Hindu socio-economic hierarchy, including those like the Hunas who allowed themselves to be recruited as mercenary soldiers by regional rulers.

Yet the fact remains that large and small groups of the pastoral nomads I have referred to, did not settle down to agriculture; they retain to this day part of their pastoral-nomadic habits of life. They can still be found in considerable number in many places of Rajasthan and Gujarat and in smaller number in Kashmir, Himachal Pradesh, Haryana and even Maharashtra, for instance.

The process of such integration and assimilation of foreign peoples and their ways of life seems to have been facilitated by Hinduism building up within itself, through experience of centuries of foreign contacts, a subtle and silent mechanism that enabled it to absorb shocks and to incorporate, slowly and steadily, as many myths and legends, words, images and symbols, gods and goddesses, items of food, dress and drink, etc. as it could without antagonizing the priestly and scholastic elements of the society, which would always make a show of resistance but accept them willy nilly in the long run by putting on all these a Hindu garb. Knowledgeable people alone know how much foreign material has gone into our lives, even into our myths and legends, our religious ideas and rituals, into the iconography of our gods and goddesses. I would only present to you one example which may not be known to many of our archaeologist. Early Indian coinage of Saka-Kushana affiliation has made our archaeologists familiar with a non-Indian goddess whose name is given on the coins as NANA. This name in its Indian version has been transformed into Nayana, Nayni, Naina, and all along the slopes and foot of the central and western Himalayas one would find small slowly temples dedicated to this goddess Nayana, Nayni, Naina and place names like Nayna, Nayni or Naina, the hill station Nainital being the most well-known amongst them. But I am not here to cite examples and give references; all that I am seeking to do is to interpret historically and sociologically, certain well-known facts of the Indian situation.

To come back to my main theme. I have referred to the *janas* of foreign origin in relation to the Hindu-Brahmanical social organizations. But what happened to the *janas* of indigenous origin that did not belong to the Indo-Aryan speaking stock and the Hindu-Brahmanical social organization of which *jati* was pivot ? To be able to find an answer to this question one must look at these *janas* from the point of view of their geographical locations and their socio-economic relations with their dominant Indo-Aryan speaking Hindu-Brahmanical neighbours.

Briefly, those *janas* who were defeated in war and taken prisoner, were immediately made economically and socially subservient altogether, being reduced to slaves, labourers and servants. Eventually, they came to be incorporated in the Hindu social organization, but seem to have been given a place at the lowest bottom, not very different from that of the *Chandalas*.

Then there were those who lived on the periphery of the Indo-Aryan speaking Hindu-Brahmanical peoples, or were surrounded by the later. These peoples were able to maintain somewhat their socio-religious, economic and cultural identity. But because of their close locational proximity and the steady pressure of the larger and techno-economically more organized communities, they were obliged to enter into social and cultural and more importantly, economic contacts, communication and exchanges with their neighbours who were socially, culturally and economically more dominant. Through this process, initiated and controlled mainly by the latter, the non-*jati janas* were slowly but eventually obliged to give up their identity and succumb to the pressure. Inevitably they were led to a position where they had to enter the Indo-Aryan folds of Hindu-Brahmanical communities who in their turn and in the process, accepted and incorporated, especially at the rural-agricultural and folk level, not a few of the ideas, images, symbols, deities, rites, rituals, beliefs, customs, myths and legends of the erstwhile non-*jati janas*. This was no generosity on the part of the Hindu-Brahmanical order, as many culture-historians would like us to believe; and anthropologist or sociologist would tell such believers that here was just a simple sociological law in operation. However, the elements which I have just referred to are known in our tradition as *lokachara* and *desachara* as distinct from *sastrachara* or the priestly and textual commandments. These elements collectively go to constitute an important aspect of our rural-agricultural life, which is called *lakayatra.* The *janas* thus incorporated in Hindu-Brahmanical society Hutton has chosen to call 'tribal castes', and others have followed him. The number of such 'tribal castes' in our traditional lists of *janas*, is considerable, and it seems they were all given a very low grade in the Hindu social organization of *jati*, side by side with or even lower than those who were called *hina-jatis* and who were allowed the profession of *hina-silpas*, or crafts that were considered low in social estimation. Slowly and steadily, through more than two millennia of history, most of the so-called 'tribal castes' have been all but completely merged into the *jati* complex of Hindu society, and a smaller number, along with the lower order of the *jati* system, into the Muslim society as well. In each case they made their impact felt, especially in the Hindu-Brahmanical society to which they gave, apart from many other things, new forms and attributes to old traditional gods and goddesses, sometimes even new ones altogether. I have said that they were placed at very low levels of the vertical *jati* structure, which is a fact, by and large, but a few of their groups and individuals, by virtue of their gaining in economic status either through accumulation of land and more control over agricultural produce and/or through transactions in trade and commerce, were able to gain in social status as well. The process is still on before our eyes. But these no longer belong to the category of what in anthropological and sociological circles and in government papers, are called 'tribes', however under-developed may be their socio-economic position. In howsoever unsatisfactory a manner it may be, they are now in the mainstream of our contemporary life and history.

Our concern at this seminar is *not* with these communities of people, but with those *janas* who have not yet been drawn into this mainstream. Who are they ? These are the *janas* who have been for centuries living far away from the mainstream, in the relatively isolated and inaccessible and less fertile and less agriculturally productive regions of forests, hills and mountains. And since they were living far away, they did not have much of an occasion to feel the pressure, until recent times, of the ever-advancing and more powerful social, economic and political forces. These were the *janas* that today lie all along our eastern and north-eastern frontiers and along the central and western Himalayas upto a height of 13,00 feet, *janas* that are predominantly of the Tibeto-Burman stock, except perhaps the Nagas and one or two other less known *janas*. This entire Himalayan area from our eastern frontiers to Ladakh is naturally a very sensitive one, and for more than one reason, a very significant one as well from the political and economic points of view. Many of these *janas* do not, for obvious reasons, make their appearance in history or in our historical geography before the nineteenth century, but many others do, for instance, the Nagas who are mentioned by Ptolemy in the second century A.D., the Khasas, the Kiratas which seem to be almost a blanket name for all the Tibeto-Burman peoples in our eastern and north-eastern frontiers, the Kinnaras of the Kinnaur district of Himachal, the Kullutas of the Kulu valley, the Bhotas of the Bhutan-Sikkim area, and a few others whose names appear in the traditional *puranic* list of the *janas*. Our medieval Assamese, Bengali, Maithili and Hindi literature too, gives us a few names of such peoples.

Then there is the whole belt along the old Paryiyatra, Vindhya and Suktimat hills which are collectively known to us as the Vindhyan ranges, stretching from almost the borders of Rajasthan to what are called Chhotanagpur and Orissa hills which are but extensions of the Vindhyan ranges. All along these ranges and their slopes and feet and in the forests and valleys nursed by them live and have been living for centuries some of the oldest *janas* known to our history and culture : the Nishadas and the Savaras, the Kollas and Bhilolas, for instance, and many other cognate and semi-cognate *janas* of whom history and historical geography have not kept any record. But we know from one of our epics, the *Ramayana*, the land *par excellence* of the *janas*. Ethnically the majority of the *janas* seem to be of proto-Austroloid origin though a few of them speak languages of Dravidian affiliation.

Besides these two geographical areas of non-*jati janas* there is also a good number of them in smaller aggregates and dispersed in relatively smaller areas in almost all the southern States, especially in Andhra Pradesh and Kerala, and in Maharashtra. But these *janas* are all more or less in close contact and communication with socially and culturally, politically and economically more powerful communities belonging to the *jati* complex. And finally there are the isolated aggregates of relatively much less developed *janas* in the Andaman and Nicobaris lands.

In relative security, seclusion and isolation these communities of people have been living for centuries in varying degrees of agglomeration, and in varying levels of primary economy. They have come to acquire through centuries, the knowledge and experience of certain technologies, weaving, use of certain metais and certain crafts, and primary low-level agriculture. They have also been living in varying levels and degrees of contacts and communications with the communities of peoples of the *jati* social organization. But whatever the levels, degrees or stages, they were on the whole able to maintain their separate *jana* identity, each *jana* its own; each maintained its own socio-religious and cultural life, its political and economic organization, and this despite occasional disturbances created by their neighbour non-*jati janas* or instrusion by the more advanced and powerful *jati janas* from the plains, hills, plateaus and valleys near or far. But since such disturbances and intrusions were occasional natural calamities that just came and went, they could and did not seriously or permanently affect the usual, normal tenor of their social life and culture and the general pattern of their political and economic organization. History has left certain evidences of kings, princes, military and other adventures including merchants, making occasional penetration in the deep non-*jati jana* areas of Chhotanagpur and Orissa, of what was once Central India and of Madhya Pradesh. Buddhist and Hindu-Brahmanical missionaries and religious leaders also did so from time to time. It would be futile to argue that these contacts and communications meant nothing at all, but since, as I have said, these were not sustained, continuous and extensive and did not bring about nor aim at any basic social and economic change, these communities were able to maintain their own identity. Parenthetically, I may point out that Islam being fundamentally a city-oriented religion and Muslim royalty, nobility and the merchant class being urban in outlook, the Turkish, Afghan and Mughal rulers and their governors and officers were, by and large , not interested in the distant, relatively inaccessible regions inhabited by the communities of people we are speaking of, unless for mining and metallurgical interest as in the Chhotanagpur area, or for political and military exigencies as in the north-western frontiers of those days. They however knew of them, particularly of the nomad-pastoral communities of Sind, Baluchistan, Rajasthan and other areas. They distinguished these people from the rest of the population and even discriminated against them.

Indeed, not until the British appeared on the stage of Indian history and consolidated their position in the country, was any stir felt and experienced in the midst of these communities of people. Such stir was slowly and steadily brought about by several factors and forces which differed from region to region not only in their intensity but differed also in accordance with the colonial needs and aspirations of the British rulers.

First, for obvious reasons they had to 'open up' the entire country to be able to rule effectively and intensively, which obliged them to evolve eventually an effective and extensive communication system including postal, telegraphic, roadways and railway services. Effective and an intensive rule called also for a common administrative system that aimed at drawing, as much as possible, the entire country within its orbit, obliging the administration to send its officers, police and army to the farthest and deepest depths of the land.

Secondly, for long, at least upto 1887, there remained the primary obligation of British mercantile colonialism, to make the military conquest as complete as possible from one end of the country to other, east to west, north to south, and if possible, to go even beyond, to Burma in the east, Nepal, Bhutan, Sikkim and Tibet in the north, and Afghanistan in the north-west. This called for, internally, a political and military police that demanded maintenance of a strong thrust into all the resistance areas including the forest depths and hill-sides and hill-tops. This policy affected the *janas* of the entire Vindhyan areas from Chhotanagpur an d Orissa to what was the until recently called Central India or Madhya Bharat. Externally, a foreign policy was also called for a policy that involved diplomacy as well as military preparedness, and military adventures wherever necessary. The aim of this policy was to make the colonial empire of India safe from any kind of aggression from any foreign power, which meant the drawing of Burma, Nepal, Sikkim, Bhutan, Tibet and Afghanistan into the orbit of its influence and protection, and if possible, within the empire itself. This policy involved the *janas* of the entire Himalayan region including the right and left arm of the mountain range, from the Assam and Chittagong ranges to Hindukush and the Karakorum, a policy that opened up the entire region to public gaze and attention. This was sharpened further by a corollary policy dictated as much by politics and military considerations as by the felt need of having hill resorts on the Himalayas that could provide for the British rulers, their wives and children an escape from the summer heat of the north Indian plains and opportunities for the education of their children. Thus came into existence the series of hill-stations from Darjeeling to Muree, all on the lower Himalayas. British policy towards the peoples of these regions was of course somewhat different from that towards the *janas* of the interior regions, but I need not go into this question.

Thirdly, it follows from what I have said that the British administration found it necessary to send their engineers and their contractors, civil and military officials of all ranks and their provision suppliers, the traders and shopkeepers into all these inaccessible regions of the interior and the frontiers. All these elements, especially a good many of the lower ranks of Indian officials, contractors, middle-men, provision suppliers, traders and shopkeepers, all of whom had to recruit their labor and servant force from amongst the local peoples but who themselves belonged to the Hindu-Brahmanical community, slowly but eventually came to develop their personal and groups vested interests in the localities wherein they were posted. Such interests were initially directed towards gaining local prestige and power which was not difficult to acquire, with the administration, police and army to support them, and finding ways and means of opening markets for consumer foods including luxury materials. Thus was money economy introduced among the communities, slowly but surely. But what came to disturb these communities most is the land hunger of these vested interests. Taking advantage of their vantage position they started acquiring lands and introducing cash-cropping through means foul rather than fair, lands that belonged to and were used by local communities. Ultimately this land-hunger affected not only their agriculture but also their hunting and fishing rights. The process has been on until very recent times, not only in the plains but also in hill areas.

Fourthly, following the administration and the army and all that these connoted, came the 'Messengers of Christ', the Christian missionaries belonging to more than one church. They penetrated deep into many of these communities, converted a good many of them, brought education and health to them and opened them up to a modern world. Naturally, these missionaries, some of them very well-meaning and pious souls determined to save these communities of people from everlasting damnation, as they saw it, were all a part of the great establishment of the foreign rulers, and consciously or unconsciously their mission was directed towards upholding and strengthening that establishment.

Fifthly, administering these areas involved more or less close and detailed objective knowledge of these communities of peoples that we are speaking of their social and religious life, their political and economic organization, their behavior pattern and so on. Thus was initiated the series of studies that led to the foundation of what today we call Indian anthropology. The authors of these studies, Dalton, Risley, Thurston, Enthoven, Crooke, Russsel, were all administrators. Their investigations involved deep probing into the lives and societies of these peoples and led eventually to the formulation of policies that were meant to guide the administration of these communities of people.

The cumulative effect of these policies was far-reaching. Within a period of one hundred years were brought about in the life of these communities what may be called tremendous changes.The very fact that they were exposed to wider life,a quicker tempo,a modern legal and administrative system,and altogether a different way of life the pressure of which was ever on the increase,meant a great disturbance in these communities,economic, social and psychological.The first signs of such stirrings and disturbance started making themselves manifest already by about the second quarter of the nineteenth century, and in the civil disturbances that followed the great 1857 rebellion, a fairly good number of these communities of people, spread over wide area, found themselves involved.Even later,in the last quarter of the nineteenth and first four decades of the twentieth century sporadic disturbances among them and classes with the authorities were frequent, and recent history has kept record of many of these. Immediate and remote causes of such clashes and disturbances included interference with their social and religious customs and practices which were considered barbarous by the authorities, introduction of cash-cropping in areas of shifting and low-level cultivation, alienation of their community lands, their exploitation by landholders, money-lenders, middle men, police and forest and other officials, and their claim for higher social status, an attempt which was not unoften resisted by the *jati* Hindus. Indeed, by about the third and fourth decades of this century, in many of these communities signs were evident enough that a new self-consciousness, consciousness of their social, political and economic rights and privileges, was introducing new and disturbing elements into their over-all life and activity.

Today, after more than a century of uneasiness and unrest amongst these communities of people when one looks back it seems curious that this new and important feature on the periphery of the mainstream of our life, was hardly ever taken note of by our social and political thinkers and leaders of even the late nineteenth and early twentieth century. Surendranath Banerjee was unaware of it, so were Aurobindo Ghosh and Bipin Chandra Pal, Mahadev Govinda Ranade, Gopal Krishna Gokhale and Bal Gangadhar Tilak. The attention of the Indian National Congress was never drawn to this growing socio-political problem, though the question of the lower and depressed orders and the untouchables of Hindu society had started receiving attention from about the 1880's. Not until Gandhiji entered the stage of Indian life was this matter looked at as a national issue. His concern for the *adivasis* or *adimjatis* as much as for the untouchables and the depressed orders of Hindu society, the institutions which he helped to bring into existence and what he himself did towards the furtherance of their cause of social and economic uplift, are all too well-known to need recounting.

But the legacy of history seems to have been pursuing us. From the number of so-called anthropological studies brought forth by the British administrators, *caste* and 'tribe' understood in the concepts with which they were familiar, seem to have been an obsession with them, and unfortunately, with us as well. They knew too well that India had been aspiring for nationhood, and the whole burden of their argument seems to have been that the people of India could not weld themselves into a nation so long as these two major hurdles were not removed. How true they were ! But it was not to their interest to remove them; by their policy they only accentuated them and encouraged the divisive forces. Some of the so-called 'tribes' they chose, for administrative reasons and in the name of 'law and order', to brand as 'criminal tribes' and as late as 1935, created the remarkable phrase and social category called 'scheduled tribes' along with another called 'scheduled castes', two phrases and concepts that we inherited from the 1935 constitution and chose, in our wisdom, to incorporate in the constitution of the independent and sovereign Republic of India !

This has been unfortunate and unwise to my mind. It has conditioned our attitude towards these communities of peoples and our approach towards the solution of their problems which are theirs as much as the rest of the Indian population.

IV

What I have sought to do is to place before you my interpretation of the historical background of what we call the 'Tribal Situation in India', how it came about and how we inherited a situation that started taking the shape and form it has assumed today.

The entire burden of my argument from the point of view of Indian nationalism, has been that

(a) The communities of peoples we are concerned with in this seminar, are not 'tribes' but *jana* or peoples, just like the peoples of other territorial and cultural regions of India, the Panjabis, the Andhras, the Keralas, the Tamils, the Bengalis etc., for instance. What distinguishes the former from the latter, is the socio-economic system of *jati* to which the latter still belong. *Jati* is not *caste* nor is it just a socio-religious system; it is also an economic system, hereditarily and hierarchically organized according to groups recruited by birth.

(b) Indian history and culture have not been unconscious of the need for integration of the communities of what we have learnt to call 'tribal' peoples, but integration in practice in history meant incorporation of these communities of people into the body social by drawing them into the fold of the *jati* system and placing them at different hierarchical levels of the system, generally at the lower levels. The process was somewhat naturalistic, and hence a slow, patient and gradual one, but through the centuries it succeeded in drawing a very large number of these communities of people into the mainstream of Indian life and culture, transforming them and being itself transformed by them. But today, because of the introduction and functioning of new production system, the *jati* system is on the first lap of disintegration. The question of drawing these communities into the *jati* system does not therefore, arise any longer. Rather the question, to my mind, seems to be as to how we draw them into the new techno-economy that is bound to replace the *jati* economy.

(c) The British colonial-imperial policy of war, conquest, internal law and order administration, land tenure, diplomacy and external security, coupled with the policy of the Christian missions in India, went eventually to expose these communities of people to the much quicker tempo of modern life, to a new legal system and administration and to a new economic system altogether, bringing about stresses and tensions amongst these peoples, which nobody seems to have known how to cope with. British administrators sought the help of anthropology as understood in the early decades of this century in the Indian context. The suggestions that came out seem to have only sharpened the stresses and tensions, which in their turn, indirectly helped to bring the entire problem into focus.

(d) Then came the Second World War, the Japanese threat in the north-eastern frontiers and the British army, a few of these communities joining the army, and going far out of their closed areas which together led to the further opening up of some of these areas.

(e) Independent India inherited the situation which had in it the implantation of divisive seeds of 'criminal tribes' and 'scheduled tribes' in association with 'scheduled caste'. The phrase 'criminal tribes' was replaced by the phrase 'denotified tribes'; it hardly changed the situation. The concept of 'scheduled tribes' is still with us.

(f) Independence, democracy and adult franchise have released many hitherto submerged social forces; these forces are now at play in many areas of our social, political, economic and cultural life. We have been witnessing some of these forces at work in the communities of people we are here concerned with. Basically it is in part a search for a sense of identity, for a sense of belonging and for self-determination in a social order. The old order of theirs is in a fast process of disintegration.

The rest is for you to study and anaylse, and to find out ways and means as to how to meet the situation. The basic problem, I would say, even at the risk of repeating myself more than once, is one of drawing these communities of people into the mainstream of our life. How should we and they do it? For, this is a two-way traffic; the task is as much ours at theirs, that is, of those who are not yet in the mainstream of life. They must also realise that it is no longer possible, nor good for them to stand aside and aloof. Are we to follow the traditional old Indian process of slow, patient, gradual integration? Time is pressing; do we have time enough to pursue such a process? Contrarily, can we telescope the process so as to be able to bring about a quick change in the situation? Would such telescoping be judicious ? The process of integration is a naturalistic one; one can only help such a natural process, by trying to cut down time. Can one eliminate time altogether, relatively speaking?

At the very outset I said that the main problem in regard to the so-called 'tribal' communities is one of drawing them into the mainstream of Indian life and society. Any sensitive and knowledgeable person is aware that among certain tribal communities, especially among those on our north-eastern borders, nourish the apprehension that being drawn into the mainstream of contemporary India life would mean their complete absorption by the larger and more dominant society of which by far the largest in number are Hindus. Looking at the whole past history of the method of 'tribal' absorption adopted by the Hindu society, this fear is perhaps not altogether without some reason, one must admit. I must therefore at once try to disabuse the minds of such 'tribal' communities of any such apprehension. When we speak of integrating such 'tribal' communities into the mainstream of our contemporary life we are not at all thinking in terms of the old traditional method of absorption. That method was primarily based on traditional mode of production and a productive organization which was built into the *jati* system, otherwise known as the 'caste' system. India today is committed to the building up of not only a 'casteless' society, but she has also given up the older mode of production and productive organization,and taken to an altogether new method and a new economic system. Any consideration in the contemporary context, of the traditional Hindu method of tribal absorption is therefore, sheer madness to my mind. In the present context this is simply anachronistic.

Integration in the present context means to my mind, four things. *First,* this sovereign independent republic of ours which is called India, is a territorial unit with well-defined boundaries. When we speak of India it is this India that we refer to, not a Hindu or a Muslim or Christian India, the India which as everyone knows, has constitution of its own with a declared policy which is secular. Within this territorial unit there were many administrative regions where for more reasons than one there may be dissatisfaction, unrest and even disturbance. While attempts can and should certainly be made to resolve and contain them, the constitution cannot certainly accommodate a situation in which any of the constituents of the Union should be demanding secession from the sovereign territorial unit. *Secondly,* India has accepted and adopted a democratic way of life the principles and procedures of which have been built into the constitution. This democracy is based on adult suffrage, which, to my mind, is the most revolutionary step that India chose to take to bring about rapid social change in our society. This democratic way of life recognises certain fundamental rights for every citizen of India and certain directive principles for the guidance of Government and administration. *Thirdly,* India has made a full commitment of the idea of secularism, a concept which may not be sought to be understood in terms of the dictionary meaning of the word. According to the intention of our Constitution, religion is altogether left to the individual and his religious community; they can do whatever they like so long as their activities do not infringe upon the duties and obligations of persons belonging to other religions and of those of the State and the government. *Fourthly,* India has also committed herself to a new economic system which is described as aiming at socialism or socialistic pattern of society. This economic system is committed to progressive industrialization as the major means of production directed towards rapid social change.

From our point of view these four, namely, the sovereign, independent territorial unity of India, democracy, secularism and planned economy of which industrialization is the means and socialism is the goal, are the four tributaries that constitute the mainstream of contemporary Indian life. Any citizen of India who is still standing aside or away from the mainstream or who is thinking in such terms, should be drawn into it with a sense of participation that would engender in him or her a sense of involvement.

This is all that I wanted to convey when I spoke in terms of drawing the so-called 'tribal' peoples into the mainstream of India life. Nothing more, but nothing less. India has launched herself on tremendous experiment towards building up a new society based on a new economic and political system. It is a challenge to all those who call themselves citizens of India. Let us all, 'tribals' and non-tribals', take up this challenge and involve ourselves in the great experiment that is taking place before our eyes. There are stresses and tensions, worries and anxieties without doubt. Let us all try to resolve them.

VIII Integrated Area approach to the Problems of the Hill Tribes of North - East India

B.K. ROY BURMAN

An integrated area approach will require careful analysis of the social situation in NEFA in its four dimensions. These are:

(A) Geo-political Perspective;
(B) Socio-cultural Perspective;
(C) Political and Administrative Perspective; and
(D) Economic Perspective

A. Geo-political Perspective

Tribal communities in North-East India are living on the fringe of three great political communities — India, China and Burma. Historically, some of them played roles of buffer communities, and others the roles of 'bridge communities' in between these three great political communities. When smaller communities play the roles of bridge or buffer, it is in the interest of the larger contact societies to maintain the separate identity of the first.

In the context of the recent changes in the political organizations of the countries concerned as well as in the modes of transport and communication the bridge and buffer roles of the small communities have become dysfunctional. Therefore a re-adjustment of relations is inevitable. If it causes strain and tensions, the same are inherent in the structure.

From a structural analysis, it is immaterial whether certain agencies like foreign missionaries or the emergent middle class have been involved in the postures of defiance on the part of the small communities concerned. Even if these agencies were not here, more or less similar developments would have taken place. It is significant to note that in Burma, one of the most important

Reprinted from K.Suresh Singh (Ed.), 1972, *The Tribal Situation in India,* Simla: Indian Institute of Advanced Study, in association with Motilal Banasrsidass, Bangalow Road, Jawahar Nagar, Delhi - 7, pp. 72-97 with the permission of the Director, IIAS.

ethnic groups defying the authority of the Central Government, is the Shans, who are Buddhists of the same denomination as the other Burmese. As regards the role of the educated middle class, it is to be noted that even some of the Naga tribes among whom educated middle class is completely absent, have been involved in the hostile activities.

Here, it is to be noted that at present in Nagaland, the revenue receipt is about 80 lakhs; on the other hand, the annual expenditure is to the tune of approximately Rs 20 crores. Had this massive aid not been injected from outside, either the standard of living in Nagaland would have remained extremely low or the emerging educated middle class would have to raise the resources by pressurizing their own people for development, so that they could get job opportunities. It follows that at this stage, in their own interest, they should be co-operating with the rest of the country. If in spite of that, a considerable number of educated middle class from among the Nagas entertain separatist ideas, it is not because of their class character, but because of other socio-cultural factors.

An important fact, in terms of geo-political perspective is the type of political organization in the countries surrounding hills of north-east India. They are China, Burma, Pakistan and of course India itself. China is a totalitarian State, Burma is an effete democracy, Pakistan is an obscurantist country; on the other hand, in spite of severe limitations India is a secular democracy. It can be assumed that the more discerning section among the hostile elements of the regions are conscious of the significance of the above facts.

B. Socio-cultural Perspective

For an appraisal of the socio-cultural perspective the following facts are important:

1. Race

There is a predominance of Mongoloid racial type among the tribal communities of the hills of North-East India as well as of the adjoining countries (Sikkim, Bhutan, Tibet, China, Burma, Thailand, Laos and Cambodia).

2. Origin

Many of the tribal communities in NEFA, Manipur and Mizo Hills have tradition of migration to their present habitats during the last two centuries only. Again most of them have traditions of originating either in Tibet or in South China from where they entered India via Burma.

3. Language

The languages mainly spoken among the hill tribals of North-East India are

(a) Austric, (b) Tai and Kachin, (c) Tibeto-Himalayan and North Assamese, (d) Bodo, (e) Naga and (f) Kuki-China. Among these languages, Bodo language is found in the plains of Assam and Tripura also. Except for a small area in Jalpaiguri and Cooch Bihar districts of West Bengal, speakers of Bodo language are not found anywhere else in India. Outside Khasi and Jaintia Hills, speakers of Austric language are found only in Chhota Nagpur and Nicobar islands in India. On the other hand, speakers of all the above languages except Bodo are found in one or more of the following countries: South China, Burma, Thailand, Laos, Cambodia, North Vietnam and South Vietnam.

4. Material Culture

Hutton, the eminent anthropologist noted in the twenties, that there is a great deal of similarity in the material of tribal communities beginning from the coast of the Pacific Ocean to the hills of Assam.

5. Political Organization

The political organization among the tribal communities of the region may be broadly put under two categories - (a) authoritarian type, (b) republican type. For instance, the authoritarian type of leadership prevails among the Garos and Lushais of Assam, Singaphos and Khamptis of Lohit District, Noctes and Wanchos of Tirpa District of NEFA, Konyaks, Changs, Eastern Semas of Nagaland, Kuki Chin Group of tribe of Manipur and Reangs, Chakmas, etc. of Tripura. The republican type of political organization prevails among the Monpas, Aksa, Sherdukpens, Tangsas and Adi group of Manipur. From an ecological perspective, it appears that the two types of political organization are distributed in alternate cycles. There are, however, a few communities whose political organizations fall in various positions in this model. For instance, the Khasis had organized states with chiefs whose position was more or less similar to that of constitutional monarch. The political organization of the Ao Nagas also falls somewhere in between authoritarian and republican type of political organization.

It is significant to note that political ferment is more noticeable among those tribes which have republican type of political organization. Perhaps it is because of the fact that this type of political organization provides broader base for the mobilization of the population concerned, for new type of political action.

In recent years, in some of the areas where the authoritarian type of political organization prevails, the chiefs were subjected to great pressure, and even challenge from the emerging educated elites. For instance, in 1953, by an Act passed on the recommendation of the District Council of Mizo Hills, chiefship was abolished. In Manipur, the educated Kukis were also challenging the authority of the chief; but it seems that they did not wield effective influence. In 1967, when the rights of the chiefs were abolished by an Act passed by the Manipur Assembly, the chiefs could rally the entire community in their support. Even the educated elites, who were opposing the chiefs in the past, staunchly stood by them, against what they thought to be an imposition from outside. In March, 1967 the special correspondent of the *Statesman,* Calcutta, observed that the Act passed by the Manipur Assembly could be one of the reasons of the revolt of the Kukis in February that year.

6. Pattern of Internal Integration

Many of the tribal communities in the region have segments which have traditions of diverse origin. They are integrated into a single community through various devices. For instance, among the Aimols of Manipur, there are two exogamous Moieties with different traditions of origin. They remain united through marriage alliances and adoption of various ritual roles. In many of the Ao Naga villages, there are two divisions, namely, Chungli and Mongsen, with slight differences in dialect and type of political organization. If a village had been established by the Mongsen people, the Mongsen language and political system would prevail, even if subsequently the Chunglis constitute the majority population in that village. Same would be the position in a Chungli village. But offices in the village organization will be held by the Mongsens and Chunglis, in the proportion of their numerical strength. Thus, integration is attained through the continuity of tradition and ideology of the original settlers, but at the same time making adjustments in the share of political power.

7. Age-group Organization

Another factor contributing to the integration of clans and phratries of diverse origin, was the institution of the bachelor's dormitory and the age-group organization. Cohorts belonging to different age groups were organized for diverse economic, social and political roles in the village community. The younger age-groups constituted the cadre for rendering various services to the community during war and peace. The bachelor's dormitory in the village was their rallying centre. It seems that the age-group organization provided the operational base for the hostile activities in many Naga areas.

8. Institution of Automatic Change of Leadership

In some Naga areas, there is a system of counting generations, in terms of a thirty years' cycle. After every thirty years, a new generation starts with a new name. The persons who were holding office at the time of the termination of one generation, would automatically be relieved of their posts and would not be able to hold the same offices during the next thirty years. Even if a young man came to occupy an office two years before the closure of one generation, he would be relieved of his post at that time.

Each generation has a name, for instance, six years ago, in an Ao Naga village, a generation which was named as "courageous generation" came to its end. The generation which followed, is named "the generation of wild rumour". It seems that the hostiles have made use of this tradition to their advantage.

Among the hostile Nagas, there is an arrangement of automatic change of leadership every three years. It seems that the old tradition has provided a precedence for this arrangement. One of the charges of the Kughato group against the Pro-Phizo group among the hostiles, is that they had violated the arrangement by General Mow's failure to hand over the charge of the army to his successor at the end of three years.

9. Modern Type of Association among the Ao Angami, Sema and Other more Advanced Nage tribes.

Among the above tribes, there are clubs and associations of the educated youths for quite some time. They are trying to modernize their society. Sometimes their intervention in the local affairs is resented by the villages. But they are allowed free hand in the affairs at the regional or state level. The Adi group of tribes in NEFA, most of the Naga and the Kuki groups of tribes in Manipur, the Lushai and other groups of tribes in Mizo Hills have their respective youth organizations. Some of them are publishing separate journals of their own.

10. Student Unrest

Among the Adi group of tribes, and Tangsas of NEFA, the Ao Nagas of Mokokchung and the Khasis of K & J Hills, student unrest has been reported from time to time. These relate to three types: (a) matters connected with their educational facilities, (b) language policy of the Government and (c) economic and political aspirations of their respective communities.

11. Educational Imbalance

In the Mizo Hills, though the level of literacy is high, a very low proportion of literates have effective education. In Nagaland also the position is more or less the same. Spread of technical education is also proportionally low in Mizo Hills and Nagaland. A corrective is to be introduced by balancing the aspiration- motivation pattern of the people and comprehensive manpower planning.

12. Medium of Instruction

In the autonomous hill districts of Assam, their respective mother-tongues constitute the media of instruction for primary education. Gauhati University has recognized Khasi as a language subject upto the post-graduate standard. In Nagaland also, the respective tribal languages are their media of instruction at the primary stage. Ao and Angami languages are taught in the secondary stage also. In Manipur a number of textbooks have been published in different languages with subsidy given by the Union Territory Administration. But in not a single school, these languages have been adopted as media of instruction or even as language subject. Until 1965, the policy followed by the NEFA Administration was to prepare special text-books for the primary schools in the tribal dialects of the area. A large number of books were prepared and printed in the Devnagari script. After the primary school stage where medium of instruction was Assamese, Hindi was introduced. In some schools, Hindi was the medium of instruction from that stage onwards. English was introduced in Class V.

After 1965, the policy has been substantially modified, on the recommendation of a committee headed by D. Erring, Deputy minister of the Union Government who hails from NEFA. The representatives of the local people wanted that either Hindi or English should be introduced as the medium of instruction at the earliest stage. In most of the areas the NEFA Administration has discontinued the teaching of the text-books prepared in the various tribal languages.

It is not clear whether the Erring committee had the benefit of consultation with educationists who have experiences of scientific research in such matters. In a report on the medium of instruction, published by UNESCO, it has been recommended that even in those cases where leaders of the local people do not want their language to be the medium of instruction, in the beginners class, an attempt should be made to persuade them to change their mind. It is based on sound pedagogical principles. Sometimes it happens that in the first stage of acculturation, the people concerned develop an inferiority complex about their own culture. They try to discard what they have and imitate cultures of the advanced communities. But at a second stage, there is a re-assertion of self-respect which leads to selective revival of the past. But in the process they are haunted by a feeling of being cheated, by the donor group. The language policy in NEFA should be reexamined keeping this also in view.

13. Absence of Creative Literature in Some Areas

A large number of text-books and gospel literature have been published in the Ao, Angami and other Naga languages, but there is hardly any creative literature among them. On the other hand, in some of the Kuki languages of Manipur, there are not only text-books but also religious literature, fiction, drama and other books. Among the *Paites*, publication of such literature is financed out of the profit earned through running a bus service. It is to be examined whether obsession with politics of coercion without creating the intellectual base for the same, is responsible for absence of creative literature among the Nagas.

14. Religion

A statement giving the distribution by religion of the more important tribal communities of the region according to 1961 Census is furnished below:

(Percentage)

TRIBE	HINDU	CHRISTIAN	BUDDHIST	TRIBAL RELIGION	OTHERS
1	2	3	4	5	6
Assam					
Garo	4.05	37.94	-	57.36	0.65
Kachari	98.63	6.16	0.21	-	-
Khasi	3.53	47.54	-	48.79	0.14
Mikir	75.11	10.28	-	14.24	0.37
Mizo	1.37	96.95	1.54	0.07	0.07
Manipur					
Angami	0.16	55.22	-	-	44.60
Gangte	0.04	87.83	-	-	12.13
Hmar	0.12	99.21	-	-	0.67
Kabui	0.28	35.53	-	-	64.19
Kacha Naga	0.01	45.16	-	-	54.83
Mizo	0.11	99.78	-	-	0.11
Mao	-	22.02	-	-	77.98
Paite	0.23	85.69	-	-	14.08
Tangkhul	0.01	88.09	-	-	14.90
Thadou	0.47	58.72	-	-	40.81
Nagland					
Dimasa	99.79	0.21	-	-	-
Garo	40.67	59.33	-	-	-
Kuki	3.55	54.65	-	14.43	27.37
Mikir	98.71	1.29	-	-	--
Naga	3.62	55.12	-	27.51	12.75
NEFA					
Aka	-	-	-	50.00	50.00
Apatani	-	-	-	87.18	12.80
Ashing	-	-	-	28.26	71.74

TRIBE	HINDU	CHRISTIAN	BUDDHIST	TRIBAL RELIGION	OTHERS
1	2	3	4	5	6
Dafla	1.18	1.77	-	91.72	5.33
Galong	1.48	4.55	-	33.11	60.86
Khampti	-	-	9.88	-	6.12
Miju	14.02	0.93	-	79.44	5.61
Miniyong	7.49	5.70	-	33.71	53.10
Miri	37.94	2.37	-	39.53	20.16
Mishami	86.67	0.48	-	3.33	9.52
Monpa	0.34	-	99.66	-	-
Nocte	46.97	1.51	4.55	36.36	10.61
Padam	10.00	2.361	-	19.18	47.21
Sherdukpen	10.00	-	90.00	-	-
Singhpo	15.38	-	69.23	15.38	-
Yagin	42.98	0.88	-	49.12	7.02
Tangsa	2.42	1.21	3.03	49.70	43.64
Wamcho	-	-	-	1000.00	-

Note: While calculating the percentage distribution of religion, tribal population of NEFA for whom All India census schedules were not canvassed have been excluded from the table.

It is to be noted that in the above statement "others" include "indefinite belief" and "religion not stated". It seems that the tribal religion and "others" can, in most cases, be included in the same category.

15. Significance of Christianity in the Hills of North-East India

In the autonomous districts of Assam and Nagaland, Christianity was introduced early. Even then the most spectacular increase in the followers of Christianity has taken place from 1940s onward. In the post-independence period, the spread of Christianity has been stimulated considerably.

In Manipur and Tripura Christianity was introduced at a comparatively late period. In these areas also the more rapid spread of Christianity has taken place after independence. In NEFA conversion to Christianity has taken place during the last few years.

It is frequently alleged that people have been converted to Christianity on a mass scale by allurements provided by the foreign missionaries. In some parts of the hills of North-East India, this may be true. But it is certainly not the whole truth. In many areas, Christianity appears to have served the role of filling up the intellectual and spiritual vacuum caused by the growing scepticism among the tribal population, about their traditional faith and world-view. It has also tended to serve as the vehicle for unification of the isolated small communities into a larger entity.

Among the Kuki tribes of Manipur, it has been found that there is a proliferation of local churches. In many cases, the different tribal communities broke away from the denomination originally adopted by them, because, the church organizations of those denominations failed to satisfy their ethnic aspirations. This seems to indicate that the tribal communities are not always the tools of the foreign missionaries, that they are made out to be.

Significance of Hinduism

The tribal communities in the hills of North-East India seem to have some amount of inhibition in the matter of adoption of Hinduism. One of the factors responsible for this is that Hinduism in their eyes is identified with the domination of the people of the plains. In the past their relations with the people of the plains were not always happy and hence their inhibition in this matter is understandable.

A New Trend

In recent years a trend towards revival of respect of and re-interpretation of old cultures and traditions of the tribal communities concerned is perceptible. In a seminar recently held at Shillong on the Sociology of the Hills, a highly educated Khasi presented an interpretation of their traditional religion at a high level of abstraction, incorporating the elements of the universal religions. It is significant that the Christian Khasis present in the seminar including the president of the council of Presbyterian church, Assam, fully agreed with this interpretation. One could feel the resurrection of a new pride in the their past. In a recent publication on Khasi History and Culture by Dr. Hamilton Bareh, a Christian Khasi intellectual, the attempt to rationalize and sublimate the traditions in the terms of modern values can be noticed again and again.

A Christian Naga lawyer in this book on 'Ao Customary Laws' has tried to interpret the Ao traditions of origin from rocks, as an allusion to the continuity of the culture of the Aos from the pelacolithic past. A pastor was made to pay a fine of Rs. 50 by his Christian followers for deriding this tradition of origin from six rocks in one of his sermons.

In course of a discussion, a non-Christian Naga professor of history stated that he did not feel it necessary to become Christian or Hindu. Their old religion has got the concept of the supreme God who is a moral being, dispenser of good, and having compassion for mankind. What was necessary was to free these elements of their religion from the grocer elements which had got mixed up, due to ignorance of the forces of nature. A Kabui Naga chief in an outlying village was also found to speak more or less on same terms.

A Christian Paite youngman stated that an intergeneration conflict was going on among the Christians belonging to his community. The educated younger persons were asserting that one need not adopt the customs of the Hebrews for becoming a good Christian. In a Naga village which adopted

Christianity about 90 years ago, it has been found that in their beliefs and world views, themes of Christianity and their ancient traditions have blended together in a new synthesis.

Strain of Secularization

While in the rest of the country reorganization of the society on the basis of secular principles of humanism is a dominant theme, acceptance of the same at this stage is difficult for some of the people of the hills. It requires a measure of maturity of the society and its culture and a capacity to look upon the past of the community with certain amount of detachment. But while a new trend in this direction is perceptible among a section of the intelligent, for the mass of the people belonging to the hills, who are trying to shake off their primitive past, it is difficult to look upon the same with detachment. Recently, an educated Naga youngman stated that his people would rather try to forget that till recently they were carrying on head-hunting or some such practices. So over-whelming is their feeling about these one or two erratic practices that it becomes difficult for them to see beyond this and appreciate that there are basic values of humanism even in the past. They tend to feel that they were wallowing in the mire of shame and were redeemed by the Christian missionaries. Thus it is necessary for them to have particularistic attachment, whereas secularism which the rest of the India is trying to attain requires universalistic detachment. There is thus a difference of levels and this difference is to be bridged through understanding, sympathy and also respect.

Inter-ethnic Relations

In the past, the various tribal communities had traditions of hostility or alliances. Generally the alliances were brought about through ceremonial friendship of individuals belonging to the different communities. Sometimes, friendships were ritually effected at group levels also.

Though the hill tribes of NEFA, Nagaland, Manipur lived in isolation as distinct entities, there were many elements of culture common among them. As a result, social boundries of the different communities were not always rigidly defined. With changes in the power structure or in the level of economy, the various communities effected different types of alignments. For instance, the Tagins of NEFA some time tended to be identified with the Daflas. Later on the dominant sections among them identified themselves with the Adis. At present some are asserting their distinct identity. This phenomenon can be described as trans-sociation. The same process is found in Manipur to a very large extent. A large number of tribes, like the Anals, Aimols, Chirus, etc., who were in the past included under the Kuki constellation, are at present claiming to be Nagas. This sort of change of alignments is possible through selective emphasis on particular elements of culture, in response to changing situations. The process of trans-sociation involving some of the tribes of the hills of

North-East India is not confined to the boundary of the country only. It takes place across the borders involving tribal communities like Kachins, Shans, Karens and Chins of Burma and Thailand.

Minority Groups in the Hill Areas

As reported by the Pataskar Commission, besides the major tribes, there are a few smaller tribes in the autonomous districts. Many of them complain of deliberate neglect of their areas for development purposes, difficulties in the matter of education of their children, incompatibility of the laws passed by the regional council with their customs and social practices and broadly describe the union of the two tribes as that of a "cat and dog tied together to one post".

The scrutiny of the proceedings of the district council of Garo Hills shows that the Hajongs have very frequently complained of discrimination against them. In the United Khasi and Jaintias Hill, the Jaintias had the same complaint. In 1963 on the recommendations of the committee appointed by the Governor of Assam, a decision was taken to establish a separate district council for the Jaintia Hills. In Mizo Hills the Smaller tribes like the Lakhers, Pawis; Chakmas are afraid of being swamped and "Lushainized" by the Lushais. Chakmas are Buddhists and even adopted Bengali script for their language. The Paites consider the Mizos too puritanical. The Maras belong to a different Mission, the Lakher Independent Mission. When the Bible Society of India did not agree to reprint the Bible in Mara language, due to low sale and asked the Maras to consider whether the Lushai version would be acceptable to them, their anger was roused as they felt that the Mizos were behind this move:

C. Political and Administrative Perspective

Typology of Political Process at Community Level

For an appreciation of the political developments in the hills of north East India a conceptual framework is necessary. The same is indicated in the diagram below:

Phases of expansion and growth		Phase of humanist resurrection	Phase of decadence
	Cosmopolitan humanism Nationalism/Transationalism	Sub-nationalism (base) Authoritarian Pseudo-internationalism	
	Proto-nationalism Infra-nationalism	Social disorganisation or anomie.	

(Base) = Tribalism

Tribalism, in the sense of complete negative attitude, towards modern technology, and contact with outside world hardly exists anywhere today. Perhaps the Jarawas of Andaman are the almost solitary instance of a community in the grip of tribalism.

The phenomenon of infra-nationalism is found among the primitive people living in comparative isolation and unaware of the symbols at the national level, who have been seized with the urge of rapidly modernizing themselves so as to avail of the benefits of advanced technology. It is not possible for small communities living in isolation to enjoy the benefits of advanced technology. To become viable, the diffused commands of the small communities over their resources including the tangible resources like land, forest etc., and intangible resources like skill, ideas and ideologies must give place to unified command of larger society. Infra-nationalism is the phase of expansion of identity of small communities.

The next phase is that of proto-nationalism, where new types of symbols have stabilized themselves and an elite has emerged out with vested interests in consolidating the expanded identities of the small communities. Proto-nationalism, however, derives its sanction mainly from the traditions, myths and legend of the primitive past.

Nationalism is the more evolved form mainly sustained by a conscious elite with economic and ideological base which cuts across local and regional loyalties. Nationalism to stabilize itself would require a minimum technological base and also certain values of rationality and pan-human world view.

The organizational base of nationism can be of two types:

(a) National society under colonial rule striving to thwart foreign domination and seize power to regulate its own affairs.
(b) National state with sovereign power.

Cosmopolitan humanism is based on the maturity of people to see the unity of mankind beyond the nation state. It requires the relentless striving of a determined elite committed to the values of rationality, freedom and human dignity. But even under the phase of cosmopolitan humanism, nation-state may continue as a convenient unit for political organization.

Nationalism may degenerate into sub-nationalism or totalitarian pseudo-internationism. In sub-nationalism the national symbols are accepted ostensibly, but at the same time, parochial loyalties become the primary consideration for political action. It generally takes place at the stage of conversion of a nation society to a nation state. In a nation society the national symbols are accepted by the different segments of the nation as a matter of grace. These are not levers to power; but in a nation state the national symbols are also levers to power and privileges for the dominant groups. There is, therefore, resistance from the subdued groups. When the nation is committed to programme of welfare state, the process of subnationalism is accentuated. Welfare state would require maximization of the outposts, which is possible only through the strategic investment of resources. This tends to increase the gulf between the more developed regions and the less developed regions. Sub-nationism is the by-product of this.

Totalitarian pseudo-inter-nationalism is the outcome of aborted social and economic revolution of the part on a nation state.

Both sub-nationalism and totalitarian pseudo-internationalism, by ensuing a process of negation of the moral consensus of the society, leads to anomie or social disintegration. It is a phase of atomization of individuals on a large scale but it is also a phase, when the best sons of men are required to fight against complete self liquidation with their back against the wall. Thus the chances of humanist resurrection are very great in this phase.

In this seven-point typology, most of the tribal communities of North East India are going through the phase of intra-nationalism or proto-nationalism or sub-nationalism. These processes are not necessarily incompatible with the processes of emergence of nationalism. It depends on geopolitical factors and the character of the nation itself, whether infra-nationalism or proto-nationalism can be incorporated in the nation structure of a particular country, or whether a new type of nationalism, transnationalism, with alignments cutting across the boundaries of more than one nation state, would emerge out.

Challenge to the Moral Greatness of Indian Nation

From the foregoing discussion, it is obvious that infra-nationalism or proto-nationalism constitutes a challenge in the Toyenbian sense to the moral greatness of a nation. The balance-sheet of the strength and weakness of Indian nationhood faced with this moral challenge is as follows:

Strength

(a) Existence of vocal elite at the national level committed to the values of democracy, secularism and liberalism.

(b) The constitution of India guaranteeing freedom of religion and culture to all sections of people and requiring the State to undertake measures for the socio-economic development of the weaker sections of the population.

(c) Tradition of diversity in unit in India.

Weakness

(a) Frequent compromises with forces of obscurantism tampering with the secular and democratic character of the nation. The examples are many. The adoption of Assamese as the State language, by the legislature in the face of the opposition of all minority groups, is always pointed out by the hill people of the regions, as an imposition. Adoption of symbols of Hinduism in state functions, for instance, performance of *puja* at the time of inauguration of the Oil Refinery in Assam, is pointed out by the non-Hindu tribals as an evidence that in spite of the formal commitment to secularism, the State has not been able to free itself from the influence of those who would like Hinduism to be adopted as the State religion. The fact that the agitation against cow slaughter has considerable support in the country is also pointed out by the tribals as a reason for their misgivings. There is also the demonstration effect of sub-nationalism in the different parts of the country. When an industry in the public sector is established in a State, because of the agitation carried on by the people of that State, and not because of economic or technological considerations, it is very difficult to convince the tribals of the hills, that more resources cannot be invested for developing their areas, on economic ground only.

Prospect of emergence of a tribal trans-nationalism cutting across the political boundaries of India, Burma and other adjoining countries. The possibility of this taking place should not be ruled out because of the following factors:

(a) Similarity of race, culture and traditions as described earlier;
(b) Involvement of some of the tribal communities in the borders of north-east India, Burma, Thailand, Laos and Vietnam in separatist movements against their respective nation states; and
(c) Organizational and ideological support provided by interested foreign powers.

It is not only China but influential lobbies in some other countries, for instance, USA, which seem to be giving moral support, if not material help to the separatist movements of the ethnic groups of north-east India and the adjoining countries. In this connection mention may be made of two articles published in June 1968, in the *London Observer,* by Crane, Director of Strategy Analysis Centre, New York, supporting the cause of the various ethnic minorities of South East Asia.

Implication of tribal trans-nationalism cutting across the boundaries of existing nation states in S.E. Asia. It is to be examined whether confrontation with the ethnic *aspocations* of the small communities constitute any moral problem for the nation- states particularly those which are committed to the values of democracy and freedom.

In this connection, the following factors are to be kept in view:

(i) Differences in the levels of technological development
The different tribal communities of South-East Asia are at different levels of technological and economic development. While many of them are shifting cultivators, there are communities like the Angamis, Karens, etc. which have advanced terraced cultivation and wet rice cultivation.

(ii) Uneven levels of education
The levels of educational development are very uneven among the communities concerned. Even in Nagaland, while the Ao and Angami Nagas have produced a large number of highly educated persons the tribes like the Konyaks, Phoms, Yimchungrs have hardly any matriculate.

(iii) Religious diversity
Religious diversity among the various ethnic groups is also quite considerable. Among the Nagas of Nagaland the Christians constitute 56.12% of the total population, the rest are followers of other religions.

In Burma and Thailand, influence of Buddhism is not insignificant among many of the tribal communities.

(iv) Linguistic differences
The linguistic differences among these ethnic groups are considerable. There are both Tibeto-Burmese speakers and Austric speakers among them.

(v) Differences in political system
As already mentioned, broadly, there are two types of political organizations among them, namely authoritarian type and republican type. There are a few communities whose political organizations fall in various intermediate positions in this bi-polar model.

(vi) Diversity in the pattern of inter-action with the neighbouring communities.
Because of the current hostile relations of many of those communities with their respective nation-states, it should not be overlooked that frequently they had cordial relations with their neighbors in the past. They have even myths and legends of descent from common ancestors. The ethos of the surrounding great traditions have also been adopted by them in varying degrees; for instance, a section of the Nocte Nagas adopted Vaishnavism centuries ago.

If inspite of the wide range of diversity in so many vital aspects, transnationalism consolidates itself among the tribal and other minor ethnic groups of South-East Asia, it can be only through digging out the highest common factor of unity from the hoary past. This will stimulate process of atavistic revival of primitive past and primitive symbols and for making the structure apparently viable, there will be an unholy marriage of primitivism and modern technology. The resultant prognosis is that of the emergence of a type of neo-fascism.

It does not require much farsight to visualise that the seeds of self-destruction will be embedded in the trans-nationalistic structure in the very process of its formation. In the first phase this may not become obvious, because the unification of the communities, whatever may be the platform, will make them viable for incorporating certain aspects of modernism. But after a phase of development, the anachronistic nature of the structure will reveal itself with disastrous effect.

If this analysis is correct, the moral constraints of the Nation, States, confronted with the ethnic aspirations of the tribals and other small communities along their borders, will not require them to make adjustments in their political jurisdictions. But it should not mean that there is no moral issue involved.

The problem of potential trans-nationalism of the tribal and small communities of South-East Asia may be considered from another point of view, namely, geo-political point of view. As pointed out earlier, a trans-nationalistic structure cutting across the various nation-states in South-East Asia can be consolidated only with the organizational and other supports of interested foreign powers. This will provide the scope either for the coming back of the erstwhile imperialists by the back-door or for the spread of totalitarianism, resulting in loss of freedom and dignity of the human individuals involved. This is certainly unacceptable to the Asian countries.

Moral Challenge

The awareness of the above geo-political implications of the infra-nationalistic movements of the tribal and other minor communities along the borders may cause two types of responses among the nation-states concerned. There is a possibility that they may try to meet the situation through sliding down to the worst form of chauvinism. There is also a possibility that they may look upon the problem as challenge to their moral greatness. In this connection, the nation-state concerned should perhaps try to look inward and examine what might be the causes of small communities turning against them, at this moment of history. In many cases, when the nation-societies of South-East Asia attained freedom from colonial rule, the small communities within their sphere of influence, opted to co-operate with the central leadership. Though there might have been some misgivings even at the initial stage, they violently turned against their respective nation-states only later on. The reasons for the same may be as follows:

(i) At the initial stage the small communities attached themselves to their great neighbours mainly because of their admiration of the fighters for freedom.

(ii) At a later stage, centrifugal tendencies gained strength because of the exhaustion of the moral fervour of the liberation struggle and also failure of the nation-states concerned to build up resilient economic and social structures.

It seems that the basic problem today in South Est Asia as a whole, and India in particular including the hill areas, is that of an aborted social revolution. The responsibility for competing it, does not lie with the tribals alone, rather it should be borne by the rest of the nation to a greater extent.

To meet the moral challenge provided by tribal infranationalism protonationalism and trans-nationalism the following ideological, political and administrative measures will have to be undertaken by India.

(a) *Shedding Complacency* It seem that Indian nation has a split personality. In spite of its avowed belief in unity in diversity, there is a good deal of intolerance in Indian society. This is manifested in agitation against cow slaughter or agitation for driving away certain language from the country and so on.

(b) *Integrated planning for development not only economic planning but also planning of political strategy and social strategy.* For integrated planning following factors in addition to the economic factors are to be kept in view:

1. Harmony of political process and the administrative and legislative processes.
2. Strategic orientation of the existing political institutions.

Harmony of political process and the administrative and legislative processes. It seems that in the matter of national integration the key strategy of the political elite in India has been that of ensuring good administration and extending basic amenities of life to the population concerned. But it should be realised that in the context of modern democracy, there is an optimum level beyond which good administration and amenities cannot be provided, without the broad based participation of the population concerned in the decision making processes at the different levels. Extension of amenities through good administration only, without stimulating the compatible political processes may even lead to alienation of the population concerned. This seems to be taking place in NEFA. Undoubtedly, amenities have been provided on an unprecedented scale, but there are also signs of discontent in different parts of NEFA. People who enjoy unlimited amenities without a sense of responsibility towards the agency providing the amenities, tend to become opportunists and entertain unlimited aspirations. This inevitably leads to unlimited frustration also.

The problem of people's participation in NEFA was examined by Erring Committee in 1965. But it seems that the committee examined the question of the people's participation in development activities and judicial activities at various levels. The question of participation of the people in the political decision- making processes at the entire territory level does not seem to have received much consideration. This requires more thought.

In this connection, another question, namely, that of continuation of the Inner Line Regulation is to be examined. A study conducted by the Institute of Applied Politics, Delhi, came to the conclusion that for the economic development of NEFA a vast army of skilled manpower will be necessary and that to facilitate their availability, the Inner Line Regulation should be abolished. While there is considerable force in this argument, the psychological reaction among the tribal population concerned cannot be ignored. It seems desirable that the Inner Line Regulation should not be abolished at this moment, but at the same time it must be very much liberalized so that the various national political parties and skilled personnel may function in NEFA freely. To safeguard the economic interests of the tribal population, restriction against alienation of their land should be continued. Besides, preferences should continue to be given to the local tribals in trade, commerce and other economic activities.

The importance of appropriate political processes for promoting development activities can be appreciated from the fact that spectacular developments have taken place in Nagaland only after it attained a Statehood.

Strategic Orientation of the Existing Political Institutions

The question of strategic orientation of the existing institutions is to be particularly examined in the matter of functioning of the Autonomous District councils of Assam. It seems that by and large the district councils are functioning as organs of local-self government with powers of legislation in the matters relating to customary laws. The different district councils have not enacted legislations in a uniform manner. The diversities are the results of inter-play of various political and other interests. A systematic and continuous analysis of the same should be made, so that the hands of the development oriented and integration oriented leadership can be strengthened. In other words, the district councils require to be politicized.

One significance of the district councils is to be appreciated here. These have functioned as cushions for extending the command of the State over the resources which traditionally remained under command of the local communities. In the context of the proposed autonomous hill state, the probable and the desirable roles of the district councils require to be examined from the points of view of administration, development and political strategy.

Creation of a Determined Cadre for National Integration

It seems that much importance has been given to creating economic interests in support of national integration. While such an approach has considerable advantage, it should be appreciated that without an ideological base, its effectiveness is very limited. The following steps seem necessary in this connection: (a) Functioning of national political parties in the hill areas, (b) getting the hill people involved in political actions confined not only to themselves but also to other people in the different parts of the country, (c) enabling the competent hill leaders to take leadership roles on all India platforms.

In the matter of creation of cadre for national integration, the problem of creation of a national lobby for the hill areas should receive due consideration. It is desirable that the various political parties and the members of Parliament should be kept well-informed about the implications of the different developments in North-East India. In this matter, in addition to mass media, a great responsibility would lie with the universities departments as well as research workers attached to different organizations. But they can fulfil their role, only if there is an atmosphere of freedom. It is known that quite frequently the research workers are not allowed to publish the results of their study for years after the completion of the studies. This is a positive disservice to the nation.

D. Economic Perspective

Nature of relationship of the tribal communities with their land resources. Prior to the promulgation of new legislation by the respective district councils, the land tenure system in the autonomous hill district of Assam was guided by Regulation V of 1986 (Chin Hills Regulation, 1986). Under this regulation, rights of the village communities as well as of the individuals over their respective territories have been recognized. After the constitution of the District Councils, the Garo Hills, Khasi and Jaintia Hills, and Mizo Hills district councils have promulgated separate legislations of their own. In the legislations promulgated by the Garo Hills District and Khasi and Jaintia Hills District Councils, with certain limitations the rights of the village communities have been recognized and also the rights of the traditional chiefs have been continued.

The legislation enacted by Mizo Hills District Council has abolished the rights of chiefs on payment of compensation; the rights of the village communities have also been curtailed considerably, causing local resistance. It is not unlikely that the District Council here is considered as an organ of the Government. In that case the alienation of the local communities by the District Council may be contributory factor to the Mizo revolt.

It is to be noted that on abolition of the rights of the chiefs, compensations have been paid, not on the basis of the extent of land under their jurisdiction, but on the number of households under them. This seems to indicate that the rights of the chiefs were political in nature rather than economic. This aspect more clearly comes out in Purim Kuki area of Manipur. There also the chiefs are recipients of certain payments from the households under them; but the office of the chief was not hereditary. By rotation people of different clans came to hold the post.

In North Cachar and Mikir Hills, the respective District Councils have extended the Assam land and Revenue Regulation, 1886. This regulation has curtailed the rights of the village communities quite considerably.

In Manipur, the Administration tended not to recognize the rights of the village communities over their respective territories. Lands were taken over for the public purposes without payment of any compensation. This question was contested in the court and a decision was obtained recognizing the rights of the village communities.

In Nagaland, the Chin Hills Regulation, 1986 operates. In NEFA, Jhum Land Regulation, 1947 is in force. The Jhum Regulation 1947 is in fact an adoption of the Chin Hills Regulation of 1986. The only significant difference is that whereas according to the Chin Hills Regulation, the customary right to Jhuming land is considered to be established in village or a community, when the village or community has enjoyed the right to cultivate or utilize such Jhum land for not less than 5 years prior to the making of the regulation, according to the Jhum Land Regulation 1947 such right is recognized if it has been enjoyed for not less than 5 years prior to the making of the regulation. Both in Nagland and NEFA the rights of the village communities over the respective territories have been recognized at least theoretically.

In some parts of NEFA, a village has large tracts of land under its jurisdiction, only parts of which can be economically utilized by it. In some cases, under the jurisdiction of a village, there are sub-villages where communities, other than the dominant community live. In such cases the people of the sub-villages are required to make some payment to the dominant community. This also seems to bring out the political relationship and not merely economic relationship of the tribal communities with their territories.

There is another aspect of the relationship of the tribal communities with their respective territories. Such territories frequently constitute symbols of the unity of the community. Any scheme of development which may cause dissociation of the tribal community from its traditional resources is therefore likely to lead to anomie or disorganization. It is desirable that development programmes should be based on continuation of the rights of the communities.

They should not be given compensation at the outset, but should be granted rehabilitation subsidy. On the other hand, for three generations they should be paid dividend, on the development of their resources in addition to compensation at a graduated scale, which for all practical purposes, will be perhaps of symbolic value only. Such an approach will provide the population concerned an opportunity to adjust to the new situation and stabilize the new way of life.

Forest Policy

In the context of the above analysis, it seems that forest policy in the different hill areas requires to be carefully examined. According to the forest law prevailing in Assam, reservation of forest can be made only through the process of acquisition. But the Chin Hills Regulation, 1986 and Jhum Land Regulation, 1947, have stipulated that no formal acquisition proceedings would be necessary; though reasonable compensation shall be paid for the lands taken over. This stipulation does not, however, appear to be implemented in a uniform manner. In Tirap district, while bringing forest areas under reservation the right of the Nocte chiefs of Namsang and Baduria as well as of the village communities have been fully recognized. The net revenue of the forest is divided in the ratio of 25 : 75 between the Government of the one hand and the chiefs and the villagers on the other. But of the latter, one third or Rs. 5,000 whichever is less, is credited to the chief's personal account, the rest being deposited in the people's fund. It is out of this fund that 40% preference share has been subscribed to a veneer mill established at Deomali (Tirap district) to process the timber produced in the Nocte area. The remaining 60% share capital has been subscribed by the Assam Railway and Transport Company (all in equity shares).

In Nagaland, recently the forest Department has come to an agreement with the Zellong Range Council, under which the forest areas of that range have been declared as Reserve Forest. The Forest Department will deduct management cost to the tune of 25% and the balance of the revenue will be handed over to the Range council.

This system of associating the local community with the development of the forest is unique in Tirap and Zellang range of Nagaland.

Some of the tribal communities in NEFA are making a demand that the same principle should be extended to their areas also. They point out that they have the same type of right as that of the Noctes in respect of the forests in their respective jurisdictions.

Development of Agriculture

In all the hill areas shifting cultivation is the prevailing from of cultivation. But in many areas wet rice cultivation and terraced cultivation have been introduced. Besides, there are orange, betel-nut, pineapple and other plantations also.

In the strategy of development of agriculture, maximum emphasis is given to the conversion of shifting cultivation to settled cultivation, but due to the nature of the terrain, it is obvious that bulk of the area will continue to be under shifting cultivation until new type of economy comes up in the region. But only in limited areas attempts are being made to introduce scientific Jhuming. By and large, this aspect of economic development has been neglected. It is to be noted that under the broad complex of shifting cultivation, there are different techniques, some of which are of more evolved form and cause soil erosion only to a limited extent. While it is not possible to completely stop shifting cultivation, much more emphasis can be given on introducing more advanced techniques of such cultivation.

Economic Dislocation as the After-Effect of the Partition of India

In Garo Hills, Khasi and Jaintia Hills and Mizo Hills, the economy was considerably disrupted due to partition. The cash crops of these regions used to be exported to the areas now belonging to East Pakistan. On the other hand they used to obtain many of their consumption needs from the markets in East Pakistan. The Assam Government had appointed three committees for these three autonomous districts to ascertain the extent of disruption of the economy and to recommend ameliorative measures. All the committees pointed out that the economic health of the districts concerned could be restored only by expanding the communication facilities several fold. They also recommended that more emphasis should be given on food crops than on cash crops. At the same time the need of establishing food processing industries was emphasized. It seems that even now communication facilities are not adequate and the capacity of the food processing industries is also limited. One of the constraints seems to be paucity of finance. Manipur has a common arrangement with Assam in the matter of obtaining finance from Industrial Development Corporation. But there are allegations from the Manipuri industrialists that the arrangement has not functioned to their complete satisfaction.

Mining and Quarrying Industries

A cement factory and a few coal mines are in operation in Khasi and Jaintia Hills. There are mineral resources in other areas also. But due to transport difficulties all these resources have not been adequately exploited. Besides, in some of the hill areas, for instance, hill areas of Manipur, exploitation of forest resources would be difficult unless co-operation of local tribals is available to greater extent.

Other Industries

In the matter of establishment of other industries in the border areas certain strategic considerations should be kept in view. They are as follows:

(1) In areas where there is availability of power or any scope for introducing any power and communication facilities, attempt should be made to introduce (a) mechanized and small industry (b) industries which would serve the needs of efficient farming (c) small industries geared to defence production (d) improved training programmes for higher technology (e) perripatetic training services and repairing facilities and (f) developing nucleus markets.

(2) In areas where power and technical communication facilities are lacking, emphasis should be given on (a) agro-industries generally directed to achieve self-reliance and self-sufficiency as far as possible (b) agro-industries which could cater to defence needs, for instance, dairy produce, poultry, meat, and preserve food (c) industries for slack agricultural seasons (handicrafts and art pieces) (d) training programmes to improve traditional skills (e) supply of raw materials and basic necessities and purchase of surplus produce (f) marketing facilities for exportable surplus products.

It is to be noted that some of the hill areas have dense bamboo forests, adequate to establish paper mills. Sugar industry also has scope in the region.

Even if heavy subsidy is required, several industries should be started in the region. This will introduce new outlook and new way of life among the people. Also this will instil a sense of confidence among them. Besides, establishment of industries which have markets in the plains, would create a stage for integration. Considering these direct and indirect benefits, the cost involved should not be a matter of highest importance. But the key to the industrial development would be the availability of power. The water power potential of the region is immense. But the same has still to be adequately tapped.

In the matter of development of industries, special attention is to be given to the maintenance of the social base of entrepreneurship. In some areas it has been found that with the expansion of marketing facilities the local enterprise tend to be replaced by imported enterprise. Besides, the indigenous qualities for which some of the products have demand outside, are also lost and ultimately the market dries up. It is only with imaginative measures that such a development can be prevented.

Two Faces of Regionalization

The strategy of development of regional economy should form a part of the overall strategy of development of the national economy. Besides regionalization has two faces. On the one hand regionalization in economic development would replace the parochial interests of the local communities by a larger interest. On the other, a tendency might grow up for lop-sided developments of the region without consideration of the total national interest. It seems that the techno-economic surveys conducted by the National Council for Applied Economic Research for the different administration units separately have not taken this aspect of the question into consideration adequately. More thought is to be given to this matter and many more studies are to be undertaken.

A Question of Basic Strategy in Economic Development

There is a tendency among some intellectuals to make a shibboleth of the programme of evolving socialistic pattern of society. While it is our national goal, it is to be approached through phased out tactics. It is to be noted that Soviet Russia introduced socialism in Central Asia, only in the late twenties. In the first phase of the revolution in that area, more attention was given to the liberation of the people from the institutions and interests imposed by Czardom.

It seems that at this stage, while steadily laying down the base for socialistic pattern of society, there should be less talk of socialism in the hills. If the nation can swallow the Birlas, it will do no harm if a few petty leaders of industry come up from among the hill people.In the caste-ridden Indian society, this will be of symbolic value. It will enable the tribal people to participate in the national life at higher level. For national integration, not only multi-dimensional, but also multi-level participation is indispensable.

It follows from the above that more emphasis should be given on removal of outside exploitation and nourishing local enterprise. In Manipur, out of about twenty forest contractors, only 3 or 4 are local people. In NEFA and Nagaland though preference is given to the local people in trade and commerce, frequently outsiders control the same, under cover. Perhaps an attempt to provide entrepreneural training may help to meet the problem to some extent.

Conclusion

From the foregoing appraisal of the various problems in the hills of North-East India, it is obvious that a mere plan of economic development would be utterly inadequate. Along with economic planning, there should be social and political planning in an integrated manner. A single agency should be entrusted with the task at the national level. At present the different aspects are being dealt with by different agencies in an uncoordinated manner. The dominant focus in dealing with the tribals of the region seems to be that they are primitive people. But this is not right. The primitive complex of the tribal communities of the country as a whole, and of North East India in particular, is rapidly giving place to the minority complex. While at the beginning of the 1950s, a welfare approach was the appropriate one, now welfare approach should only play a subsidiary role, to the integrated political, economic and social approach.

IX Pluralist Basis of Indian Politics

A.H. HAQQI

India has a civilization which is as old as that of China with a similar unbroken continuity. But, as against China, India has never been effectively governed from one political centre with the result that, till the recent integration of the princely states, India has been more or less a geographical expression with a wide variety of political systems and cultures sprouting and co-existing in different parts of the country. History and geography may or may not be regarded as determinants of political behaviour but they are certainly important variables, and, in the case of India, history and geography have played an important role by the creation of cleavages and the strengthening of the differences between the inhabitants of the different regions of the country, especially the peripheral regions in certain areas.

Notwithstanding the separation of the territories that now constitute Pakistan and Bangladesh India is still too large with an area of 32,80,483 sq. km. and too populous with a population of over 600 millions which is greater than that of Africa and Latin America combined; extremes of climate, 9.4°C being the lowest average minimum temperature in Kashmir and 48.9°C being the highest average maximum temperature in Rajasthan, and rainfall ranging from 200cm on the west coast form Trivandrum to Daman and Assam, 100 to 200 cm. in West Bengal, Bihar, East Uttar Pradesh, to under 50 cm. in Kutch, Western Rajasthan, South Punjab; the nature and dimensions of Indian topography are still subcontinental.

Physiographically, India consists of seven major regions: northern mountains, including the Himalayas and the North -East ranges, the great Indo-Gangetic plain, the central highlands, peninsular plateaux, east coast, bordering seas and islands. The distances that separate the different regions (North-South 3,218 km., East-West 2,977 km) and the obstacles to inter-communication have been and are still too formidable to permit, much less encourage, mobility. Hence, the overwhelming ecological and cultural diversity that is India.

Reprinted from *Punjab Journal of Politics*, 1984, Vol. 8, No, 11 pp. 129-133 with the permission of the editor.

Indian political geography is, moreover, neither neat nor tidy. The political sub-divisions designated states and union territories are the product and by-product of 'wisdom and chance'. Inhabited by persons belonging to various religious and linguistic/cultural groups, the states are of all shapes and sizes, and have been constituted together with language on the basis of not one but several factors: historic and political factors (UP, Bihar, Jammu and Kashmir), language and socio-cultural distinctiveness (Assam, West Bengal, Orissa, Karnataka, Andhra Pradesh, Tamil Nadu, and Kerala), ethnic and economic considerations (Nagaland, Meghalaya, Manipur and Tripura), religion, script and communal sentiments (Punjab, Haryana and Himachal Pradesh), and integration of princely states and viable grouping (Rajasthan and Madhya Pradesh). With an area of 2,94,413 sq. km., Uttar Pradesh is twice as large as Czechoslovakia (1,27,876 sq. km) and is more populous (88,341,144) than the two Germanies (GDR and FGR: 78,729,00) combined. Madhya Pradesh sprawls over an area of 4,42,841 sq. km. which is greater than that of Britain (230,609 sq. km), Bulgaria (110,912 sq. km) and Portugal (91,631 sq. km) combined; while with a population of 41,654,114, it is almost as populous as Spain (34,490,0000) and Sweden (8,143,000) taken together. There are, on the other hand, small units like Nagaland (area: 16,527 sq. km; population: 516,4490), Tripura (area: 10,477 sq. km; population: 1,556,342), and Sikkim (area: 7,298 sq. km; population: 208, 609), which appear to be dwarf, if not pygmies, in the company of giants.

The same 'diversity' and 'disparity' is to be found in the religious and linguistic composition of the population. The major religious communities are Hindus (84%), Muslims (11.4%), Christians (2.1%), Sikhs (2%), Buddhists and Jains and they are to be found in almost all the states but in varying strength. The Muslims, the largest religious minority otherwise, are a majority in Jammu and Kashmir.

There is a bewildering variety of languages and dialects, and India is a veritable 'tower of Babel'. There are, at least, twenty five languages with more than a million speakers, and no less than fifty other Indian and tribal languages and dialects with more than a hundred thousand speakers, though all of them are not literary languages. Only fifteen languages (Assamese, Bengali, Gujarati, Hindi, Punjabi, Kannada, Kashmiri, Malyalam, Marathi, Oriya, Sanskrit, Sindhi, Tamil, Telugu, and Urdu) have been recognized as the principal or 'national' languages of India (though according to the Linguistic Survey of India, there are no less than 179 languages and 544 dialects).

Recognizing the nature and extent of the social segmentation, the dimensions of ecological and cultural diversity, and the different levels of development of the various regions and socio- cultural groups, and also fully cognisant of the suspicions, fears and prejudices subsisting between the different castes, communities and regions, the fathers of the Indian republic, therefore, sought to preserve and promote 'unity in diversity'. This was attempted by:

(a) incorporating a detailed list of Fundamental Rights, and Directive Principles of State Policy to assure the dignity of the individual and higher standards of living in larger freedom to all the citizens without any distinction or discrimination.

(b) abolishing untouchability and making special provisions to ameliorate socio-economic conditions of socially, educationally and culturally backward communities and to obligate the state to promote the educational and economic interests of the weaker sections and, particularly, the Scheduled Castes and Tribes;

(c) ensuring representation of the Scheduled Castes and Tribes in parliament and the state legislatures by reservation of seats for them on the basis of population; and of the Anglo-Indians (a religious, social and linguistic minority) by authorizing the President and the Governor to nominate, if necessary, two members for the Lok Sabha, and one member for the state assembly concerned respectively; and

(d) assuring specific educational and cultural rights to the minorities and guaranteeing the right of every minority "having a distinct language, script or culture of its own" to "conserve the same".

Without going into the controversy about the "true nature" of Indian federalism, it may not be unreasonable to say that the constitution does reflect and seek to ensure the federal character of Indian polity through some of its provisions such as those relating to: (a) the procedure for amendment of 'federal' features of the constitution; (b) the schedule listing fifteen languages as national languages; (c) the educational and cultural rights of minorities; (d) the directives requiring the Union Government to take adequate steps for the encouragement and development of non-Hindi languages worthy of their status as 'national languages', and (e) to encourage and foster the growth and development of Hindi, "so that it may serve as a medium of expression for all the elements of the composite culture of India" (Article 351).

It may, perhaps, be not out of place to make in this context a specific reference about the Indian President. Though the constitution is based upon the 'Westminster model' with a council of ministers responsible to the Lok Sabha 'to aid and advise' him, the President is elected by an electoral college consisting of the elected members of Parliament and the state assemblies by proportional representation through single transferable vote, which, in effect, means that the states take part in the election directly through their assemblies and indirectly through their representatives in the Rajya Sabha, elected by their assemblies. It can, therefore, be inferred that he represents as nobody else can the 'unity in diversity' that is India, and, perhaps, that is why the Vice-President does not succeed him in case of resignation or death (as in the US), but a fresh election has to be held.

In the case of political/discretionary appointments, one finds a definite pluralistic approach to give representation (may be not always adequate or to the entire satisfaction of the interests concerned) to the various groups and regions both at the central and state levels. There are, however, complaints that at the lower or 'grass-roots' level, the minorities and weaker sections are not getting their share of 'loaves and fishes' in non-competitive appointments nor in the social welfare benefits and expenditures.

Interestingly, by coincidence or definite design, it has so happened that the highest office has rotated between south and north; the first Indian Governor-General was C. Rajagopalachari (Tamil Nadu) and he was succeeded by Rajendra Prasad (Bihar) as the first President of the republic. The other distinguished occupants of the Rashtrapati Bhavan have been S. Radhakrisnan (Madras), Zakir Husain (UP), V.V. Giri (Tamil Nadu). Fakhruddin Ali Ahmed (Assam), N. Sanjiva Reddy (Andhra Pradesh), and currently Zail Singh (Punjab).

The Prime Ministership, however, has seemed to be a monopoly of Hindi-speaking north, and Shastri that too of Uttar Pradesh and of the Nehru family. Right from 1947, with a short interlude of about 18 months when Lal Bahdur Shastri was Prime Minister (1964-65), and again during the short-lived Janata experiment (1977-79), it is the Nehrus (father and daughter) who have ruled India. But with Morarji Desai (Gujarat) taking over as Prime Minister (1977), there were two deputy Prime Ministers—Jagjivan Ram (Bihar) and Charan Singh (UP), providing an excellent foil to and balances between tradition and modernity, between Hindi and non-Hindi speaking regions. But all said and done against them, it cannot be gainsaid that the Nehrus 'belong' to the whole country and they enjoy an ample measure the affection, respect and confidence of the masses, irrespective of their mutual antagonisms or conflicting claims. If the Congress has been a monopolist in Indian politics, it is because of them and, if India was/is a 'soft state' as asserted by Myrdal, it was/is because of the Nehruvian commitment to pluralism and the doctrine of 'concurrent majority'. With whole-hearted acceptance of variety and heterogeneity, there was and is no alternative to self-restraint and a pluralist approach. May, however, one add that a short dose of authoritarianism during the internal emergency (June 1975 to March 1977) made the 'democrats' sorry and worry about the Indian masses and the future of 'the largest democracy in the world', that the attempts towards a single-dimensional political discipline ended in an unprecedented electoral/political disaster for the Congress (led by Mrs. Gandhi) and that the formation, functioning and failure, of the Janata government has been an unconscious endorsement of the Nehruvian model of development?

Not only the framers of the constitution but also Nehru and his successors had, thus, consciously opted for a pluralistic model of development for the solution of the problems of a highly pluralist community: reshaping of the polity and the economy, modernization of the socio-institutional framework and the cultural milieu, removal of poverty and ignorance, and that of the extreme inequalities between different regions and different subsisting between the different communities, religious and regional, linguistic and cultural, about their future and role in independent India.

X Cultural Pluralism, National Integration and Democracy in India

IQBAL NARAIN

Societies characterized by cultural pluralism have a built-in tendency for conflicts but are not necessarily prone to balkanization. In fact, balkanization of a country is not essentially a function of cultural pluralism as of failures at the managerial level, both economic and political. Conflicts in culturally plural societies may appear to be acute, particularly in situations beset with economic scarcity and exposed to exploitation by a political elite in quest of a support structure in a democratic polity. But, in fact, these conflicts may basically be only aspects of bargaining pressure politics, working toward a reasonable politico-economic deal rather than secession from the national mainstream. If the politico-economic system shows enough resilience to be firm and yet accommodative to reasonable demands, it may succeed not merely in containing the conflicts that aim at unreasonable and extravagant bargaining but also convert them into healthy partnership deals in the interest of nation-building. For, after all, political development is "a process of meeting new goals and demands in a flexible manner."[1]

Cultural pluralism and the conflicts emanating from it, thus may not necessarily impede the processes of national integration and the emergence of a democratic political community. In fact, the phenomenon of cultural pluralism may provide for firmer traditional moorings[2] which may, in turn, serve as a safeguard against conversion of the essentially modernizing process of rising expectations into what Daniel Lerner calls a "revolution of rising frustration."[3] This is not a plea for rationalizing all traditional moorings per se, let alone a socio-economic *status quo*. These are, finally, to be subsumed, even modified but not perhaps replaced entirely, by a secular, essentially modern loyalty structure which, in fact, can only be created around a new sense of community, emanating from a consciousness of common destiny,

Reprinted from *Asian Survey,* Vol. 16, No. 10, 1976, pp. 903-917, by permission of the Regents of the University of California.

shared goals and values. The creation of a secular loyalty structure may also be the result of dialectical interaction between "tradition" and "modernity", each infiltrating and transforming the other. The argument about safeguards has been advanced, however, because political modernization itself is the result of a protracted process, as it

> involves the extension of political consciousness to new social groups and the mobilization of these groups into politics. Political development involves the creation of political institutions sufficiently adaptable, complex, autonomous and coherent to absorb and order the participation of these new groups and to promote social and economic change in the society.[4]

It is proposed to study in the specific Indian context the pattern of interaction between cultural pluralism on the one hand and national integration and democracy on the other against the theoretical perspective elaborated above. By way of a preface to this discussion it is worthwhile to point out that cultural pluralism manifests itself, among others, in regionalism (inclusive of linguistic regionalism), religion and caste. These three factors are being used, therefore, as the key foci in the operational definition of cultural pluralism. It may also be added that cultural pluralism in the specific context of its operational definition constitutes the central theme and national integration and democracy are being treated as contextual sub themes. The effort is to just pose the central theme against the sub-themes and bring out the relationship between them, as well as the impact of the one on the other.

Regionalism, National Integration and Democracy

The central argument being developed here is that regionalism need not *necessarily* have a negative correlation with national integration and democracy, as has sometimes been argued in the literature on Indian politics. The elaboration of the argument requires a close look at the nature of regionalism itself.

The Phenomenon of Regionalism —Notes towards a Characterization: Regionalism in the Indian context, as elsewhere, is a nebulons phenomenon. While it embodies in positive terms a quest for self-identity and self-fulfilment on the part of an area, it reflects, negatively speaking, a psyche of relative deprivation on its part—not always viable in terms of rational economic analysis, let alone prone to rationalization. It is worth emphasizing here that for a balanced relational perspective on regionalism, national integration and democracy, it is necessary to remember that regionalism has a positive content which has, more often than not, been ignored by political analysis of the Indian situation. It may be recalled here that the 1965 report of the States Reorganisation Commission had hoped that regionalism would be imbued with a positive content if the States were reorganized on the basis of language:

> It is obviously an advantage that constituent units of a federation should have a minimum measure of internal cohesion. Likewise, a regional consciousness, not merely in the sense of negative awareness of absence of repression or exploitation but *also in the sense of scope for positive expression of the collective personality of a people inhabiting a state or a region* [emphasis added] may be conducive to the contentment and well-being of the community. Common language may not only promote the growth of such regional consciousness but also make for administrative convenience. Indeed, in a democracy the people can legitimately claim and the government have a duty to ensure that the administration is conducted in a language which the people can understand.[5]

A closer look at the stages of growth, components and types of regionalism is necessary to bring its nature into sharp focus in the Indian context.

(1) *Stages of Growth:* It is wrong to assume that there was no regionalism in the pre-independence period, though it was much more dormant than what it came to be in the post-independence period. This was the case largely because of scarce opportunities for economic development, limited avenues for political participation, obsession with communal politics, and the primacy of a common national objective — liberation from colonial rule. Regional simmerings however, had started in the wake of the implementation of constitutional reforms under the Acts of 1909, 1919 and 1935. The establishment and role of the Justice Party in Madras and, to a lesser extent, of the Akali Dal in the Punjab in the pre-independence period are concrete examples of emerging regionalism in India.

As far as the post-independence period is concerned, there are three major landmarks in the development of regionalism. First, there was independence and the establishment of a democratic polity, committed to nation building on the principles of democracy, secularism, national unity and social justice. All parts of the country thus came to develop aspirations for a fair deal and began to compete for it; any crisis in expectations or lag in competition in any quarter could lead to regionalism.

Secondly, there was the integration of the princely states, resulting in the merger of relatively small identities into new and bigger identities. The new states thus came to develop for the time being split personalities with the loyalties of their citizens (and even of their political elites and the civil servants) torn between old territorial boundaries and new territorial structures. The split in state personalities would have one of its significant manifestations in the regional sentiment which would develop around loyalties to old territorial units.[6] This is the primary factor explaining the success of princes in elections. The princes often received overwhelming support in their former territories in the newly created states and relatively much less in other parts of the same state.

Finally, there was the process of linguistic reorganization, as a result of which 28 states were reshaped and reduced to 14 states and 6 centrally administered territories. To these, new states were added when, for example, Bombay was divided into Gujarat and Maharashtra and Punjab into Punjab and Haryana. But, as Rasheed- ud-din Khan has correctly pointed out:

> let it be remembered that, even as it is, the states are not constituted, despite popular misgivings, entirely on the basis of linguistic homogeneity. A closer scrutiny will reveal that together with language many variables and critical factors like ethnic-cum-economic considerations (Nagaland, Meghalaya, Manipur and Tripura); religion, script and sentiments (Haryana and Punjab); language-cum-culture (Maharashtra and Gujarat); historical and political factors (U.P. and Bihar); integration of princely states and the need for viable groupings (M.P. and Rajasthan); and, of course, language-cum-social distinctiveness (Tamil Nadu, Kerala, Mysore, Bengal and Orissa) have played a decisive role in the composition of the Indian federation.[7]

And, indeed, the terms of reference of the States Re-organisation Commission had enjoined it to take note of four broad principles: (i) the preservation of the unity and security of India: (ii) linguistic and cultural homogeneity: (iii) financial, economic and administrative considerations: and (iv) the successful working of the national plan. It cannot be denied, however, that language has been the critical factor in the reorganization of states, which became a force to be reckoned with in the context of regionalism, so much so that the epithet "linguistic regionalism" gained currency in the literature.

(2) *Components:* The foregoing discussion of the stages of growth should serve as a background against which a meaningful discussion of the components of regionalism can be attempted. It is a multi-dimensional phenomenon, at once geo-cultural, politico-economic and, above all, psychic. The components of regionalism also overlap each other. It is, therefore, difficult to say which of them play a *seminal* role and which of them are important as a *booster.* Still for purposes of analytical clarity these can be grouped and discussed under these general headings.

It will just be emphasizing the obvious to point out at the outset that the geo-cultural component is the heart of India's historic cultural pluralism, and the bedrock of the phenomenon of regionalism. It consists of several factors, which are not only important *individual* but also in *conjunction* with each other as with other groups of components. A conjunctional perspective alone will, in fact, bring out their real import.

First, there is the factor of *geographical boundaries* to which the people of an area usually relate their identity. One, however, is at times tempted to hypothesize that the territorial orientation is not central to its being, let alone to its becoming. The geographical boundaries, though still clung to, are, historically speaking, symbolic, at least in the Indian context, of overlapping and vanishing footprints on the sands of time, which need economic and political props for their retrieval from time to time in the increasingly hazy memories of succeeding generations. For example, memories of old geographical boundaries of princely states still haunt and are exploited but it can hardly be denied that they are yielding place to new and bigger territorial identities — eg. Rajasthan, Madhya Pradesh, or Orissa. While there is an obvious element of exaggeration here, this has been deliberately introduced to put across the point more sharply.

Secondly, there is the factor of *history,* which also buttresses regionalism by way of cultural heritage, folk lore, myths and even symbolism. Nothing perhaps illustrates this better than the story of the rise of the Dravida Kazhgam (DK) and the Dravida Munnetra Kazhgam (DMK) in Tamil Nadu. States like Maharashtra and Andhra Pradesh also had cultural identities predating the formation of present state boundaries which would, in turn, split them on regional lines even after the creation of new states.[8] But history cannot always withstand the onslaught of rugged politico-economic realities as is again borne out from the change in the stand of the DMK from secession to one of autonomy within the federal constitutional framework.

Thirdly, there is the factor of *caste.* This, *per se*, is not so important except when combined with dominance (as in Maharashtra) or when working in conjunction with linguistic preponderance (as, again in Maharashtra or Tamil Nadu) or religion (as in the Punjab or, negatively speaking, in Tamil Nadu in the sense of Anti-Brahmanism) and so on.

Finally, there is the factor of *language,* which "is perhaps the most important mark of group identification. But it is more precisely discriminating than either colour or race, for whereas the latter may be overridden by other forces so that they cease to refer to any real interacting community of persons, language as it is used in working link.[9]

Linguistic homogeneity strengthens regionalism both in positive and negative senses; in the former in terms of strength in unity and in the latter through emotional frenzy. Language is "an expression of shared life, thought structures and value patterns" and as such has the potential to unite people emotionally and make them work to improve their common destiny as also to add to their bargaining strength.[10] The story of the creation of Punjabi Suba is one case in point. But common language is no bar to intra-state regionalism. Nothing perhaps illustrates this better than the Telengana issue where economic grievances and demands triumph over considerations of linguistic homogeneity. The very fact that language helps a group to forge an emotional identity also tends to make it highly inflammable, as shown by the language riots in Assam or in Tamil Nadu. The language issue in terms of regional language *vs* Hindi *vs* English, as distinguished from other problems created in the wake of linguistic reorganization of states, led at times to such disturbing questions as "can different states in India communicate with each other if English is withdrawn and Hindi does not emerge as a link language?"[11] But these fears were exaggerated : English is continuing and resistance to Hindi is weakening.

Among the problems specifically related to linguistic reorganization of states is the issue of the relationship of minority linguistic groups with the majority linguistic group in a state. There are two types of minority linguistic groups: (i) linguistic minorities belonging to linguistic groups which are majorities elsewhere — e.g. Bengalis in Assam — and (ii) minorities which are not majorities anywhere. After identifying these two types, Myron Weiner has posed the problem in these words.

> Whether the minority group is a tribe or a religious or linguistic group, it is often concerned with preserving its cultural identity and resisting assimilation into a larger regional culture. Regionalism, like its earlier counterpart, nationalism, has become a powerful loyalty within most of the states and has resulted in a degree of assimilation of related social groups, but minorities are often even more protective of their status in the midst of growing regional sentiments. This protectiveness takes many forms; there is often a fight over representation in the governing Congress Party and in the state assembly, insistence upon a share of appointments in the state administration or a proportion of admissions and scholarships into colleges, technical schools and universities, and where the minority group has its own language, pressure for the use of the mother tongue as the medium of instruction of local schools and in the publication of state laws and administrative regulations.[12]

Experience has shown, however, that the thrust of linguistic regionalism can be contained, if not neutralized altogether, with a relatively equitable sharing of the economic cake, tactful political and administrative management and firm handling by a nationally-oriented leadership. "It is, therefore, inappropriate to feel (or affect) shock and despair whenever the problem manifests itself as, for instance, in the question of language. For it is rather the relation of the problem to the 'resources' for dealing with it that merits attention.[13]

Indeed, the economic component is the crux of regionalism. It is common knowledge that India is underdeveloped economically, in spite of the many achievements registered since independence. The resources are scarce and demands disproportionately heavy and overgrowing in the wake of the continued population explosion. There is thus acute competition among individuals, groups and regions within a state. That at heart is the story of regional economic imbalances to which the political elite in general, and the ruling elite in particular, have contributed in no small measure. Demands for distributive justice thus are generated and assume the form of movements. It may be recalled here that most of the demands for constituting new states were primarily based allegedly on unfair and unequal distribution of developmental benefits and expenditures in the multi-lingual states.[14] The Telengana movement is a classic instance in point. The simmerings of discontent in the depressed regions of Bihar, Orissa, Madhya Pradesh, Maharashtra, Gujarat, U.P. and Rajasthan are other examples. The scramble for the location of steel plants and other heavy industries and charges of partisanship levied against the center in this regard is another instance in point. The tensions between migrants and "sons of the soil" are another manifestation of the economic component. The conflict between Assamese and the Bengalis in Assam and the Shiv Sena's slogan of "Maharashtra of Maharashtrians" illustrate the point.

Seling Harrison noted that:

> As population mounts, densely populated regions cry *lebensraum* at their neighbours. Bihari labourers flock to jobs in the mining and industrial towns bordering Bengal and overcrowded Bengal protests in return that the towns should be the preserve of its own jobless. The tea plantations of Assam become a battleground between rival groups of Assamese and Bengali unemployed. Tamilnad demands the border plantation land of Malayalam-speaking Travancore-Cochin to protect Tamil coolies from Malayalee competition. With each year that more and more people- made more and more group conscious by scarcity compete ever more feverishly for limited economic spoils, the hope for ultimate escape from India's vicious circle grows more remote.[15]

One would, however, take Seling S. Harrison's concluding observation with a grain of salt without becoming complacent about the gravity of the situation. In order to set the perspective right, it should be noted here that, simultaneous to these 'natural pulls and pressures for distributive justice, the processes of economic integration have also been at work in the wake of nationally planned economic development. As a result of these processes, even relatively "have-not" regions have come to realize that their ultimate salvation lies in remaining a part of the Union and not in seceding from it. Thus, while regions within a state try to forge a separate identity as a state, they do not want to leave the Union. The demands for Punjabi Suba, the separation of Gujarat from Maharashtra, or of Haryana form Punjab illustrate the point. In fact more often than not, these movements want the Union to arbitrate.

Moreover, the thrusts of technological and scientific developments are also in the direction of unity and collaboration, as is the secularization process in the wake of industrialization, migration and urbanization. Finally — and most important of all — the economy, particularly since the declaration of the emergency in 1975, is showing up in good measure. One is, therefore, inclined to argue that regional pressures emanating from economic compulsions are more bargaining counter than secessionist moves.

The political component of regionalism is important even though politics does not so much create regionalism as accentuate it. Politicians, in their not always so enlightened self-interest, exploit situations of regional deprivation and unrest and convert them into movements to forge and strengthen their individual and factional support bases. It is common knowledge, for instance, that the in-fighting within the Congress Party gave a lease of life to the Telengana agitation, and that the Shiv Sena thrives in Maharashtra with the connivance of the Congress bosses there. Regional political parties like the DMK (Tamil Nadu), Akali Dal (Punjab), Jharkhand Party (Bihar), in fact build up and survive on the accentuation of the regional sentiment. The same is true of the Senas, such as the Shiv Sena in Maharashtra or the Gopal Sena in Kerala. Border disputes, like the one between Maharashtra and Karnataka, also thrive by fomenting regional sentiment.

The fact that the opposition political parties have regional pockets of influence almost compels them to swim on the tide of regionalism.[16] The same is true of Congress factions at the state level, as the party cannot ignore the local and regional power configurations even though in the broader context one-party dominance, with its national leadership, has been a unifying force in the country:

> In its effort to win, Congress adapts itself to the local power structure. It recruits from among those who have local power and influence. .... The result is a political system with considerable tension between a government concerned with modernizing the society and economy and a party seeking to adapt itself to the local environment in order to win elections.[17]

The Congress has to adjust not merely at national but also at state levels, as indicated by the stance of the Congress on the language issue in Tamil Nadu or the border dispute between Maharashtra and Karnataka. Again, power and patronage are important media for fulfillment of regional aspirations that are at times deliberately used by the ruling elite in favour of one region and at the cost of others. Thus regionalism always thrives on real or fancied charges of political discrimination which has, in fact, become an important facet of politics of regionalism itself.

Finally, regionalism in India, as elsewhere, is a psychic phenomenon, and in the ultimate analysis its roots lie in the minds of man. Each individual, by and large, carries a split personality; he is in part sub-nationalistic and in part nationalistic. There is always a natural tendency towards the primacy of the sub-nationalistic over the nationalistic though he does not usually pose the two as an either/or situation. "To be Indian" is not necessarily "not to be a Maharashtrian."[18] The situation is best brought out in an observation of Lokenath Misra:

> My first ambition is the glory of Mother India. I know it in my heart of hearts that I am an Indian first and an Indian last. But when you say you are a Bihari, I say I am an Oriya. When you say you are a Bengali, I say I am an Oriya. Otherwise, I am an Indian.[19]

Thus, it should not be surprising if emotional overtones get easily attached to regionalism which, in turn, can be traced to its cultural moorings and economic roots. Patterns of socialization which, more often than not, tend to be more parochial than national also contribute to it.

(3) *Major Types:* Like the components, the types of regionalism also seem to overlap. Though regionalism is not always conterminous with present-day state boundaries, a state can still be treated as a unit for evolving a typology of regionalism. One could thus identify three major types.

First, *supra-state regionalism* is built around, and is an expression of group identity of several states which join hands to take a common stand on an issue of mutual interest *vis-a-vis* another group of states. The group identity thus forged is usually *negative* in character: it is usually against some other groups identity. It is also *issue-specific* in that it is confined to certain matters on which the group would like to take a common and joint stand. It is not at all a case of total and permanent merger of state identities in the group identity; in fact, intra-group rivalries, tensions and even conflicts continue to take place at times even simultaneously with group postures. South *vs* north in India on such issues as language or the location of steel plants illustrates the point.

Second, *inter-state regionalism* is coterminous with state boundaries and involves juxtaposing of one or more state identities against another on specific issues which threaten their interest. River water disputes in general and the Maharashtra - Karnataka border dispute in particular can be cited as examples.

Third, *intra-state regionalism* embodies the quest of part within a state for self-identity and self-development in positive terms and negatively speaking, reflects a psyche of deprivation and/or exploitation of a part in relation to other parts of the same state. This type of regionalism is most rampant, typified by a Vidarbha in Maharashtra, a Saurastra in Gujarat, a Telengana in Andhra Pradesh, an East U.P. in Uttar Pradesh, and an East Rajasthan in Rajasthan.

Regionalism in India, thus is a complex amalgam of geo-cultural politico-economic and psychic factors. It is difficult to decipher which component has ascendency, though it will not be surprising if the economic factor has an edge over the others in the ultimate analysis. It is also interesting to note that regionalism in India seems to manifest itself in overlapping circles in terms of its major types which as such are not immune to fusing into each other and also being diffused, if an overarching national loyalty structure emerges to transcend them.

Regionalism and National Integration: After a rather protracted discussion of the nature of regionalism in India, we turn briefly to the vexing issue of its relationship with national integration. But before this is done, a word on national integration is appropriate.

The problem of national integration is not peculiar to India. In fact, it concerns both developed and developing countries, though it is more acute in the latter owing to the phenomenon of the politics of scarcity. National integration as a concept is at once holistic and multidimensional. It is holistic because it takes into account society, culture, economy, policy, administration and even education by way of tributaries. It is multi-dimensional because it is a question of cultural, politico-administrative, economic and emotional integration. And its multi-dimensionality makes it at once a goal and a process. The three more important facts of national integration are: (a) culture-group integration; (b) elite-mass integration; and (c) integration of sub-nationalistic and national identities. We are concerned here more with the first and the third to which we shall turn when we talk of regionalism and democracy. This digression apart, the quest of national integration in the specific context of India is a search for unity in diversity. Plurality is accepted and prized and as such the path chosen for national integration is one of unity in diversity and not of unity out of diversity. The country also opted for *democractic-participatory,* as opposed to *elitist-authoritarian* instrumentalities to accomplish the task of national integration.

As far as the relationship between regionalism and national integration is concerned, the main thrust of the argument so far has been that in India, as elsewhere also,[20] the former is neither necessarily antithetical to the latter nor should it be surprising if both show enough resilience not merely to coexist but also to develop a mutually reinforcing and creative partnership. The two are not basically antithetical because both are pro-development, one aiming at the development of regional identity and the other of national identity. The development of the country is a total phenomenon and as such there is a natural and mutually reinforcing interdependence between the development of regional and national identities. The prospects of coexistence and creative partnerships are also there for several reasons. First, India's approach to unity in diversity is conducive to culture- group integration. Second, the processes of economic integration are also rending to buttress relationships along this pattern. Third, with improvements in the economy, there are also prospects for a relatively equitable distribution of the economic cake. Finally, an overarching sense of national identity seems to be emerging, as is obvious from the fact that in periods of national crises regional identities have voluntarily merged themselves with the national identity. The overlapping manifestations of regionalism, as stated elsewhere, makes it flexible enough to adjust to emerging national identity. As Rasheed-ud- din Khan has stated in a somewhat different context:

> Every urge for autonomy is not a divisive, but most probably a complementary force; it would not lead to balkanization but to the restructuring of national identity; it is not a fissiparous but a normal centrifugal tendency in a federation; it should not be taken as a call for disintegration of the national sovereignty, but its re-integration.[21]

Regionalism and Democracy: The relational pattern obtaining between democratic politics and regionalism in India is similar to that pertaining between regionalism and national integration. Democratic politics is based upon bargaining and competition within the confines of a one-party dominant system. The competitive and bargaining character of politics has, in turn, encouraged regionalism which, in the process, has tended to become a bargaining counter through which the deprived groups press for access to and accommodation in the political system (particularly in terms of developmental benefits). It is true that regionalism can give a violent orientation to the style of bargaining or pose demands far in excess of the capabilities of the political system to cope with. But until now, at any rate, it has also practiced self-denial and retreated when it perceived that the limits of bargaining had been reached and would yield no further results. So long as this happens for processes of democratic politics, regionalism and national integration can coexist without necessarily weakening each other in any fundamental sense. Democratic politics has also led to elite-mass interaction through elections and the phenomenon of linkage politics developing in the country in the wake of democratic decentralization, popularly known as *panchayati raj*, which should ultimately help the process of elite-mass integration by bridging the gulf between the center and the periphery. Thus, for the reconciliation of the competing claims of regionalism and national identity the political system of the country should continue to remain both federal and democratic.[22]

Religion, National Integration and Democracy

We will now turn very briefly to religion, which is another important component of cultural pluralism in India. We are taking a rather delimited view of the phenomenon by focussing on relations between Hindus (who constitute about 83% of the population) and Muslims (who, in spite of partition, still number about 65 million, making India the third most populous Muslim nation in the world after Indonesia and Bangladesh). There are other religious groups in India but it is the relationship between Hindus and Muslims which is at once the crux of the problem of national integration and basic challenge to the democratic polity in the country. Partition, far from solving the problem, left lingering, poignant shadows behind, which only made the problem psychologically complex, particularly in the initial stages of the country's independence.

India is, however, committed to be both a secular state and a secular society. The process is underway and the country has made much headway. And yet it has a long way to go. Communal riots still take place on flimsy grounds and the sufferers are usually the Muslims — e.g., the Gujarat and Delhi riots of 1969 and 1974. The country still has communal organizations such as the Hindu Mahasabha, the Ram Rajya Parishad, Rastriya Swayamsevak Sangh and the Jana Sangh as bulwarks of Hindu communalism and the Jamaat-e-Islami and Muslim League of Muslim communalism.

It is, however, noteworthy that Muslims are occupying high places in educational, administrative and political echelons and that the state does not discriminate on religious grounds. Muslims enjoy freedom of religion under the constitution. What is most important to note is that each election has tended to make religious minorities, particularly the Muslims, conscious of their political importance, thus drawing them gradually into the national mainstream. This softens the communal edges of political parties and strengthens the secularization process.

Secularization is also a function of economic development, particularly industrialization. Thus the various religious communities are also learning to live and work together for their economic rights within the trade union movement, where communal political parties have not had much success. The bannings of some of the communal organizations under the emergency should further weaken this ingress. It cannot be denied, however, that by its very nature secularization has to be a slow process in the Indian context, for:

> Political analysis must recognize that in a country which is characterized by glaring inequalities, appalling poverty, a highly vitiated social and political climate, wherein traditional loyalties are critical and where the political system has encouraged the jockeying for power caste by caste and community by community, the political secularisation of minority communities cannot take place at a rate faster than that of the whole society. The whole society must grow and change at an even pace, and that is the only course for the successful socio-political integration of an otherwise diverse and plural society.[23]

And the hope of the change lies in the urge of the religious minorities to find a place in the system through participation in democratic politics, thereby sharing developmental benefits and in the process ultimately becoming part and parcel of the national mainstream.

What the religious minorities are aspiring to is already an empirical reality in the context of the interaction of caste and democratic politics in India which has, in fact, passed through three phases. First, there was the *stage of arousal* through democratic confrontation between traditionally high but numerically minority castes and traditionally deprived but numerically majority castes. The traditionally dominant castes continued to holdsway but the deprived castes became conscious of their numerical strength. Secondly, there was the *stage of displacement* when numerically strong middle castes (mostly peasant castes) took over from the traditionally dominant castes. Thirdly, a *stage of second displacement* seems to be in the offing when the lower castes will replace, where possible, the middle castes or at any rate successfully bargain with them for coalitions on respectable (if not their own) terms. Thus almost a bloodless social revolution may be accomplished through the democratic process, under which the deprived strata of society are being brought into the national mainstream and also obtaining a share of the economic cake, thus giving them a stake both in the political system and nation- building activities.

Other facets of the interaction of caste and democratic politics do not require detailed description. It should suffice here to note that social stratification has tended to yield place to political stratification, transforming the near exclusive role of caste (already neutralized with each political party exploiting it) to at best a conjunctional role, particularly as one moved from micro to macro-levels of democratic politics, where political arithmetic requires political alignments cutting across caste, thus secularising this institution in the process.

The descriptive hypothesis emerging from the foregoing analysis is that cultural pluralism, national integration and democratic politics are to be treated as living, interactive and mutually reinforcing parts of a continuum. Developed as individual entities and equally strong in unison with the national identity, this is the "promised land" towards which regions, religious groups and castes seem to be led by the dynamics of democratic politics as it operates in India. If the destination is reached, the descriptive hypothesis of today may become the empirical reality of tomorrow. Let us, therefore, remember what Michael Novak had to say — in a different context — in this regard:

> Political unity depends in part upon cultural pluralism. Ethnic consciousness and expressiveness do provide one answer to the desperate search for a sense of belonging in America. There can be no strong individuals without strong loving families. To be strong and loving, families need to be rooted in and supported by communities.......If it were not for three of their institutions — religious freedom, universal suffrage,

and the right to assemble and petition — the rest of us would be a lot worse off, and, in some ways even less ethnic.[24]

REFERENCES

This article is based on the paper presented at the first National Conference on the Asian studies Association of Australia in May 1976.

1. Alfred Diamant, "The Nature of Political Development" In Jason L. Finkle and Richard Gable (eds.). *Political Development and Social Change* (New York Wiley: 1966) p. 92
2. This is because of competition for survival among different cultural identities which constitute cumulatively the phenomenon of cultural pluralism.
3. Daniel Lerner, "Toward a Communication Theory of Modernization", in Lucian Pye (ed.), *Communications and Political Development* (Princeton: Princeton University Press, 1963), pp. 327-50.
4. Samuel P. Huntington, *Political Order in Changing Societies* (New Haven, Conn: Yale University Press, 1968), p. 5.
5. *Report of the States Reorganisation Commission* (New Delhi: Government of India, 1955), p. 46.
6. For elaboration of this issue Iqbal Narain see, *et.al.* (eds.) *State Politics in India* (Meerut: Meenakshi Prakashan, 1967), particularly the Introduction, pp. XXI-XXIV.
7. For the development of the argument, see Rasheed-ud-din Khan (i) "Territorial Reorganization" Seminar, 137, Jan. 1971, pp. 28-32; and (ii) " The regional Dimension," *Seminar* , 164, April 1973, pp. 35-45. [The extracts here and elsewhere are from (ii) pp. 35-36].
   In fact, Khan argues that "the process of the infrastructuring of the Indian federation is not yet over". He makes a plea for further restructuring in terms of smaller and medium sized states, as 'the problem is really not of numbers but of the recomposition of the Indian federation based on objective criteria of socio-economic homogeneity and felt group identify in order to provide a more respective infrastructure for an industrial breakthrough and modernization." (p. 36)
8. There are four sub-regions in Maharashtra — Western Maharashtra, Marathvada Vidharba and Bombay city. Andhra Pradesh also has two subdivisions — Andhra and Telengana.
9. W. H. Morris-Jones, "Language and Religion within the Indian Union" in Philip Mason (ed.), *India and Ceylon: Unity and Diversity* (Oxford: Oxford University Press, 1967). pp. 51-52.
10. Rohit Dave has dealt with the advantages of remapping of stares on the basis of language in terms of better elite-mass communication, resulting in better socialization of the people in democratic values; attention to social problems peculiar to a linguistic groups, and sharing of "membership of a regional society and possession of a distinctive cultural, literary and historical tradition." (Rohit Dave "Regionalism in Maharashtra." *Janata,* (15) March 23, 1958.
11. Selig S. Harrison. *India: the Most Dangerous Decades* (Madras: Oxford University Press, 1960). p. 9.
12. Myron Weiner, "Political Development in the Indian States" in *State Politics in India* (Princeton: Princeton University Press, 1968). pp. 26-27.

13. Compare an observation of W. H. Morris-Jones: "The difficulties of India's language problem, in other words, are not unevenly matched by the political remedies she can bring to their handling — not least a body of shrewd and seasoned politicians trained in the states but conditioned, in layers as they move up the party ladder, to think in all-India terms." Morris-Jones, p. 66.
14. *Report of the States Reorganisation Commission,* p. 225.
15. Selig Harrison, p. 333.
16. Rajni Kothari has described the opposition political parties in India as regional in character because of their regional pockets of influence. But in view of their national aspirations, it is not entirely true to describe them in this manner. Rajni Kothari, "The Congress System in India," *Asian Survey* 6:12 (December 1964). pp. 1161-1173.
17. Myron Weiner, *Party Building in a New Nation (Chicago: University of Chicago Press, 1967). p. 15.*
18. Lokenath Misra, M.P., *Debates on the Report of the States Reorganisation Commission* (New Delhi: Lok Sabha Secretariat, December 1955, Vol. 11, col. 647).
19. Rasheed-ud-din Khan, p. 35.
20. In the U.S. example, "there appears to be a development toward both greater regionalism and greater nationalization". Ira Sharkansky, *Regionalism in American Politics,* (New York Bobbs-Merrill Co. 1970). p. 67.
21. Rasheed-ud-din Khan, p. 36. He is, however, not oblivious of the fact that "it is evident that like national chauvinism, unhealthy regional or sub-regional patriotism is equally cancerous and disruptive both of creative regionalism as also of syncretic nationalism." (p.35)
22. The author does not agree with the view "Is the resolution of intense but conflicting preferences in the plural society manageable in a democratic framework? We think not." Alvin Rabushka and Kenneth A. Shepsle, *Politics in Plural Societies : A theory of Democratic Instability* (Columbus, Ohio, Charles E. Merrill Publishing Co. 1972). p. 217
23. Imtiaz Ahmed, "Secularisation," *Seminar,* 144 (August 1971) p. 26.
24. Michael Novak, *The Rise of the Unmeltable Ethnics: The new Political Force of the Seventies,* (New York: Macmillan and Co., 1972). p. 321.

XI Geographical Bases of India's Federal Polity

SANDEEPTA ADHIKARI

Geography of Federalism

Federalism, like most institutional forms, is a solution of, or an attempt to solve, a certain kind of problem of political organization. The essence of federalism lies in the society itself. It is a device by which the federal qualities of the society are articulated and protected (6). A federal system of government is one in which there is a division of power between one general and several regional authorities, each of which in its own sphere, is coordinate with the others, and each of which acts directly on the people through its own administrative agencies (1). This spatial political system is born when a number of usually separate or autonomous political units (or units with some pretensions to autonomy) mutually agree to merge together to create a state with a single sovereign central government, while retaining for themselves some degree of guaranteed regional autonomy (3).

A federation is a geographical phenomenon and it develops out of interaction between the spatial structure on the one hand, and the political process and the system on the other hand. It is a device by which the state is adjusted to the nation and vice-versa (10). A federation is the most geographically expressive of all political systems. It is based on the existence of regional differences and recognizes the claims of the component areas to perpetuate their individual characters (11). The distribution of political powers, as between the central or federal government, and the government of the states or provinces, may reflect geographical conditions and the strength of local feeling. The federation framework is geographically suitable for large states. It is a means by which a state is able to cope with the problems of administering a very large area with many and sometimes conflicting sectional interests (2). Countries of large size with small population or large population concentrated in widely scattered areas, are obviously suitable for federal form of government (12). Presence of empty areas within a large state, usually gives

Reprinted from *The Indian Geographical Journal* Vol. 58, No. 2, December, 1983, pp. 162-171, with permission of the Editor.

rise to the feeling of separateness and that leads to centrifugal tendencies (5). Presence of isolated population clusters and urban foci does create a condition for the adoption of federal polity.

A federal structure is geographical requirement or necessity for multi-core states with different politico-religious ideologies. It is also a geopolitical solution for territories occupied by people of widely different ethnic origin, languages and religions and cultures as well. This is especially true in those cases where these differences have regional expression. Federal frameworks by their flexibility, have been able to accommodate expanding territories. Provisions are made for areas which have not yet been incorporated and the status for such territories shall have to be adjusted accordingly(2). Uneven distribution of resources and economic imbalances within the state do favour a federal polity. The ultimate goal of federal polity is national cohesion within the larger political framework . Federation does not create unity out of diversity, rather it enables the two to coexist(12).

Histogenesis of India's Federal Polity

The history of evolution of India's federal polity appears to have been consistent with the basic principles of geographical size, shape and contiguity of the territory and vice-versa. The idea of provincial autonomy was given a fuller expression in the Report on the Indian constitutional Reforms (the Montague Chelmsford Report) of 1918 which envisaged the future of India as a sisterhood of States, self-governing in all matters of purely local or provincial interest. This report was the basis of the Government of India Act of 1919 which introduced in the country a 'federal in embryo'(18). Wide geographical distribution of the British Provinces and the nature of their discontinuities separated by the intervening territories of the Princely States seemed to have formed the geographical basis for the growth of federalism in India before independence.

The Princely States of that time were the sources for centrifugal forces and especially threatened the integrity of the British Provinces. The federal system was thought to be means of containing the rising trend of centrifugal tendencies in marginal or peripheral areas of British Provinces and vice-versa. Besides the spatial distribution of various ethnic and religious communities was a potential factor for the British authorities to think over introducing the federal system. The geographical distribution of the Muslims in certain geographically contiguous areas accentuated them to demand for a federal political organization so that they could preserve their cultural and religious aspects and achieve regional autonomy in their majority areas. The Montford Reforms contributed towards the development of federal polity as a possible solution to the growing geopolitical problems arising out of the geographical distribution of ethnic and religious communities. The Montford Reforms helped in the linguistic reorganization of provincial boundaries and provided safeguards to the Muslims to ensure their own future as a district religious community. Federalism, with its promise of regional autonomy, seemed to offer the solution for the problems of Muslims, the Princes, the linguistic patriots and the imperial interest of Britain (3).

An Indian Statutory Commission, under the chairmanship of Sir John Simon, was set up to study the complex nature of spatial political problem of India. Many of the recommendations of the Simon Commission were incorporated into the Government of India Act of 1935, which may be regarded as blue print for the federal constitution for India. The Government of India Act of 1935 provided a considerable measure of autonomy and responsible government, instead of the system of dyarchy introduced by the Government of India Act of 1919 (11). The federation envisaged in the Act of 1935 represented the concessions which that Government was willing to make (8). The Act, in fact, created a kind of equilibrium between the centripetal and centrifugal forces.

Choice of New Delhi as a Federal Capital

The choice for a location of the federal capital has often been a matter of vigorous dispute at the time of the formulation of the constitution. The existence of local rivalries, in a sense, is implicit in the existence of a federal constitution (10). The location of the capital must represent and symbolize compromise and territorial adjustment among the various constituent units of the federation. It must have a "federal raison d'etre" of its own to manifest the centripetal idea.

Calcutta was an arbitrary capital. It was created to represent a preconceived geographical framework. As the process of the territorial integration, initiated by the British, came to an end by 1910, it was found that Calcutta would be unfit to maintain the integrity of the newly established British empire.

The move to shift the capital from Calcutta to Delhi in 1911, was in part a conscious effort to harmonize the alien rule with the historic tradition and in part a recognition that the location of Delhi would serve the purpose of geographical function of territorial compromise and adjustment. Delhi had always been a natural capital because of its separate existence for the last few centuries (12). In fact, it is an organic capital because it has grown up around a nucleus.

Lying in a narrow passage where deserts on the west, mountains on the north, hills to the southwest give way to the productive valley of the Ganga and to easy routes to the south, Delhi possessed of some compelling advantages which no ruler aiming at a tight control over the subcontinent could fail to discern (4). Delhi is said to possess a conveniently central administrative location and has persisted as the major control point of India since at least the tenth century A.D. and was prominent for a millenium before that (7).

New Delhi occupies a 'neutral site' relative to Calcutta, Bombay and Madras, the three original control points of the British administration and penetration in India. It is a product of politico-geographical forces of federal bargain(12). It is the focus of the centrepetal forces radiating in all directions toward the peripheral areas and represents the 'state idea'(5), a manifestation of the "raison d'etre" of the federal polity of India.

Geographical Base of Indian Federation

The spatial extent of India is one of the conditioning factors which has greatly helped in the rise of federalism in the country. Because of its size, it has a large population with heterogeneous ethnographic religious structure, and equally bestowed with regional diversities. Because of its large geographical size, India is bestowed with physiographic and climatic diversities which have largely impeded the process of cultural and linguistic integration. Mountains, plateaus, forests and deserts have long been an obstruction to the growth of 'circulation field' for India. Instead of a single circulation field, there are a number of small circulation fields widely separated from each other by relief features. The traditional center of political power which lies near Delhi, finds it extremely difficult to exercise durable control over the peripheral areas(15).

Another conditioning factor which has helped in the rise of federalism in India, is the wide and unequal distribution of resources and specialisation in varied nature of economic activities. There are regions which are rather specialized in agricultural activities or industrial activities. There are intermediate regions also. Such an economic spatial pattern has given rise to regionally grouped economic diversities and unequal economic growth.

India has always been a multicore country. A core area is defined as an area in which or about which a state originates as a spatial organization(17). It would be an area of immense strategic economic and human significance. North of the Vindhyan lie the Aryan Hindu Core-area and south of it, specially in the coastal tracts of the Godavari, Krishna and the Cauvery lies the Dravidian Hindu Core area. The Ganga valley from Gaya to Hardwar is the nucleus of the Hindu Region(9) and the Krishna-Godavari delta formed a dominant nucleus with a subsidiary nucleus in the Cauvery delta of the Dravidian Peninsular India(16). One more secondary core area seemed to have been crystallized in the upper Godavari Basin. Muslim core area is found located in North-West and East India. A Sikh-core area has almost grown up in the Punjab Region with a separate iconography. These core areas have been the foci for separate political-religious ideas operating in the opposite direction. Indian federation has been able to successfully contain the problem created by the spatial distribution of the core-areas. Diversities in races and religions in India too necessitate the federal set-up.

India's geography is thus closely linked up with its federal polity. Most of the geographical factors do form a negative base for India's territorial growth and integration and jeopardise the process for achieving the national cohesion. Federalism, in fact, is a centrepetal force for India, that it keeps the centrefugal tendencies under control.

Reogranization of Territories

The Indian Independence Act of 1947 created a two-tier spatial political structure for Indian dominion. They were, British Provinces and the territories of the Princely States. An option was provided in the Act for the Princely States that they could either remain independent or acceded to any of the Dominions. There were as many as 526 Princely States. However, the Princely States were acceded to the Indian Union by peaceful negotiations before August 15, 1947, though police action and referendum had to be resorted in some cases, like Junagadh, Jammu and Kashmir and Hyderabad (Fig 2). However, in a little over two years after independence the political geography of India was rationalised by the merger, or the consolidation and integration of states. The process of integration took into account factors like ethnic and linguistic homogeneity and historical traditions(13).

In 1950, just on the eve of the inauguration of the Constitution, there were three categories of states each with a pattern and status of its own. 'A', 'B' and 'C' were the main categories of the three-tier spatial political structure of India (Fig. 3). The territorial arrangement of the different categories of states was geographically inconsistent and required a rational reorganization of territories on the linguistic pattern. Keeping in view the growing demand for realignment of territories on the basis of linguistic pattern, the Government of India set up the States Reorganization Commission by an Act in 1956, followed by an amendment (Seventh) of the constitution in the same year, for this purpose.

The States Reorganization Commission considered the problems of rearrangement of political boundaries of the units of the federation and on the basis of the recommendations of the States Reorganization Commission, the three-tier state system disappeared and fourteen new federal provinces along with seven federally administered union territories were created (Fig. 4).

However, the SRC overlooked the tribal demands. Sikhs in the Punjab demanded the Punjabi Suba and some of the Union territories gradually acquired the statehood. There was a need to make the federation more viable and flexible to accommodate aspiration of those people whose demands were not included in the report of the SRC. Keeping in view these developments a few more provinces have been created especially in the northeastern region: Tripura, Meghalaya and Manipur. Haryana has been taken out of Punjab and the Union territory, Himachal Pradesh, became a federal Province. Sikkim became a federal province in 1975, through a referendum (Fig. 5). At present there are twenty two federal provinces and nine federally administered union territories. These provinces have been created by a series of amendments of the constitution.

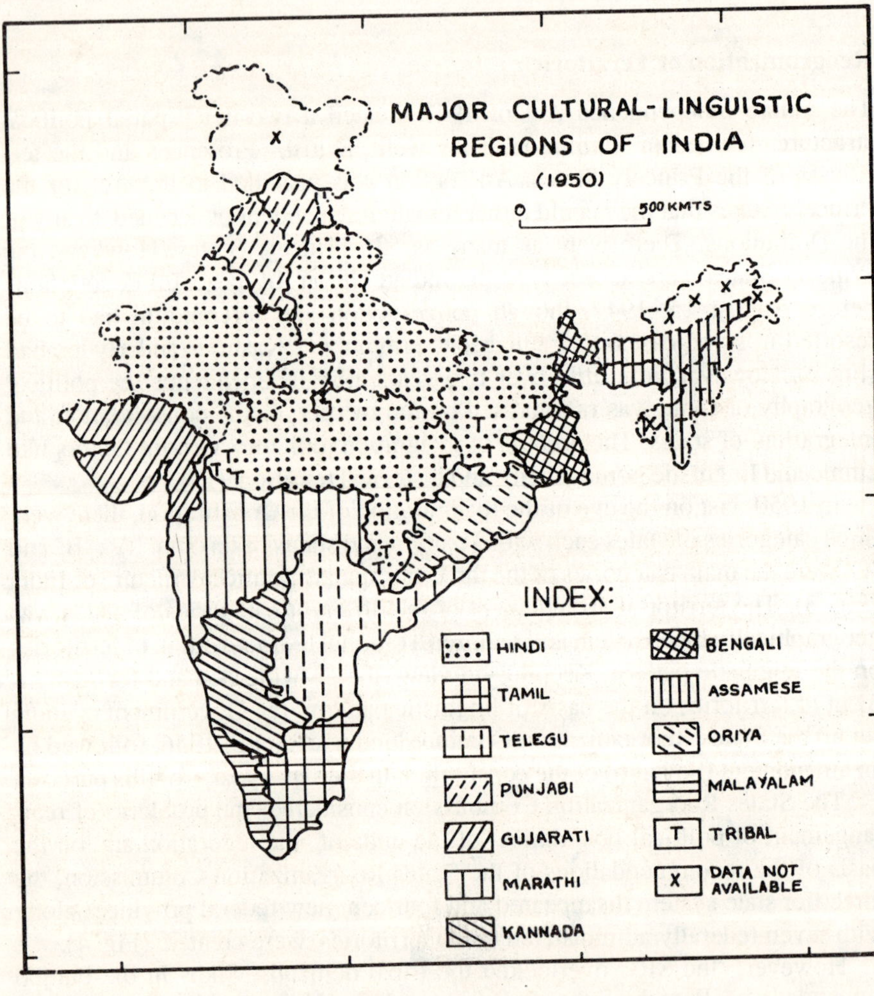

Fig. 1 Major Cultural-linguistic Regions of India

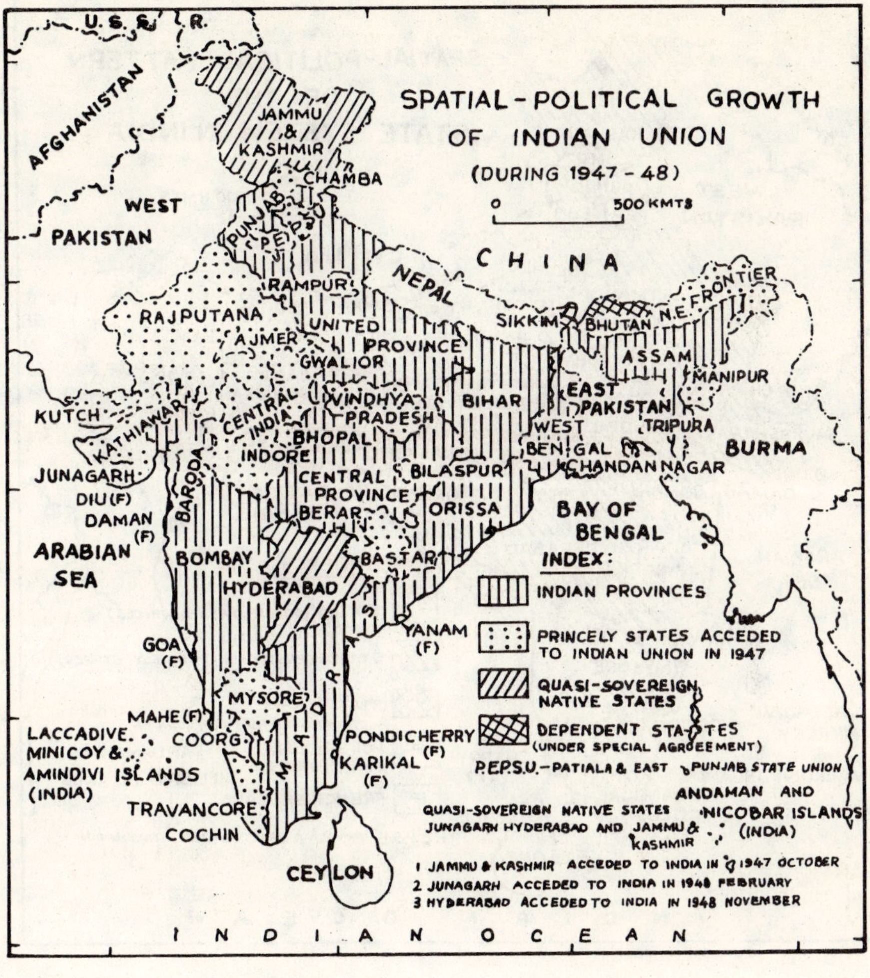

Fig. 2 Spatial-Political Growth of Indian Union.

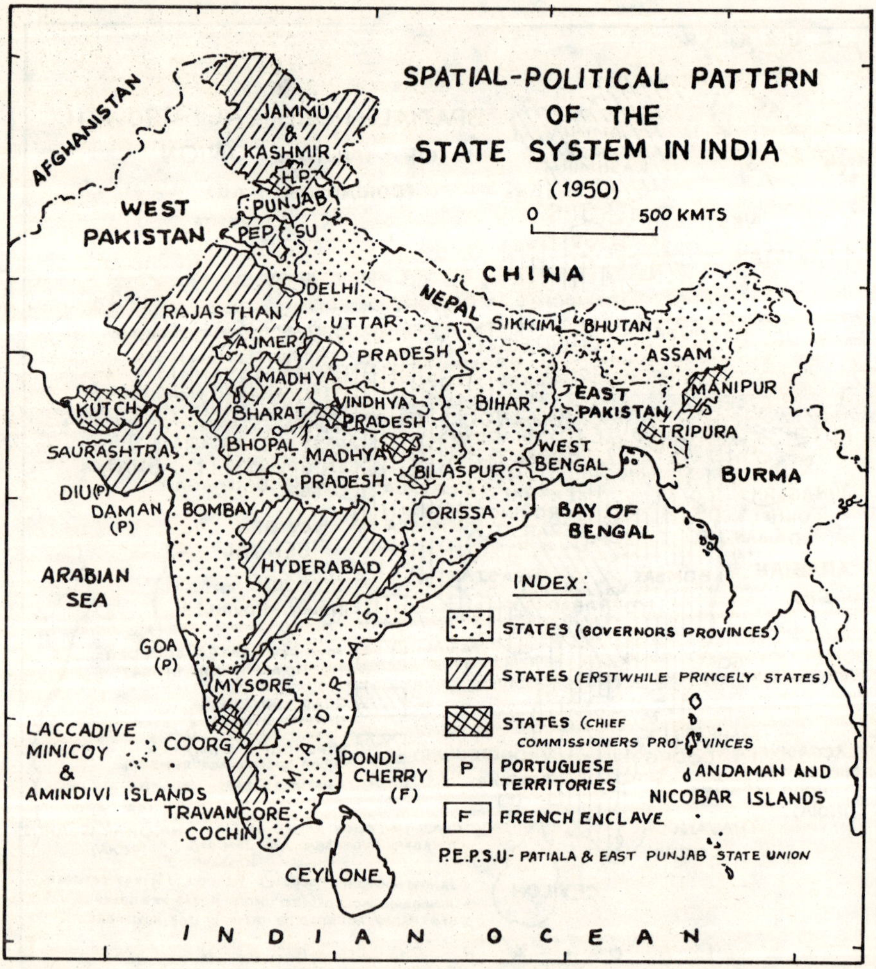

Fig. 3 Spatial-Political Pattern of the State System in India.

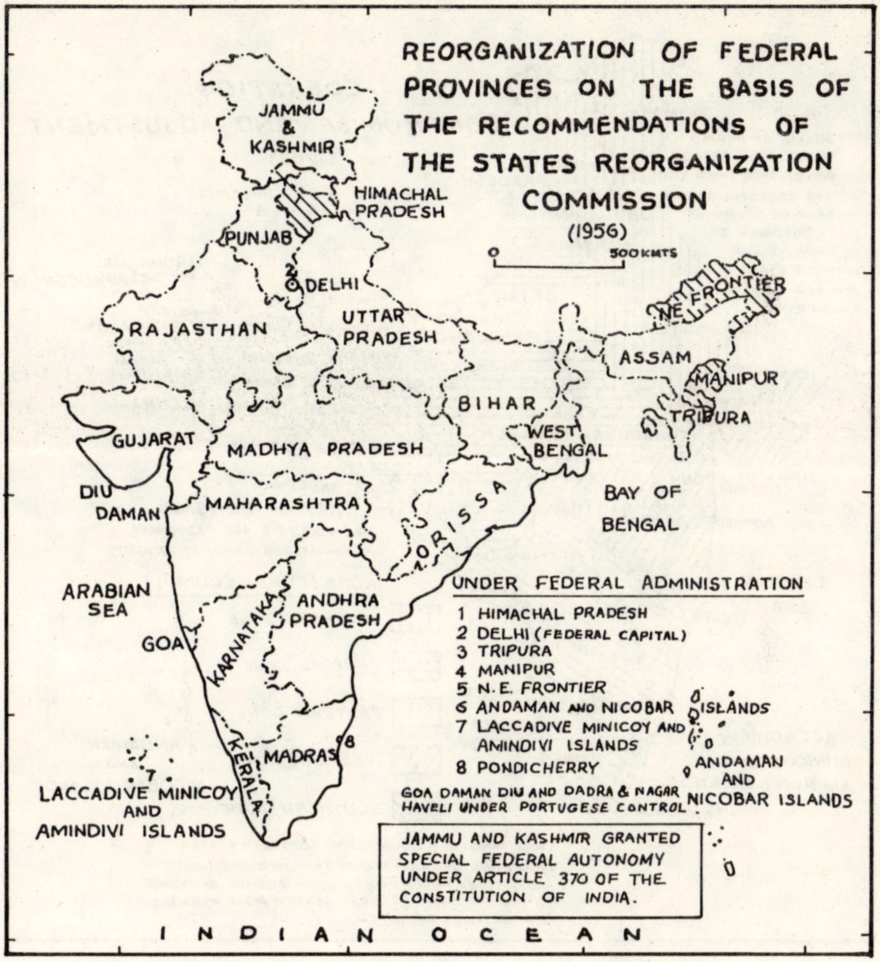

Fig. 4 Reorganisation of federal Provinces on the Basis of the States Reorganisation Commission.

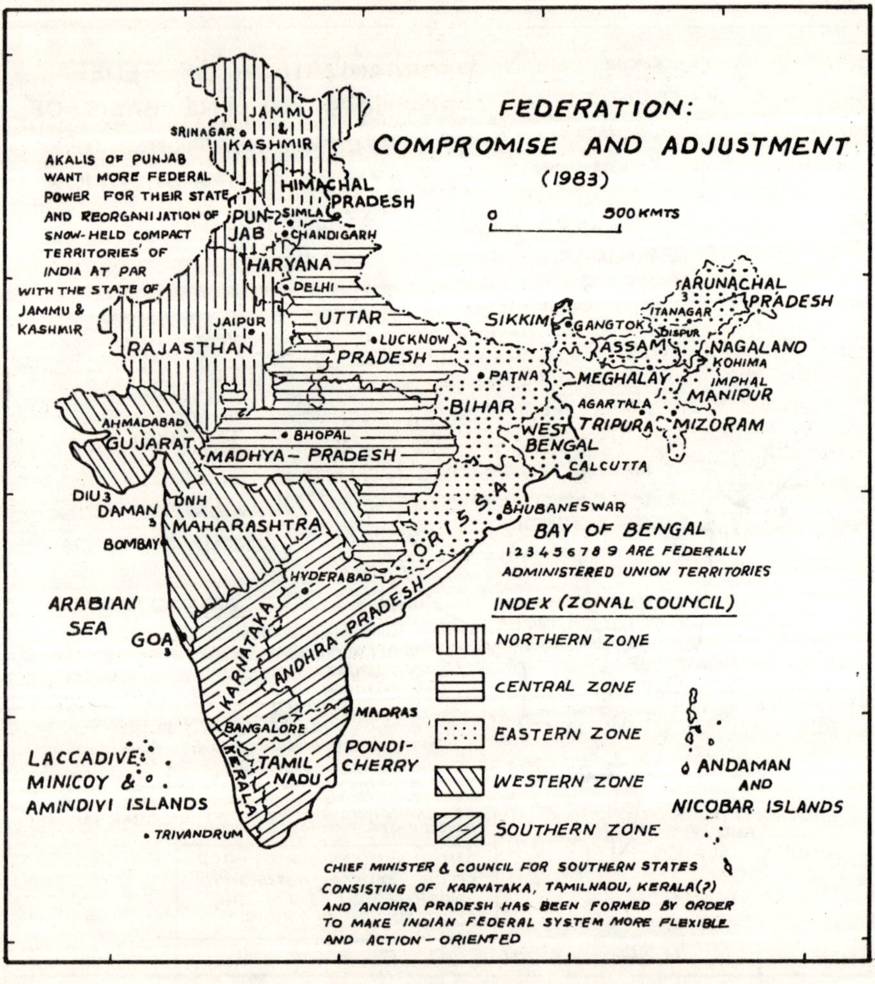

Fig. 5 Federation : Compromise and adjustment

Nature of Indian Federation

Federalism in India has been developed as a result of the willingness of the participating constituent units to give up some or their individual advantages in the interests of a greater state. It required a great deal of compromise and adjustment on the part of many people. The degree of compromise required to render the state a permanent and stable entity is quickly recognized when it is found that millions of Muslims and Sikhs remain within this Hindu-dominated country and that the people of diverse race, religion and language have been thrown together not only within the confines of India as a whole but within the individual provinces of India(2).

Federal provinces in India are unequal in their spatial extent, economic structure and interest, plural societies and character. Recently, one more factor has added to their unequal disposition and character, that is the political ideology and the federal system in India, is highly asymmetrical in the true sense of the term(14).

Conclusion

Federalism is best suited in Indian geographical conditions. Indian federation belongs to compromise and adjustment category because of its asymmetrical nature. The polity does not create unity out of diversity, but, it enables the two to coexist. However there is a growing demand for a settlement of bilateral and multilateral issues and establishment of cooperative federations in the fine spirit of the constitution with understanding between the provinces and the Centre. And this requires a re- examination of the Centre-State relations.

REFERENCES

1. Birch A.H. (1966). Approaches to the Study of Federalism. *Political Studies.* 14. pp. 15-33.
2. Blij Harmde (1972). *Systematic Political Geography.* New York: John Wiley.
3. Dikshit R.D. (1975). *The Political Geography of Federalism: An inquiry into Origin and Stability.* Delhi: Macmillan.
4. Dikshit R.D. (1982). *Political Geography: A Contemporary Perspective.* New Delhi: Tata-McGraw Hill.
5. Hartshorne R. (1950). ''The Functional Approach in Political Geography', *Annals of the Association of American Geographers.* 40(2) pp. 95-130.
6. Livingstone W.S. (1952) 'A Note on the Nature of Federalism'. *Political Science Quarterly.* 67(1) pp. 81-95.
7. Murphey R. (1956) 'New Capitals of Asia', *Economic Development and Cultural Change. 5, pp. 216-225.*
8. Palmer N.D. (1961). *The Indian Political System.* Boston. Houghton Mifflin.
9. Panikkar K.M. (1955). *Geographical Introduction to Indian History:* Bombay Bhawan's Book University.

10. Pounds N.J.G. (1973). *Political Geography*. 'New York': McGraw Hill Book Co.
11. Robinson K.W. (1964) 'Sixty Years of Federation in Australia'. *Geographical Review*. 50, pp. 1-20.
12. Spate O.K.K.(1956). 'Two Federal Capitals: New Delhi and Canberra' *Geographical Outlook*. 1, pp 1-8.
13. Srinivasan N. (1956). *Democratic Government in India*. Calcutta. Allied.
14. Tarlton C.D. (1965). Symmetry and Asymmetry on Elements of Federalism : A Theoretical Speculation, *Journal of Politics* 27, pp. 861-874
15. Tayyeb A. (1966). *Pakistan. A Political Geography: London: Oxford University Press*.
16. Thapar R. (1974). *A History of India (Part I)*. London: Penguin.
17. Whittlessey D. (1939). *The Earth and the State*. New York: Henry Holt & Co.
18. Whyte F. (1926). *India : A Federation*. Simla: Government of India Press.

XII The Centrist Future of Indian Politics

SUSANNE HOEBER RUDOLPH & LLOYD I. RUDOLPH

The most striking feature of Indian politics is its persistent centrism. During the thirty years and five parliamentary elections in which the Indian National Congress was the dominant party among India's several parties, it benefited from three interdependent characteristics: first, a national apex organization with national goals, leadership and strategy: second, a centrist (for India) ideology and centrist policies; and third, a pluralist basis of support. Despite the country's notorious social, economic, and cultural diversity, the characteristics of the Janata party, victor in the sixth parliamentary election of 1980, and the several parties that split from Congress and Janata between 1978 and 1980 resemble those of the Indian National Congress.

Why are India's national parties centrist and pluralist rather than left or right class parties? Why, in a country with prominent religious communities, are confessional parties like those of the Middle East or Western Europe that appeal to religious sentiment and identity ineffective and defensive?

There are several reasons why we expect national centrist parties to constitute the mainstream of party politics over the next decade: (1) the tenuous quality of national class consciousness; (2) the fragmentation of the confessional and social minorities; (3) the electoral strength of disadvantaged confessional and social minorities; (4) the increasing political consciousness and strength of "bullock capitalists" and "backward classes"; (5) the imperatives of capturing power in Delhi; (6) the constraints imposed by cultural diversity and social pluralism in the federal system; and (7) the advantages that accrue to a centrist national party when parliamentary seats are won by pluralities in single member constituencies.

Class Politics

A leading reason for centrism is the tenuous quality of national class con-

Reprinted from *Asian Survey*, Vol. 20, No. 6, 1980, pp. 575-594, by permission of the Regents of the University of California.

sciousness. To many it may seem plausible, even inevitable, that the politics of a country with as much inequality,. poverty, and injustice as exist in India will take the form of class conflict. But class has not been and may not become the principal medium for representation. Consciousness and organization with respect to inequality, poverty, and injustice have been as much or more the provenance of other kinds of social groupings such as status groups and cultural communities as of class. Organizations representing language, caste, and territorial (sons of the soil)[1] interests and those of disadvantaged minorities(backward classes, scheduled castes and tribes, Muslims) have been more successful than class-based organizations in creating consciousness and guiding political action to deal with the economic interests of the poor and oppressed.[2]

If class is to become a stronger determinant of political action, suitable objective conditions are required. Conventionally, this has meant heightened conflict between capital and labour as capital accumulation accelerates and larger proportions of the work force enter wage employment in the industrial sector of the organized economy. The future of class politics in India, however, is likely to remain as tenuous as its past.

Although India is a leading industrial country (fifteenth in world rankings), it has no national classes. Its industrial work force is proportionately small and stationary and poorly organized. India's five million factory workers in 1971 constituted 3% of its total work force of 180 million and 28% of the eighteen million employed in the organized sector of the economy (10% of all workers)[3]. The proportion of workers employed in the organized sector (like those employed in agriculture and services) has remained constant between 1960 and 1977 and more or less constant form 1901. About 54% of organized sector workers in 1978 belonged to India's fragmented, competing, and locally oriented trade unions[4]. Unions, and leadership positions in unions, have proliferated much faster than membership. Between 1927-1932 and 1974, average membership per union steadily declined from 2618 to 737[5].

Organized labour's apex organizations have little control over subordinate units or influence on national policy or politics. Their fragmentation along party lines and competitive relationship make it difficult for them to pursue common policy or political objectives.[6] There is no equivalent in India to Britain's or Germany's national trade union congresses, France's or Italy's rival but national apex labour organizations, or even America's AFICIO.

The closest approximations to national labour organizations in India are unions representing workers in public sector industries and government departments that span the subcontinent. These include the railways (1.4 million employees in 1975), insurance (800,000 employees in 1974), posts and telegraphs (650,000 in 1975), banks (100,000 in 1975), and smaller but strategically placed industries such as oil, coal, steel, and airlines. Sixty-five per cent of the workers employed in the organized economy in 1974 were public employees. Federal and state government employees, including the police, have acted nationally on the basis of emulation and contagion rather than all-India organization. The railway strike of May 1974, which Mrs. Gandhi broke through massive repression and the arrest of union leaders, was an impressive demonstration of organized labour's potential for national action and may be a harbinger of the future. But working class consciousness of state employees is compromised by their status as public employees, the nominally non-antagonistic nature of their relationship to owner- managers who are "socialist," and the common interests that many white collar and skilled blue collar workers share with management. When the state controls the means of production and administration, conflict arises between worker's claims based on performance and those based on equity. As wards of nominally socialist state that aspires to be a "model employer," India's 13 million state employees tend to be dependent or rebellious rather than class conscious. Nor are they likely to make common cause with agricultural labourers and poor cultivators. Not only do they live as races apart but also, as beneficiaries of state exactions and investments, state employees can become rivals rather than brothers of their fellow workers in the unorganized economy.

The limited and episodic capacity of India's blue and white collar workers to engage in class politics is paralleled by the marginal capacity of owners and managers of capital in the private sector to engage in class politics or evenly to influence policy on their behalf. India's large private firms rarely approximate the national scope (in markets, employment, investment, and influence on consumption and political consciousness) that national firms in advanced industrial economics do. Furthermore, the conglomerates that replaced the British managing agencies have been placed on the defensive by the widely accepted ideological position that the profit motive produces antisocial consequences. Legislation and licensing procedures that regulate the kind and amount of commodities produced as well as state control of credit and investment reflect their limited capacity to insure their own freedom of action by influencing politics and policy. The frequent statements and policy proposals of spokesmen for private capitalism find little public favour with party leaders and elected and appointed officials because they want to avoid the political costs associated with legitimizing their views. Those who accept campaign contributions from private sector donors and support their interests governmentally find it convenient to attack them in public.

In marked contrast to private capitalism, state capitalism under the name of the public sector is honoured and well endowed. Planned investment over the past thirty years has created a public sector that, as Jawaharlal Nehru intended, occupies the commanding heights of the organized economy. From the third plan to the sixth, the ratio of planned investment between the private and public sectors has been 40:60.[7] Public sector industries and financial institutions more often than not are national in ways comparable to their counterparts in advanced industrial economies. Of India's twenty largest firms, the top ten are in the public sector. The sales of the largest public sector firm,. Indian Oil Corporation, exceeded the combined sales of the top ten firms in the private sector and the combined sales of the top ten public sector firms exceeded the sales of the top 101 private sector firms.[8]

The existence of state capitalism on this scale severely limits the objective possibilities for rightist parties grounded in as politics of industrial and finance capitalism. The public sector is of and within the state. Its apex industry organizations, firms, and multifarious credit institutions *ab initio* engage in bureaucratic rather than class politics. The organizational interests of state bodies are pursued more in the rooms and corridors of central secretariat buildings and in bargaining sessions with agents of competitive apex bodies and interest groups than in parliament, on the hustings, or in the streets. The politicians and bureaucrats who articulate and pursue the interests of state capitalism do their best to avoid or escape class or interest groups politics by making ideological claims that place agencies of the state above or outside the political struggle.

As the nation's largest employer, the state's claim to be above or outside class or interest group politics affects its relations with its own workers. On the one hand, state employees who provide public or essential service de jure or de facto are not expected to use strikes to make demands or bargain collectively. On the other hand, when state capitalist firms try to live up to the state's paternalistic claim to be a model employer or, like large corporations in the advanced industrial economics, increase prices in order to pass on the costs of buying industrial peace that was achieved by granting increased wages and benefits, they not only contribute to inflation but also, in a poor developing country like India sacrifice the country's long term collective interest by dissipating investment funds in present consumption.

The domination of the means and relations of production by state capitalism in the organized sector severely limits the objective conditions for industrial and financial capitalism in India to become a conscious, organized force in politics. State capitalism's ambiguous dual role as the representative of the public or people's interest and as the nation's largest and most powerful holder of capital and employer of labour radically diminishes the means available for the conduct of class politics.

Confessional Politics

If there are reasons why India does not have national class parties, why not confessional parties ? Why not a national Hindu party ? Moderate Christian parties play an important role in the politics of western European states where "rationalist" culture and "secular" norms are more advanced than in India. Why has not India been affected by the religious revivalism that has swept the

Islamic world or by confessional parties that, earlier, played an important part in the politics of the Middle East and Southeast Asia ? Why does "centrism" in India include secularism as much as it does "socialism" and democracy ?

The support base for national confessional party, the Hindu majority, is illusory even while confessional minorities play an important role in national and regional politics. The overwhelming Hindu majority, 83% according to the 1971 census, is in part an artifact of categorization. In addition to the 17% (95) million) of the population who are not Hindus, another 15% (80 million) are members of scheduled castes. Their categorization as Hindus is as much a result of census enumeration as of their choosing. Most are not susceptible to political appeals based on a Hindu identity or Hindu interests. Similarly, it is questionable whether the 30 million out of 38 million tribals classified as Hindus in the 1971 census share a Hindu identity. They too are not available for party appeals to Hindu nationalism and interests. Indeed, many are actively engaged in asserting a variety of cultural or subnational identities and defending their interests against "Hindu" encroachment and appropriation. Together, Muslims, other non-Hindus, and scheduled castes and tribes counted as Hindus (11, 6, 15, and 5% respectively) constitute 37% of India's population.[9] Of the 63% of the population that is left, only a fraction of uncertain magnitude shares a Hindu identity that has political saliency. More important, the Hindu majority is more fragmented and competitive along class, caste, and regional lines of cleavage than are India's minorities.

The potential support base for a Hindu confessional politics is located, for the most part, among the traditionally literate, spiritually initiated ("twice born") upper castes in the relatively backward Hindi-speaking heartland states of northern India. In the nine large and more advanced states of southern, eastern, and western India, the kind of Hindu consciousness and social structure needed to support confessional party politics is marginal to nonexistent.

Prior to the 1975 emergency regime, it was the erstwhile Jan Sangh— which became a fraction in the Janata Party — and its sister organization, the RSS, that articulated the ideology of a Hindu Rashtra (Nation) and symbolic, cultural, and social policies that addressed Hindu interests and values. A rough measure of the potential support base for a Hindu confessional party is the proportion of the vote captured by the Jan Sangh in elections before 1977— e.g., 9% in 1967 and 7.4% in 1971.

Minority Politics

By contrast India's important minorities— the scheduled castes, Muslims, scheduled tribes, and Christians— like American Blacks and Spanish speakers, have had a greater propensity, now declining, than the Hindu majority to vote en bloc.[10] Among the minorities, the scheduled tribes vote together more than scheduled castes and Muslims. The minorities, community identity is more pervasive and politically salient than the Hindu majority's, and they share common economic and social conditions to a much greater degree than Hindus. Again, like American blacks and Spanish speakers, they suffer simultaneously from poverty, discrimination, and powerlessness, objective circumstances that have been translated into subjective awareness of economic exploitation, status deprivation, and political repression. The objective conditions and consciousness of India's minorities have been less subject than the Hindu majority's to the cross cutting or divisive effects of class, caste and region.

Muslim and scheduled caste and tribe support is particularly vital for electoral success in the five Hindi heartland states of northern India where the undivided Janata enjoyed its greatest electoral successes, and where in 1980 the opposition to Congress remained strongest. In other words, minorities are most significant numerically in just those states where elections since 1967 have been most closely contested. In Uttar Pradesh, Bihar, Madhya Pradesh, Haryana, and Rajasthan, which include India's two most populous states and elect 40% of parliament's 542 members, the three minorities together constitute 37% of the votes in the first three, 34% in Rajasthan, and 23% in Haryana (see Table 1).

TABLE 1 – *Parliamentary seats and minority population proportions in Hindi Heartland States, 1977.*

State	No. of Seats	Percentage of Population			
		Minority Muslim	Scheduled Castes	Scheduled Tribes	Total of Three Minorities
(All India)	(542)	(11)	(15)	(7)	(33)
Uttar Pradesh	85	16	21	Negligible	37
Bihar	54	14	14	9	37
Madhya Pradesh	40	4	13	20	37
Rajasthan	25	7	15	12	34
Haryana	10	4	19	Negligible	23

SOURCE : India, A Reference Annual, 1979, p. 10

Because minority support is vital to electoral success, gaining or holding it has become a common feature of party strategy and ideology. Parties that rely primarily or exclusively on class appeals to reach the poor and disadvantaged have been less successful than those that seek to represent minority interests and identities. The electoral successes of the Congress party under Jawaharlal Nehru and Indira Gandhi were in large measure a result of support from India's largest minorities, Muslims and the scheduled castes, who regarded Congress as their friend and protector. Janata's success in the March 1977 election followed the alienation of these minorities from Congress and their support of Janata. All three national parties contesting in 1980 tried to occupy the centre by avowing that they were secular and socialist— i.e., that they recognized the minorities' need for protection and help and the state's obligation to eliminate destitution (or poverty) and provide economic opportunities to the poor and disadvantaged. Each made symbolic and policy claims designed to gain the support of India's minorities: Janata's scheduled caste leader Jagjivan Ram and its Gandhian commitment to eliminate destitution and promote employment; Lok Dal's son-of-a-*kisan* (cultivator) leader Charan Singh and its commitment to promote secularism and the interests of the agrarian sector where most of the disadvantaged minorities live; and Congress I's Nehru legacy of friendship and protection for Muslims and the Scheduled castes personified in his daughter Indira and its effort to eliminate poverty by promoting industrialization.

The nominal reason for Janata's three splits in July 1979 and March and April 1980 was the credibility of its commitment to secularism, a code word for the party's attitude towards the minorities and Hindu communalism. The Charan Singh, Jagjivan Ram, and Chandra Shekhar-led fractions alleged that the former Jan Sangh fraction's "dual membership" in the Janata party and in the upper caste dominated RSS (Rashtriya Swayamsevak Sangh) was incompatible with the party's secular commitment. The RSS's ideology of Hindu Rashtra, they argued, would exclude the minorities from the equal benefits of citizenship and inhibit policy initiatives on their behalf. Leaders of the former Jan Sangh depicted Hindu Rashtra as an inclusive term that encompassed India's composite Hindu culture and plural society. They stressed that the Janata party's founder, the late Jayaprakash Narayan, found no incompatibility between his commitment to secularism and to Gandhi's social philosophy and Jan Sangh and RSS support prior to and after the formation of the Janata party in May 1977. In April 1980, at the time of Janata's third split, Atal Behari Vajpayee, former Jan Sangh leader and President of the newly formed Bharatiya Janata party, reaffirmed the new party's commitment to Jayaprakash Narayan's philosophy. Emphasizing that secularism and unity in diversity were the cornerstones of Indian nationalism, he said that his party would give the foremost place to promoting the interest of the minorities and the backward sections.[11]

By advocating secularism, the Janata party and its successors could continue to bid for the support of scheduled caste and Muslim voters, many of whom had deserted Congress to vote for Janata in 1977 as a result of emergency excesses. Indeed, Janata in 1977 captured all 38 scheduled caste seats in the five Hindi heartland states (Table 2), contrary to the impression conveyed by the 1980 outcome in seats when Congress-I won 62% of the seats in Hindi heartland states (Table 2), the 1980 outcome in party vote shares there makes clear that opposition inroads on Congress support from minority communities persisted at diminished levels. In 1971, Congress-I won 65% of the scheduled caste seats in the seventeen large states, in 1977 Janata almost reversed the result by capturing 58%, but in 1980 Congress-I apparently regained its 1971 hold on scheduled caste support by winning 61% (Table 2). This aggregate result is misleading, however, because scheduled caste votes in Hindi heartland states were more closely divided in 1980 than they were in 1971 and 1977 when we infer the minorities voted decisively for Congress and then for Janata. In U.P., not only India's most populous state but also the state with the highest proportion (21%) of scheduled caste voters, Janata and Lok Dal candidates in 1980 gained 58% of the vote in scheduled caste constituencies, 6% higher than their average vote in U.P. (52%) and 22% higher than the Congress-I's vote in U.P.'s scheduled caste constituencies, and in the state as a whole, both of which averaged 36%.[12]

TABLE 2 – *Scheduled caste constituencies, seats by party, in the 1971, 1977, and 1980 Lok Sabha Elections.*

17 Large States								
1971			1977			1980		
Congress No. %	Janata No. %	other No. %	Congress No. %	Janata No. %	Other No. %	Congress No. %	Janata No. %	Other No. %
48(65)	7(10)	18(25)	15(11)	45(58)	18(23)	46(61)	11(18)	16(21)
Total Seats 73			Total Seats 78			Total Seats 76		

Hindi Heartland

	Congress 1971	Janata 1971	Congress 1977	Janata 1977	Congress 1980	Janata 1980
Bihar	5	1		8	5	2
Haryana	2			3	1	1
Madhya Pradesh	3	2		5	4	2
Uttar Pradesh	17			18	8	9
Rajasthan	2			4	4	
	29(85%)	5(15%)	0(0%)	38(100%)	23(62%)	11(38%)

SOURCES : Government of India, Election Commission, *Report on the General Election to the Lok Sabha, 1971, Vol. II : Report on the General Election to the Lok Sabha, 1977, vol II For 1980 G.G. Mirchandani, The People's Verdict* (New Delhi: Vikas, 1980), p. 135.

NOTE: The number of scheduled caste seats has increased over time. Congress refers to Congress(R) in 1971; Congress in 1977; Congress(I) in 1980. Janata refers to the components that united in 1977; 1971 counts the premerger parties and 1980 counts Janata, Lok Dal, and Congress U.

In Bihar, the home state of the then Janata leader and potential prime minister, Jagjivan Ram, himself a member of a scheduled caste community,

Congress-I candidates in scheduled caste constituencies polled a higher percentage of the vote than in U.P. and higher than their Janata and Lok Dal rivals, but still fell short of majority. Mrs. Gandhi's well-publicized efforts, starting as early as May 1977 at Belchi in Bihar, to recapture the confidence and loyalty of scheduled caste voters by rushing to the scene of atrocities against them, as well as the backlash among upper as well as scheduled caste voters against Bihar's Janata Chief Minister's successful effort to reserve 22% of government jobs for the backward "classes" (i.e. castes), may help to account for the fact that Congress-I candidates in scheduled caste constituencies won 48% while Janata/Lok Dal candidates won 11% of the vote. Both sets of candidates did better in scheduled caste constituencies than in the state as whole, Congress-I polling an impressive 12% and Janata/Lok Dal a more modest 4% higher than their statewide average.

Muslim seats and votes in 1980 were closely divided. Congress-I captured 29(45%) of the 64 seats with 20% to 50% Muslim voters, a result 22% below its national average (67%) of seats won. Similarly, Congress-I won only 36% of the vote in the 64 "Muslim" constituencies, 7% less than its national average of 43%. Clearly, its special relationship with the Muslim community had not been reestablished. Of the 32 of 64 "Muslim" seats captured by other parties, 12 were won by the Lok Dal in U.P. and 18 by the Communist Party-Marxist in West Bengal — from June 1977 the governing party in that state Congress-I victories, by contrast, were spread over the country[13] mainly in states where Muslims were smaller fractions of statewide population and thus more isolated than they are in U.P. and West Bengal, which rank first and second in proportion of Muslim population. Because a significant proportion of U.P. Muslims are small or middle size landholders and many Bengal Muslims are poor laboureres, there is reason to believe that Muslims in these two states who voted for the Lok Dal and the CPI-M respectively interpreted their interest more in class than in community terms, while Muslims who voted for Congress-I more often understood their interest in communal terms.[14]

Among the minorities, only the scheduled tribe voters in 1980 returned to the Congress-I fold in overwhelming numbers.[15] In 1971, Congress-I won 68% of the scheduled tribe seats, in 1977 Janata won 61% and in 1980 Congress-I improved substantially on its impressive 1971 margin by capturing 82% (Table 3). Voting patterns confirm that tribal voters, unlike scheduled caste and Muslim voters in 1980, engaged in bloc voting for Congress-I, but only at the state level. In Madhya Pradesh, the state with the largest scheduled tribe population and the most scheduled tribe seats (9), Congress-I in 1980 won comfortable majorities or high pluralities in all but one. Its overall average in Madhya Pradesh tribal reserved seats was 55%, eight points higher than its average in the state and 12% above its national average. Scheduled caste and Muslim voters rarely exceed 30% of the population even in reserved constituencies and, as a result, often reflect by choice or intimidation the voting pattern of their constituency. Tribal voters, however, are more densely concentrated, typically constituting 50% or more in reserved constituencies,[16] and vote more independently. Unlike exuntouchables and Muslims, whose level of consciousness and organization are higher and whose constituency circumstances are different, tribals, as the most oppressed and disadvantaged minorities, turned to the Congress - I in 1980, and to a lesser but significant extent, to tribal parties.

TABLE 3 – *Scheduled Tribe Constituency seats by party, 1971 to 1980, Lok Sabha*

	1971		1977		1980	
	No.	%	No.	%	No.	%
Congress	21	(68)	9	(22)	27	(82)
Janta	5	(16)	19	(61)	2	(6)
Communist Parties					1	
Other	5	(16)	3	(10)	3	(3)
	31	(100)	31	(100)	33	(99)

SOURCE : Same as Table 2.
NOTE : Congress means Congress (R) in 1971. Congress in 1977, Congress - I in 1980. Janata in 1971 means the components of the 1977 Janata; in 1980 it means Janata, Lok Dal and Congress-I.

The 1980 election results confirm that the need to win a substantial share of the minority vote is one of the essential determinants of centrism in Indian politics. Parties whose ideology, policies, and electoral strategy do not attend to representing minority interests and identities cannot compete for power at the national level. Centrist parties, in the coded language of Indian politics, espouse secularism and socialism in order to signal their regard and concern for the 37% of the electorate who are minority voters.

Bullock Capitalists and Backward Classes

The centrist character of India's national parties is also determined by the balance of forces in the agricultural sector. Polarization and left and right class politics in the countryside are unlikely because the first wave of land reform broke the power of the neo-feudal landlord class (*Zamindars* and *Jagirdars*) and because a centrist political class composed of bullock capitalists and backward classes has emerged as the leading representative of agrarian interests. The power of neo-feudal notables was initially limited by the first and largely successful first wave of land reforms in the 1950s that abolished "intermediaries", India's landlord class. Without the first wave of land reform, neo-feudal landlords using wage labour might have provided, as they have in Pakistan, the basis for a viable conservative politics.[17] The Swatantra party for a time created an alliance between neo-feudal and capitalist agricultural interests and private industrial capitalism. Its demise ended India's only experiment with class politics of the right.

At the bottom of the agrarian pyramid, except in few districts such as those in Kerala, the level of consciousness and organization among agricultural workers is too low to enable them to represent themselves effectively. India's "proletariat" in the organized economy is in fact a labour aristocracy composed of white and skilled blue collar workers. It is sufficiently separated by interest, life style, and aspirations from unskilled rural wage labour to prevent urban and rural workers in the foreseeable future from becoming allies, much less comrades. This urban-rural gap may close as agriculture workers become more conscious and organized, more members of their households are employed in the city, or, as rural producers, they become beneficiaries of the green and white (milk) revolutions. Over the next decade such linkages seem remote. By contrast with the objective circumstances and subjective orientation of the top and bottom of the agrarian pyramid that constrain class parties of the right and left, those of bullock capitalists are ascendant and promote centrism.

In the 1970s, the political influence of large landholders was successfully challenged by a new rural class we call bullock capitalists. Their political strategy was to challenge Congress alleged urban and industrial bias by generalizing the interests of independent agrarian producers to the agricultural sector and rural society as a whole. The victory of the Janata party and the formation of a Janata government in 1977 marked the ascendance of a political coalition in which agrarian interests were the senior partner. Charan Singh's March proagrarian budget, like. This six month (July-December 1979) Lok Dal-led coalition government and 1980 campaign themes (e.g. his attacks on organized labour as privileged and Nehru's urban-industrial bias)[18] created a backlash whose strength, inter alia, is a measure of his success. No party, including those on the left, has ignored Charan Singh's mass base among agrarian producers.

Eighty percent of India's population is rural, 72% of its labour force is agricultural, and 45% of its national income comes from its agricultural sector. Independent agrarian producers constitute a plurality of India's work force but are even less effectively organized and represented at the state and national level than industrial labour. Prior to 1977 they partially overcame this apparent handicap for influencing policy by direct representative in state legislature and cabinets. After Janata's victory in 1977 they were for the first time directly represented at the national level by being the largest faction in the Janata Parliamentary Party and by their influence in the cabinet, planning commission, and ministries.

The emergence of agrarian interest on the national politicai scene undid the political settlement of the Nehru era, which was based on a coalition of urban and rural interests united behind an essentially urban-oriented industrial strategy. Its senior partners were India's proportionately small (about 3%) but politically powerful administrative, managerial, and professional English-educated middle clases and big industrialists who welcomed the freedom from foreign competition of the import substitution and industrial self-reliance strategies of the second and third five-year plans. The English-educated middle classes manned the bureaucracy, built and managed the state capitalist (public) sector, and staffed large firms in the modern private sector. The junior partners in the Nehru settlement were rural notables mostly of the large landowner class. They consented to the import substitution and industrial self-reliance strategies, urban control of the central government, and the advantages that accrued to urban elites and organized workers on condition that they themselves remain in charge of the state governments. Being in control of state governments enabled them to allocate resources and control policy implementation for the agricultural sector and rural society in ways that protected their interests. One of the consequences of the 1977 election, only partially intended, was that for the first time since independence, agricultural interests were powerfully represented at the centre as well as in the states. At the same time, bullock capitalists partially displaced large landowners in the agrarian power constellation. As a result, the Nehru era policies favouring cities, centralization, bigness, and capital intensive industrialization were displaced by policies that favoured the agricultural sector and rural society, decentralization of government and the economy, and small producers using labour intensive technologies that promote employment.

The Gandhi-led Congress-I government formed after the January 1980 mid-term parliamentary election has given indication that it intends to revive the Nehru settlement. Industry rather than agriculture is to be the principal means of the growth and employment generation. Tenants, small holders, and laboureres led by neo-feudal notables and large landowners are likely to play a larger part than bullock capitalists and backward classes in Congress-I's rural support, pointing toward a variety to Tory socialism.[19] If local notables and large landowners become junior partners in Congress-I's rebuilt Nehru-style urban-rural coalition, they may be willing to support the return of an urban-oriented industrial strategy in return for a partial restoration of their status and power in rural society and state politics. At the same time Mrs. Gandhi, whose party ridiculed Janata's agriculture-led employment strategy has not ignored the policy needs of middle-level agricultural producers, no doubt in the expectation that some support can be obtained from that quarter.[20]

Middle-level agricultural producers are one component of the ascendant political class in India's countryside. This component encompasses two overlapping categories: a producer group, the middle-level peasants we call bullock capitalists, and a status group, the backward classes, an administrative euphemism for backward castes. They constitute about 25% of the population located in the lower reaches of the caste order but above the untouchables. Backward castes began to mobilize politically in the south and west after world War I, but only recently have come to be a political force in the north.[21] The centre pieces of Janata's economic and social policies articulated the political class interests, an agriculture-led employment strategy, and a new wave of reservation of seats, this time for backward castes in government jobs and educational institutions.

Bullock capitalists are advantageously placed by their objective circumstances to be the hegemonic agrarian class. As of 1971-1972 they constituted a large proportion of agricultural households and controlled more land (35% controlling 51%) than any of the other three agrarian classes — landless (27%), small holders (33% controlling 10%), or large landholders (6% controlling 39%). Large landholders, whose average size holding increased marginally, lost ground to bullock capitalists between 1954 and 1972 in terms of proportion of households and are controlled, and we infer, in capacity to practice vertical mobilization. Along with the end of pro-Congress block voting by minorities (except tribals) in the Hindi heartland states, the absolute and proportional decline in the largely pro-Congress large landholder category and the more effective political mobilization of bullock capitalists and backward castes have contributed to Congress progressive loss of support, particularly in U.P., Bihar, and Haryana. Congress decline began in 1967, touched bottom in 1977, and made only a weak recovery in 1980. Congress decline paralleled the rise of Charan Singh's BKD (a Janata faction and later the Lok Dal), the party *par excellence* of the bullock capitalists. This pattern may be interpreted in light of Paul Brass investigations in U.P., which show that Congress' strength lay especially among operators of 30 acres or more and one acre or less, while the BKD's lay among the two to twenty acre holders.[23]

Bullock capitalists resemble "yeoman" farmers in that they are independent agricultural producers. Self-employed and self-funded, their holdings are large enough to support the use of a pair of bullocks and the new inputs associated with the"green revolution". Typically, they operate between 2.5 and 15 acres. At the same time, their assets are not large enough to enable them to engage in capital intensive agricultural production based on extensive use of machinery or to require them to rely wholly or mainly on wage labour. We prefer the term "bullock capitalist" to "middle peasants" because of the mix of capitalist, preindustrial, and non-capitalist features that characterizes their economic circumstances. They operate family farms; the family broadly defined is the primary if not the exclusive source of capital, management, and labour. Bullock capitalists produce enough for the market to be oriented to and constrained by it but not so much that consumption of their own products or non-monetized exchanges have ceased to matter. The costs of their inputs, like the prices of their products, are powerfully but not wholly determined by market forces; state actions also affect agricultural prices and factor costs.

They sometimes employ but do not depend on attached or casual wage labour. Because they are self-employed, they benefit from their own high quality, committed labour. While they use and pay for capital and participate in markets, they are "non-capitalist" producers in the sense that, being self-employed, their relations of production remain relatively undifferentiated and non-antagonistic.

Most important for consciousness and politics, the economic circumstances of bullock capitalists unite the interests of capital, management, and labour. Bullock capitalists own the means of production, manage the productive unit, and themselves provide most if not all of the labour. If there is exploitation involved in their relations of production, it is self-exploitation — e.g., they benefit from the "surplus value" of their labour. Their ideological propensities are individualistic rather than collectivist or capitalist. Their objective interests do not place them in a necessarily antagonistic relationship with other agrarian classes. At the same time, to the degree that they supplement family labour with wage labour or compete for scarce land with tenants or landless laboureres, their mainly non-capitalist relations of production are no bar to severe conflict with other agrarian classes. Nonetheless, they are more free than any other agrarian class to seek hegemony by appealing to the common interest of all agrarian classes and, by making upward and /or downward alliances, to become the fulcrum of political centrism.

All these features conspire to make bullock capitalists not only a centrist class but also the agent of a centrist politics equally opposed to collectivist and capitalist agriculture. As the bullock capitalists most articulate ideologue put it:

> A system of agriculture based on small enterprises, where the worker himself is the owner of the land under this plough, will foster democracy. For, it creates a population of independent outlook and action in the social and political fields. The peasant is an incorrigible individualist; his vocation, season in and season out, can be carried on with a pair of bullocks or a small machine in the solitude of nature without the necessity of having to give orders to or take order from anybody. That is why the peasant class everywhere is the only class which is equally democratic without mental reservations. Further, the system of family sized farms or peasant proprietorship ensures stability because the operator or the peasant has a stake in his farm and would lose by instability.[24]

The political coming of age of bullock capitalists runs parallel with and is strengthened by the first wave in the northern Hindi heartland states and the second wave in the southern and western states of Karnataka and Maharashtra of the so-called backward classes movement. While the objective overlap is hardly prefect, the backward castes that were and are now being mobilized present the status face of the economic category represented by bullock capitalism. Bullock capitalists are an economic category grounded in the means of production; the backward castes are defined by the traditional ritual ranking of caste, modified by the British and Indian state's "official" sociological ranking of India's disadvantaged.[25]

Both the British government and the government of India's Backward Classes Commission designated as "backward classes" castes whose ritual rank and occupational status were above the untouchables but in the lower reaches of the caste order of traditional society. The fact that in 1977 both "backward classes" and bullock capitalists caused and benefitted from the electoral victory of the Janata coalition bears witness to their substantial overlap. The chief ministers of Janata governed northern states were themselves of or spoke for Yadavas (Ahir), Kurmi, Koeris, and other prominent backward classes who have traditionally worked the land but now aspire to higher social status and improved occupational opportunities.[26] The demand by "backward classes" for reservation of posts in government service and seats in educational institutions is partly an expression of upwardly mobile lower caste cultivators' efforts to challenge the hold on clerical positions of traditionally literate upper castes. They hope to place family members in white collar service jobs and thus improve the family's social standing, income, and security. In Bihar, U.P., and Karnataka, and later at the centre, the demands of "backward classes" became a divisive political issue by generating a backlash among upper caste Hindus and untouchables. Just prior to the 1980 election, abortive efforts were made by the Charan Singh caretaker government to have the central government follow Bihar and U.P. in reserving jobs in government service for "backward class" members.[27]

The interests and ideology of the political class formed by bullock capitalists and backward castes strengthens centrism in Indian politics. This composite class encompasses more voters than any other agrarian class taken by itself (34% of households) and any other status group taken by itself (about 25% of the total population). While imperfectly mobilized and far from homogeneous in political outlook and behaviour, on both counts this group compares favourably with other agrarian classes and status groups. As self-employed agrarian producers, their relations with capitalist farmers and agricultural labourers are less antagonistic than the wide appeal of its agrarian ideology and policies.

The prospects of bullock capitalist leadership depend on recognition in their alliance strategy and policy agenda of the mixed class character of India's rural economy. If the parties under bullock capitalist influence make downward class and status group alliances and credibly pursue polices that promote redistribution, growth, and the collective good of the agrarian sector, bullock capitalists will improve their chances for playing a leadership role. Alternative strategies such as forming alliances upward with large landowners or pursuing only its own class interests are viable but foreclose the possibility of hegemony. In the world of day to day political struggles, where personal ambition, group self-interest, and short-term calculations often govern the course of events, all three strategies are evident at different times and in different contexts.

The Imperatives of National Power

The goal of capturing power in Delhi or wielding political influence there also strengthens centrism. Power in Delhi requires a party majority or opposition standing in parliament. As a result, the powerful tendency toward party fragmentation is met by countervailing tendency toward party consolidation. Factions and personalities motivated by short run calculations designed to exploit the main chance deflect from and split parties even while India's political elites seek homes in national parties in order to exercise power and influence at the national level. Governing in Delhi means access to the advantages of incumbency. India's federal system gives the centre initial or residual control over most of the resources and rules that affect state governments and leaves a substantial portion under the centre's direct authority. The governing party gains bargaining advantages with its own and rival state level party units seeking to influence policy choice and implementation. Even though a national party may lose an election, belonging to it may be more attractive possibility than going into the political wilderness at the head of a minor party.

The Constraints of Pluralism

India's cultural diversity and social pluralism provide another reason for parties seeking national power to adopt centrist ideologies and policies. Issues and symbols highly salient for one or a few states are irrelevant to others. Madhaya Pradesh, one of India's more backward states because of its marginal agricultural economy, predominantly rural society, large and backward tribal population, and residual princely state loyalities, is very different from its neighbour Maharashtra whose productive, market-oriented agricultural economy, numerous industrial complexes, large urban centres and advanced and well-organized ex-untouchable caste make it one of India's most advanced national parties must adopt ideologies and advocate policies that articulate and represent the formidable range of economic, cultural,and social differences that exist among and within India's twenty-two states and eight union territories. The result has bedevilled efforts by scholars to establish significant correlations between socioeconomic variables and party voting.[28] For example, in 1977 the Janata in Bihar may have gotten a substantial portion of its support from poor and uneducated labourers, in Delhi from upper income and educated commercial and professional classes, and in Rajasthan from middle income traders and cultivators. Associations between socioeconomic status and party preferences have been found at the district and, to an extent at the state level, but they wash out at the national (all-India) level. In consequence, electoral pluralities and parliamentary majorities depend on articulating policies that reconcile contradictions among conflicting interests yet avoid consensual formulas that, by speaking to everybody, speak to nobody.

The Electoral System and Centrism

Finally, the formal rule governing electoral competition among parties has favoured centrism. Regardless of the number of candidates and the division of the vote among them, the candidate first past the post (plurality) wins the seat.Given the absence of national classes, communities, and interest groups and the presence of cultural diversity and social fragmentation, it is not surprising that no party has won a majority of vote in seven parliamentary elections over a thirty-year period. What may be surprising is that (with exception of the November 1969 to February 1971 and the July-December 1980 periods) parties with comfortable parliamentary majorities have governed the country since the first election in 1952. Without a majority party India has regularly produced majority governments.

The first past the post rule rewards the party with the highest plurality by inflating its seat over its vote percentage by 20% to 25%. Conversely, the rule penalizes competing national centrist parties by deflating their seat as against their vote percentage. Given the high proportion (46% in 1980) of plurality victories and the predominance of multi-cornered contests (518 of 525)[29], two phenomena that reflect the attempts by some candidates to represent narrow confessional, caste, tribe, and other diversities, the first past the post rule has given the advantage to parties appealing to a wide spectrum of classes, communities, regions, and interests.

The centre may not hold. There is a limit to how many parties can aspire to be national and centrist. Multi-cornered contests involving several centrist parties open the way to levels of vote fragmentation that can advantage parties with more focussed appeals to class, community, or special interests. If three centrist parties (or electoral coalitions), having failed to agree on seat adjustments, face each other in 300 to 400 of parliament's 542 constituencies, their average vote per candidate could decline sufficiently for the first past the post rule to reward candidates of narrow rather than broad gauged parties.

A centrist multiparty system may self-destruct by failing to produce a parliamentary majority. The possibility was implicit in the 1980 election. One hundred seats won by Congress-I pluralities would have gone to Janata or Lok Dal if they had made seat adjustments. For the first time, an Indian national election would have failed to produce a parliamentary majority. But Mrs.

Gandhi's Congress - I won 358 rather than 353 seats, a two-thirds majority with less than 43% of the vote. For the seventh time a minority vote produced a majority government. (Janata and the Lok Dal/Congress-U alliance, with 33% of the vote, won 15% of the seats.) The first past the post (plurality wins) rule once again supported centrism.

Despite the de facto regime changes of the 1975-1977 period, when the Indian state changed from democratic to authoritarian and back again, the long-run character of political ideology, electoral politics, and public policy has remained centrist. More recently, since 1977, the transformation of the party system by alternation of party control of the national government has led to a perceptible decline in the level of procedural consensus. The result has been increased conflict over state issues such as the nature of the federal system and the independence (or "commitment") with which judges and higher civil servants interpret the law and carry on the administration.

Significant as the changes since 1975 have been, they are not likely to change the centrist character of Indian politics in the foreseeable future. In the absence of state fragmentation, revolution, or international cataclysm, which we consider unlikely, the social and institutional determinants of centrism are likely to prove more powerful and durable than those supporting class or confessional politics at the center. Even if India returns to a one party dominant system or comes under dynastic one party rule, the determinants of centrism will shape party strategy and policy.

REFERENCES

1. Myron Weiner. *Sons of the Soil, Migration and Ethnic Conflict in India* (Princeton: Princeton University Press, 1978).
2. See R. A. Schermerhorn. *Ethnic Plurality in India* (Tucson: University of Arizona Press, 1978). and Loyd I. Rudolph and Susanne Hoever Rudolph, *The Modernity of Tradition: Political Development in India* (Chicago: University of Chicago Press. 1967).
3. For a good summary of organized sector figures, see A.N. Agrawal. *Indian Economy,* 5th revised edition (New Delhi: Vikas, 1979), Chapter 32. Standard sources include government of India Central Statistical Organizations Department of Statistics, Ministry of Planning, *Statistical Abstract, New Series, various numbers;* Government of India, Planning Commission, *Economic Survey,* annual, Delhi. The latest figures, which do not allow the range of comparison of the 1971 figures, are 5.7 million factory workers in 1975, total work force 212 million in 1980 (applying a 2% annual growth rate to achieve an estimate), and 21 million organized sector employees, *Economic Survey*, 1978-9, pp. 87-88.
4. Alak Ghosh. *Indian Economy: Its nature and Problems,* 22nd Revised Edition (Calcutta: World Press, 1979-80), p. 429. Ghosh's figure of 34% for the proportion of union members among workers in the organized economy has to be an estimate. The Labour bureaus of many state governments do not submit any returns (presumably because they must rely on union cooperation, which is not forthcoming) and some of

those that do submit returns report incomplete coverage. *The Statistical Pocketbook, 1979, p. 171* relying on such state Labour Bureau figures, gives five million as the number of union members in 1971. Ghosh's 34% works out to 6.1 million.

5. Agrawal. *Indian Economy,* p. 584.
6. See Rakhahari Chattopadhyay. " The Political Role of Labour Unions in India: An Interstate Study of Labour Unions in West Bengal, Karnataka and Rajasthan," Ph. D. thesis, Department of Political Science, University of Chicago, 1975.
7. Alak Ghosh, *Indian Economy,* p. 312.
8. *India Today,* May 16-31, 1979, p. 83.
9. These percentages are based on the table given in Ministry of Information and Broadcasting, *India, 1979 A Reference Annual* (Delhi: Government of India, 1979), p. 10 Tribal Christians are included among "other non-Hindus."
10. For a careful examination of Muslim voting in Bihar, which shows, interalia, increasing Muslim electoral cohesiveness, see Harry W. Blair, *Voting, Caste, Community, Society: Explorations in Aggregate Data Analysis in India and Bengla Desh* (New Delhi and Stockholm: Young Asia Publications, 1979).
11. *Times of India,* April 7, 1980.
12. Constituency reports, January 6 to 15, *Hindustan Times, Times of India, Indian Express, Statesman.* Of the eight seats Congress - I won in U.P. all were by minority votes ranging from 35% to 45%. None would have been won had Lok Dal and Janata voted together. Of the 6 seats Congress-I won in Bihar, it would have lost two had those parties voted together.
13. G.G. Mirchandani. *The People's Verdict* (New Delhi: Vikas, 1980), p. 134; *Times of India,* January 26, 1980.
14. For an examination of long-term social factors at work shaping electoral behaviour in U.P. see Francine Frankel, "Problems of Correlating Electoral and Economic Variables: An Analysis of Voting Behaviour & Agrarian Modernization in Uttar Pradesh," in John O. Jield et al., *Electoral Politics in the Indian States: The Impact of Modernization* (New Delhi: Manohar, 1977). Frankel (p. 180) shows a modest positive correlation (A1) of BKD vote with the percentage of Muslim population in the West Plain (18 western districts) of U.P.
15. The Janata/Lok Dal average vote in tribal constituencies declined from 61% to 34%. an enormous 27 point decline. Constituency reports, *Hindustan Times, Times of India, Indiana Express, Statesman,* January 6 to January 15, 1980.
16. See Gopal Krishna, "Electoral Participation and Political Integration," In Center of the Study of Developing Societies, *Context of Electoral Change in India* (Delhi: Academic Books, 1969), p. 37. He shows that in 80% of reserved constituencies the Scheduled Castes constituted less than 30% while Scheduled Tribes constituted 50% of the population in two-thirds of constituencies reserved for them.
17. Ronald Herring and Charles Kennedy. "The Political Economy of Farm Mechanization Policy: Tractors in Pakistan," in Raymond Hopkins (ed.). *Politics of Agricultural Development* (Boulder, Colorado: Westview Press, forthcoming), and (with M.Ghaffar Chaudhry) "The 1972 land Reforms in Pakistan and Their Economic Implications: A Preliminary Analysis," *Pakistan Development Review,* Autumn 1974.
18. Charan Singh articulated these themes while campaigning in Andhra Pradesh in October 1979. See the *Deccan Herald, Hindu, and Indian Express* (Bangalore) for October 23 and 24, 1979. We are grateful to V.K. Narasimhan, editor of the *Deccan Herald,* for providing these cuttings.

19. In anticipation of assembly elections in nine states projected for late May or early June 1980, ex-Swatantara neo-feudal notables (mainly Rajput) in Rajasthan previously associated with the Janata party but badly treated by Janata Chief Minister Bhairon Singh Shekhawat and adversely affected by his redistributionist policies were being encouraged by Congress-I state and national leaders to believe that they would be allocated a substantial number of tickets (nominations). Rajputs in Gujarat and to a lesser extent in Madhya Pradesh already had moved into top leadership positions in the Congress-I. His Highness Baria was coordinating Rajput-Congress-I negotiations. In U.P. Rajputs played a significant part in 1980 Congress-I Parliamentary campaigns. In Amethi, the son of the Raja of Amethi, Sanjay Singh, managed Sanjay Gandhi's successful campaign. The Rajmata of Gwalior, Vijaya Raje Scindia, Mrs. Ghandhi's Janata opponent in Rae Bareilly constituency, fared badly in part because Rajputs whose support she expected voted instead for Mrs. Gandhi.
    In the 1980 state assembly elections in U.P., 115 of the 334 Congress (I) candidates for general seats were Rajputs. Brahmans for the first time ranked second with 95. Brahmans, like Muslims and Harijans, are not likely to take well to Sanjay Gandhi's reliance on Rajput leadership. For details, see S.K. Tripathi's column in the *Indian Express,* May 16, 1980.
20. The Congress-I minister for agriculture, Rao Birendara Singh, has pushed vigorously for more favourable prices for agricultural products and inputs. "Appeasing the Rich Farm Lobby", *Economic and Political Weekly,* 15-9, March I. 1980. Finance Minister R. Venkataraman's temporary budget of March 1, 1980, which had virtually no relief for any category of taxpayer, did have one for the tax paying farmers who are by definition rich, allowing them to take into account the previous year's losses in the current year's declaration. *Times of India,* March 12, 1980. The two moves suggest appeals to both bullock capitalists and rich farmers.
21. See our *Modernity of Tradition,* Part I: Robert L. Hardgrave Jr., *The Nadars of Tamilnad* (Berkeley: University of California Press, 1979): Eugene Irschick, *Politics and Social Conflict in South India: The Non-Brahman Movement and Tamil Separatism* (Bombay: Oxford 1969).
22. The Congress percent of votes over three elections is as follows: U.P. — 1971, 49%; 1977, 25% ; 1980, 36% ; Bihar — 1971, 34% ; 1977, 23% ; 1980, 36% ; Haryana — 1971, 53% ; 1977, 18% ; 1980, 30%.
23. "The Politicization of the Peasantry in a North Indian State," mimeo, 1978, p. 66.
24. Charan Singh, *India's Economic Policy: The Gandhian Blueprint* (New Delhi: Vikas, 1978), p. 16.
25. For an account of the complex ideological and legal aspects of the backward classes movement, see Mare Galanter. "Who are the 'Other Backward Classes' An Introduction to a Constitutional Puzzle," *Economic and Political Weekly,* 13:43 and 44, October 28, 1978. Also Lelah Dushkin, "Backward Class Benefits and Social Class in India, 1920-1970," *Economic and Political Weekly,* 14:4, April 7, 1980. Today the term backward classes loosely refers to lower "Shudras," about half the castes located below the upper "Shudras"— e.g., Jats and the traditionally literate upper "twice born" castes — eg. Brahmans, Kshatriyas, Vaishya, and para-twice born such as Kayasthas and Khatris — but above the untouchables.
26. Ram Naresh Yadav and Karpoori Thakur, the Chief Ministers of U.P. and Bihar, were from backward classes. Studies that deal with specific "backward classes" include Chandrasekhar Bhatt, "The Reform Movement among the Waddors of Karnataka."

in M.S.A. Rao, ed., *Social Movements in India,* Vol. I (New Delhi, Manohar, 1978): Mark Franda, *Small is Politics* (New Delhi: Wiley Eastern, 1979), which deals with backward "classes" in Bihar; M.S.A. Rao, "Political Elite and Caste Association, *Economic and Political Weekly,* Vol. 3 May 18, 1968, an account of the Yadavs: and K.K. Verman. *Changing Role of Caste Associations* (New Delhi: National Publishing House, 1979) a study of the Kurmis.

27. These efforts reflected the commitment in the Janata manifesto of 1971 and the Lok Dal manifesto of 1979, which provided that " at least 25 per cent of Groups A and B jobs in the Central Government services will ... be reserved for young men and women coming from .... [the socially and educationally backward classes . .... both Hindu and Muslim] as recommended by the Backward Classes Commission appointed in the fifties by the Union Government itself, under Article 340 of the Constitution," *Lok Dal Election Manifesto, 1979* (New Delhi: Shyamal Basu, Publicity Secretary, 1979) p. 29.
28. Biplab Das Gupta and W.H. Morris-Jones, *Patterns and Trends in Indian Politics: An Ecological Analysis of Aggregate Data on Society and Elections* (New Delhi: Allied Books, 1976).
29. Both figures are for 1980. See Mirchandani, *People's Verdict,* p. 127 for plurality victories in 1980 and p. 103 for number of multi-cornered contests, 1980 and 1977. In 1977, 438 of 540 contests were multi-cornered but 473 of 540 contests were won by a majority, a result reflecting Janata's sweep in the Northern States. For the 1977 majority/plurality win figures see Election Commission. *Report on the Sixth General Election to the House of the People ...1977. Vol. II (Statistical) p. 16.*

XIII State and Spatial Aspects of Industrialization in Post-Independence India

VIJAY K. SETH

Once the firm foundations of industrial revolution were laid in England, the spatial spread of this powerful force operated in three different directions: (a) within the European core region;(b) to distant lands where European settlers dominated the economy (in this sense, the spatial spread of industrialization to the economies of America,Canada and Australia was a mere extension of the European core region); and (c) to other inhabited regions, where old social orders were invariably pre-industrial in type and were integrated into the European core region as colonies.

The spatial spread of industrialization in the third direction registered distinct changes after World War II. The process of decolonialization which set in during this period gave birth to independent nation-states. The newly independent states had the option to leave the process of industrialization to the free play of market forces, or to evolve certain new institutions,regulations and policies to give a definite direction to the process of industrialization.

An important question which emerges with regard to the spread of industrialization in these underdeveloped countries is whether industrialization retains the basic characteristics of its spatial structure in all spatial units (nations) where it encounters or assumes characteristics which are specifically emphasized by the respective national governments.

The former alternative is based on the premise that spatial differences in the forces that have bearing on the process of industrialization do not really matter because it is an autonomous process,which has its own logic, independent of socio political systems. The other alternative highlights the variety of forms that industrialization takes on when it enters different spatial units, depending upon the differences in physical features, social and cultural institutions—history and polity. These diversities are reflected in the specific nature of the collective entity called 'State'. Therefore, dissimilarities which produce spatial

Reprinted from *Political Geography Querterly* Vol. 5, No. 4, October 1986, pp. 331-350 with the permission of Butterworth-Heinemann, Oxford, U.K.

peculiarities are clearly manifest in the State policies.

India achieved Independence in 1947. The political independence of India signified that the process concerned with the formulation of government policies also got internalized. After Independence, India assumed the characteristics of a mixed economy and politically of democracy and federalism.[1] These characteristics are quite relevant to an analysis of the role of State in influencing the spatial pattern of industrialization. Here the nature of a mixed economy will provide us with knowledge regarding the extent and nature of the role of State; the nature of federalism will give us an understanding about the spatial distribution of political power and resulting spatial pressures emanating from different spatial units of the national economy.

For the purposes of our study, we have divided this paper into four sections. Section I provides a brief description of the nature of a mixed economy and the strategy of development and regional planning that has emerged after Independence in India. Section II analyses the nature of federalism and resulting regional pulls and pressurse Section III deals with the role of two important policy instruments used by the State in India to affect the spatial spread of industrialization. And finally, in Section IV, broad conclusions of the study are reported.

I. The Nature of a Mixed Economy and the Strategy of Regional Development

After World War II, most of the countries outside the socialist bloc assumed the characteristics of a mixed economy. This had happened in response to the Great Depression of the 1930s which revealed the dangers associated with an economic system that depended on the free play of market forces to solve its central problems. It was also in response to the rise of Keynesian economics that explained the pathology of free market economies. The needs of post-World War II reconstruction and development also gave rise to regulated market economies. With this the old classic model of capitalism came to an end and the system which emerged thereafter is termed a mixed economy.

There are marked variations among different mixed economies due to: (a) the extent of the role that State assumes—in other words, how far the deliberative processes as against market processes are given importance in influencing the working of the economy; and (b) whether the role of the State is merely to moderate the market forces, e. g. policies to counter business cycles, or to achieve certain broad social objectives which cannot be achieved through market mechanism. The latter consideration paves the way to systematic planning for growth and equity. Therefore, it is essential to know the nature of a mixed economy in order to identify the policy instruments being used by the State to influence the spatial spread of industrialization.

It was after World War I that the Indian economy started to acquire the characteristics of mixed economy. Wartime needs had forced the British government to set up certain industries within India under State ownership. The number of public enterprises further increased during World War II for the same reason. Historical circumstances also forced the British government to provide protection for Indian industries and pursue policies of import substitution (Seth,1983).

When India was on the threshold of achieving Independence, all those who gave thought to the future course of development accepted the importance that needed to be attached to the role of the State. The acceptance of the role of the State was partly in response to prevailing international thinking regarding the dangers as free market economies,and partly due to the need for development of infrastructural facilities and to provide and alternative path of development. However,there were differences in the opinions of various organized groups and national leaders regarding the nature and extent of the role of the State. Their opinions varied,depending upon their ideology, class interest or the interest of a particular stratum they represented. Broadly speaking,there were four strands of thought, associated respectively with Gandhi, Nehru, M. N. Roy and a group of Bombay industrialists. [2]

After Independence,the Indian National Congress assumed power with Pandit Jawaharlal Nehru as Prime Minister. Since it was the Congress Party that formed the government,the views held by other organized groups and political parties did not affect official policies. The approach of the Congress Party became the official approach.

The framework of the Indian Constitution (which came into effect on 26 January 1950) provided legal sanctity to the mixed character of the Indian economy when it incorporated, on the one hand, the 'Fundamental Right to Property' (which is an important characteristic of a free market economy)[3] and on the other, the 'Directive Principles of State Policy' (which envisage controls on the free play of market forces). [4]

To translate the goals of social and economic policy prescribed in the Directive Principles of the Constitution into national programmes of development, the Planning Commission was established on 15 March 1950. [5] Along with the establishment of the Planning Commission, provision was made for industrial licensing under the Industries (Development and Regulation) Act of 1951 to regulate private investment in accordance with the plan objectives. This also necessitated a clear-cut differentiation between small-scale and large-scale industries.

The political philosophy which was embodied in the Indian Constitution was made economically operational in the Industrial Policy Resolution (IPR) 1956. This policy resolution made it clear for the first time what precisely would be the role of the State in India's mixed economy. The Industrial policy Resolution 1956 divided industries into three categories:

1. Schedule 'A' industries, consisting of defence products, railways, airways, shipping and heavy basic industries, where the private sector could not enter.
2. Schedule 'B' industries, consisting of light capital goods industries, where the private sector could coexist with the public sector.
3. Rest of the industries, mainly consumer goods, which were left for the private sector.

Since then, there have been several changes in the Industrial Policy. Most of them have been concerned with increasing or decreasing the exemption limit for the investment which needs licensing (thus redefining the distinction between a small-scale and large-scale industry) and regarding redemarcation of the areas for the participation of the private sector. Another significant change in the industrial policy has been the introduction of a new sector, called the joint sector. It is intended to combine private management with capital provided from the public exchequer (Wadhwa, 1977). In spite of all these changes, the basic industrial policy has remained the same.

Under the administrative arrangement that has come to exist the Planning Commission sets the target (physical targets) for various industries for both the private and public sector, and outlays only for the public sector. Investment in the private sector is understandably enough left to be determined by entrepreneurs themselves, though subject to influence by various promotional and regulatory measures. Among the promotional measures one can mention the role of development banks. Among the regulations one can mention the role of licensing policy. Licensing authorities issue licences to the private sector units which fall within the purview of the Industries (Development and Regulation) Act, according to the plan targets. Apart from licensing regulations, private investment is subjected to various kinds of price controls.

The process of planning in India began in 1951. By now,therefore, the Indian economy has experienced more than three decades of planning. It is obvious that, during this time period,India should have completed six five-year plans. However,the Sixth Plan ended only in March 1985,because a period of Plan Holiday or Annual Plans (1966-1969)[7] delayed the formulation of the Fourth Plan (1969-1973). Thereafter,the Fifth Plan was abandoned in 1978-1879, after it had run for four years, following a change in political power at the centre. The Janata Party drafted the Sixth Plan (1978-1983). With the return of the Congress Party at the centre,the Sixth Plan was redrafted and scheduled for the period 1980-1985.

Systematic model-building for the purpose of evolving development strategy began with the Second Five-Year Plan. It was in this plan that Mahalanobis (1963) evolved a model to explain the basic strategy of industrialization. In the initial stages,planners used a two-sector consistency model, but with the availability of inter-industry tables, they used linear-programming techniques to evolve multisectoral consistency as well as optimality models (Bhagwati and Chakravarty, 1971; Rudra, 1975).

The basic strategy of development which is implicit in these plan models suggests that,in an economy with stagnant exports, the most appropriate policy for maximizing the rate of growth and self-reliance would be to invest a relatively large proportion of resources to the building up of 'heavy'/'basic' or capital goods/machine-building industries. The appropriateness of this strategy depended on the postulate that the emphasis on investment goods increases the ration of savings and investment to national income in the long run. This strategy was, in as sense,continuation and further consolidation of the process of import substitution which began during the interwar period. In this sense it helped to speed up the process of internalizing the forces of industrialization.

In analysing the role of planning for India, it is important to recognize that India is a mixed economy, one in which the private sector is a dominant sector. Therefore,in major segments of economic life, government has to depend on the price mechanism and other forms of regulations and incentives to realize its objective (Chaudhuri, 1971; Gadgil, 1972).

Indian plans, in general, have the following operational goals:

(a) to achieve an increase in the national income;
(b) to achieve a desired investment/income ratio;
(c) to reduce inequalities of income and concentration of economic power;
(d) to generate additional employment opportunities.

This list does not contain a consistent set of goals. In spite of an inherent conflict in the set of objectives planners do not specify 'trade-off' among them (Marglin, 1967). It is to be noted that balanced regional development is not considered as an important objective. At best, it can be said that the objective of balanced regional development is subsumed under the criterion of social justice. But the importance attached to asset transfer measures, such as land reforms and the minimum needs programme,is far more than is considered for balanced regional development.

The Third Five-Year Plan for the first time devoted an entire chapter to the problem of regional imbalances (Planning Commission,1961: chapter 9). Emphasis was laid on the systematic identification of underdeveloped regions as well as the promotion of studies of economic trends and rates of growth of different regions. It also realized the importance of large projects as nuclei of regional development (Planning Commission, 1961: 148- 149). To achieve regional dispersal of industries,the Third Plan suggested the policy of charging a uniform price for steel at all the rail heads in India.

Earlier the First and Second Plans documents, though positing the objective of balanced regional development, believed that the extent to which the pattern of industrial location in the country could be changed within a short time span was limited. This was so because for any industrial undertaking to operate profitably it must have easy access to raw materials, labour,power and markets. Hence,planners believed that the creation of infrastructural facilities during the plan period would significantly alter the pattern of industry location in the long run.

The Second Plan, which laid emphasis on the creation of heavy basic industries, relied on the small-scale industries to achieve regional balances. To achieve a spatial spread of small-scale industries, an 'industrial estates' programme was initiated to provide small-scale units with the advantage of common services and such other facilities as electricity, water, gas, railway sidings, etc. (Planning Commission, 1956:452-453). This suggests that the First and Second Five-Year Plans had no explicit policy of balanced regional development other than the promotion of small-scale industries.

The rationale of this strategy was that balanced regional development could be achieved by supporting two separate types of industries to serve two distinct social and economic objectives. The first was the category of capital-intensive large-scale industries concerned with the manufacture of producer and intermediate goods. The second category encompassed a whole variety of labour-intensive small-scale industries. Assuming that the location decision in the first category would be based on efficient use of resources, with a consequent tendency to spatial concentration, the second location decision would favour regional dispersal. Planners felt that this strategy would correct regional imbalances (Lefeber and Datta Chaudhuri, 1971).

This strategy failed miserably in India. It was bound to fail because of its erroneous assumptions regarding the locational pattern of small-scale industries. It is now an accepted fact that modern small-scale industries need similar types of infrastructural facilities to the ones required by large-scale industries. As a small-scale unit is not capable of internalizing the external economies,its tendency towards location in the big cities is quite obvious (Dhar and Lydall,1961; Staley and Morse,1965). For this reason too the 'industrial estates' schemes were successful only when they were located near a large urban agglomeration (Alexander, 1963).

It was perhaps the experience of earlier plans which helped the Fourth Five-Year Plan(1969-1974) give overriding importance to infrastructural facilities in attempting to achieve regional spread of industries. Planners observed that as there existed intrinsic linkages between the infrastructural facilities on the one hand, and growth and diversification of economic activities on the other, there was therefore a need for large-scale expansion of the infrastructural facilities to achieve balanced regional development (Planning Commission, 1969). This plan noted that the development of backward areas in different states was essentially the responsibility of individual state governments.

In the Fifth Five-Year Plan (1974-1979), for the first time planners accepted that there need not be a trade-off between national growth and reduction of regional inequalities. In particular,the development of backward areas was assigned great importance in view of the emphasis laid on improving the living standards of the lowest 30 per cent of the population (Planning Commission, 1973).

The Janata Party government drafted the Sixth Five-Year Plan (1978-1983), which again had an entire chapter on patterns of regional development (Planning Commission, 1978). To achieve industrial dispersal through small-scale industries,a District Industries Centre (DIC) scheme was introduced in 1978. Under this scheme,it was suggested that various facilities should be provided to groups of entrepreneurs on preselected sites for setting up small village industries.

This plan accepted that setting up large capital intensive industrial projects is not sufficient in itself to encourage the development of backwared areas (Planning Commission, 1978: 195). In order to evolve a consistent development strategy, the Planning Commission set up a high-level National Committee for the Development of Backward Areas on 30 November 1978, under the chairmanship of B. Sivaraman.

As explained earlier, due to changes in political power at the centre, the Sixth Plan was redrafted. The revised Sixth Plan (1980-1985) does not contain a separate chapter on regional development. In contrast to its catalytic role in the early phase of planning, the State is now called upon to concern itself with problems of a short-term nature. Therefore, in the new plan strategy, greater emphasis is attached to problems associated with demand management and the elimination of certain bottlenecks from the supply side, such as transportation and energy. This is supplemented by a drive to remove technological gaps (Planning Commission,1980).

It is quite evident from this brief evaluation of the regional development strategy pursued in the five-year plans that concern with balanced regional development was constrained by three important postulates: (a) removal of regional imbalances is not possible in the short run; (b) even in the long run, it is difficult to combine a high rate of growth of national income with removal of regional imbalances; and (c) the development of infrastructural facilities will *pari pasu* result in spatial spread of industrialization. The acceptance of these postulates resulted in a situation where planners concentrated their attention more on the intersectoral balances than on regional imbalances. Therefore, to translate the regional development policy into practice several *ad hoc* groups and committees were appointed.

The first working group was set up during the Third Five-Year Plan in 1964 for evolving a suitable method for identifying backward regions. What emerged from their deliberations was a set of 15 social and economic indicators of backwardness.[8] In 1968, two separate working groups were set up: one is popularly known as the Pandey Committee, and the other as the Wanchoo Group.

The Pandey Committee used certain indicators for identifying industrially backward states.[9] The Wanchoo Group was set up to suggest suitable strategies for the industrialization of backward regions, and recommended that incentives be provided for attracting industries to such regions. The proposed incentives consisted of: (a) capital subsidy during the construction stage of the project; (b) interest subsidy during initial stages of production; and (c) tax subsidy within the successful operation of the project.

In 1969, following the recommendations of the Pandey and Wanchoo working groups, a scheme known as the Central Scheme of Investment Subsidy was introduced. This subsidy was fixed at 10 per cent of the fixed capital investment subject to a ceiling of Rs. 5 lakhs. From March 1973, the rate of subsidy was increased from 10 to 15 per cent of the fixed capital, subject to a ceiling of Rs. 15 lakhs (National Committee on the Development of Backward Areas, 1980). Apart from the capital subsidy, concessions on corporate tax were provided to units located in the industrially backward regions under Section 80 HH of the Income Tax Act. Since 1972, 20 per cent of the taxable profits are allowed as tax rebates.

The National Committee, while reviewing the working of these incentives, observed that most of the incentives given by the centre had been appropriated by backward districts located in the industrially developed states. The Committee discovered that out of the total subsidy given under the Central Investment Subsidy Scheme, over 66 per cent had gone to those states which were *not* considered industrially backward by the Pandey Committee (National Committee on the Development of Backward Areas, 1980; Table 4.1, p. 21). Even out of the financial concessions given by such nationwide financial institutions as the Industrial Development Bank of India, the Industrial Credit and Investment Corporation of India and the Industrial Finance Corporation of India to different states until December 1979, 55 per cent of the concessional finance flowed to units located in states *not* considered to be industrially backward regions (National Committee on the Development of Backward Areas, 1980: 23).

The foregoing analysis suggests that the strategy to influence the spatial pattern of industrialization which has been followed by the Indian government consists of such policy instruments as: (a) location of public sector plants; (b) the use of a licensing policy to influence the location of private sector investment; and (c) use of input subsidies and other financial incentives to attract industrial investment in the backward regions. It is also evident that for implementing this strategy the role of federating units, i.e. states, is very important. This necessitates an understanding of the nature of federalism as it has evolved in India.

II The Nature of Federalism and Spatial Pressures

It has been said that the federal state

> ......is the most geographically expressive of all political systems. It is based on the existence of regional differences, and recognises the claims of component areas to perpetuate their individual characters—federalism does not create unity out of diversity, rather it enables the two to co-exist (Robinson, 1961).

Livingston makes the spatial dimension more explicit when he states 'federalism becomes nothing if it has to embrace diversities that are not territorially grouped' (Livingston, 1952; Dikshit, 1975). For the existence of federalism in reality, the spatially organized diversities should be accompanied by forms of government which have the characteristics usually associated with democracy or free government.

Depending upon the degree of homogeneity and heterogeneity of federal States, political power is distributed among various spatial units. Here the term political power refers to the ability of a spatial unit to materialize its objective function and internationally influence the decision field of other spatial units. There are some studies which have explored the possible conditions under which political power will be evenly distributed as against those which would lead to its uneven distribution over space. These studies classify federal States according to the characteristics of *symmetry* and *asymmetry.*[10]

In India, it would seem that the characteristics of assymetry would be prevalent, due to the existence of regional diversities in terms of culture, history, language and levels of development. These characteristics of asymmetry gained strength with the linguistic reorganization of the states after Independence.

The seeds of the formal structure of federalism were sown in India with the Government of India Act of 1915, Section 32 and the Government of India Act 1935, Article 176 (i)(Dikshit, 1971). Therefore, the characteristics of the federalism were inherent in the system at the time of Independence. The cultural diversities and related geographical factors made federalism inevitable in India. India after Independence emerged as a federation 'in which the paramountcy which British Government had over Indian princely states has been taken over by Union Government and applied to all its units' (Santhanam, 1961: 13).

It must also be noted that, in contrast to other federations, the Indian federation emerged through a process of 'disaggregation rather than aggregation' (Hicks, 1961: 152). This means that political power is concentrated at the Union (centre) level to start with and gets devolved over a period of time. In this kind of federalism, there is demand for more and more autonomy for states (federating units), whereas in other federations, there is demand for more and more transfer of power to the centre.

The Constitution of India which came into force after Independence also proclaimed it as a federal State. However, it seems that the Constitution has given greater powers to the centre. This bias in the Indian Constitution is related to the fact that: (a) the main objective of the Congress Party, prior to Independence, was to provide an alternative to the centralized British power; and (b) the partition of the country on religious grounds also strengthened the idea of a strong centre. The national leaders held the belief that the infinite segmentation of their society along linguistic, communal or religious lines, and the primordial attachments that inevitably accompany such a segmentation, would tend to negate the secular democratic norms as well as the development process itself.

The powers are more centralized with regard to financial aspects. The capacity of central government to exert influence on the quantum and pattern of states' expenditure is much larger in Indian than in other federal States. The net central transfers financed more than one-third of the aggregate net budgetary expenditure and more than one-half of the plan outlay of states in India during the two decades beginning with the Second Five Year Plan (George, 1982). The extent of dependence became more serious in view of the fact that in practice about 71 per cent of the central grants are given *ad hoc.* This leads to arbitrariness in the distribution of funds (Narang, 1982). This process makes the federal polity a bargaining process between central and states leaders.

Moreover, due to the presence of institutions like the Planning Commission and such instruments as licensing policy, the centre has assumed more powers with regard to the industrial sector in India. It seems that due to the existence of a nation-wide market based on common laws and fiscal policy, the centre will continue to have importance in the management of the Indian economy.

In spite of awareness on the part of planners concerning spatial aspects of economic development, the core of planning exercises and the setting of plan targets were based on the implicit assumption of a unitary State. In the beginning it did appear that Indian polity had the basic characteristics of unitary State. The Congress Party was ruling at the centre, as well as in the states. Due to this outward manifestation of the unitary characteristics of State and Nehru's personal commitment to the idea of planning, the Planning Commission was a very powerful organization.

However, in the framework of planning, there was scope left for regional pressures to influence central planning—the National Development Council (NDC) was constituted for this purpose.[11] But to start with, NDC was a weak organization (due to lack of expertise with the State Departments of Planning and Statistics, and the centre having control over the formulae for central financial assistance) (Hanson, 1966, 1966). Nehru's voice, as Chairman of the NDC, was ultimately decisive (Frankel, 1978). But after successive general elections and the formation of new states on the basis of language, new regional elites emerged, which started influencing power at the centre.[12] Hence, slowly the forces of regional pressure started showing their face on the surface and the spatial organization of political power played a larger role (Hardgrave Jr. 1970). Meetings of the NDC were the sounding board for exploring the political feasibility of the plans (Bhagwati and Desai, 1970). Some scholars describe it 'as the most important organised gathering where plans undergo adjustments in the light of the needs, pressures, prejudices and capacities of the states' (Hanson, 1966: 113).

The emerging forces of regional pressure added strength to the NDC, and pressures from the states on the process of planning at the centre necessitated the constitution of the Administrative Reforms Commission in 1968. The Commission suggested the attribution of primacy to the NDC. The function of the NDC should be the formulation of the basis for development policy, and the role of the Planning Commission should be subsidiary. The Commission also advised the establishment of Planning Boards in each state (Administrative Reforms Commission, 1968: Frankel, 1978).

The increasing power of the forces of regional pulls and pressures were even recorded by the Study Team on Centre-State Relations. This committee showed that it was much easier to settle issues between the centre and the states when the Congress Party was ruling simultaneously at both levels. But the emergence of non-Congress state assemblies increased the tensions. The committee cited the examples of Kerala, Tamil Madu and West Bengal (Report of Study Team on Centre State Relations, Vol. I, 1968). There are two very important instances to reveal the changes in the centre-state relations in the post-Nehru period. For example, the Government of Tamil Nadu Challenged the constitutionality of the Planning Commission and the powers of the centre to issue industrial licences (Rajamannar, 1971: 109). The other example is that of Kerala, which formulated an alternative national plan(Hanson and Doughlas, 1972: 175).

In spite of the fact that forces of regional pulls and pressures were gaining strength, planning in a federal State continued to be spatial in character[13]. The absence of spatial distribution of targets led to a free play of the pull and push factors within the spatial organization of political power, represented in meetings of the NDC. This resulted in allocation of plants with an uneconomic scale to as many states as was politically necessary (Bhagwati, 1962; Bhagwati and Desai, 1970).

III. Policy Instruments and Spatial Aspects of Industrialization

As has been observed above (Section I), the role of the centre is quite significant with regard to the spatial spread of industries. Even regional pressure groups try to achieve their development objectives by influencing the decisions of the centre (Crawley, 1972). In the administrative setup explained above, central government can influence the spread of the industrial sector in three different ways: (i) some changes in location can be achieved through governmental regulations, incentives and policies (industrial policy, fiscal policy and price controls); (ii) changes in spatial patterns may result from state ownership and controls of industries (public sector); and (iii) there could be locational changes which occur as by-products of government policies not directly intended to change industrial location.

Manufacturing, of course, is affected to a larger extent by investment incentives which are offered in a variety of ways. These incentives can be in the form of long-term finance at subsidized rates, through various central and state financial institutions (e.g. the Industrial Finance Corporation of India, the Industrial Credit and Investment Corporation of India, the Life Insurance Corporation, nationalized banks and various state finance corporations), tax rebates and subsidized land.

The influence of these incentives on location should be studied in the broader context of the total investment decision, of which the choice of appropriate location is only one component. Therefore, it is possible that producers incorporate investment inducement into their investment plans for reasons which are quite independent of the choice of location. Thus, the real impact of inducements may not involve locational adjustments (Cannon, 1975), and in reality it is very difficult to identify the precise relationship between inducement to location and other investment decisions (Halliwell, 1968). Moreover, policies which subsidize a particular factor of production can result in a bias in the adoption of technologies in favour of the subsidized factors themselves rather than towards influencing the location decision (Harcourt, 1968).

The Indian government followed the policy of charging uniform prices of some of the important industrial inputs—cement, steel and coal. Spatial spread of these inputs is highly localized. As these inputs are quite important in the process of industrialization, the government intends to provide regional equity so that the states blessed with these inputs should not gain disproportionately. Moreover, by charging uniform prices, the government has minimized the pull of these materials on firms' location decisions,and strengthened the pull of the market.

There are certain government policies which as such are not intended to affect the location decisions of industry, but still location decisions may be influenced by them. In this category, one can mention the impact of location of administrative headquarters and cantonments, etc., on the decision of the firms regarding choice of location. An important example of this is the role of Indian planning in the development of agriculture. Agriculture by its nature is widely spread. Agricultural prosperity has several kinds of impact on the industry: (i) in the form of supply of wage goods; (ii) in the form of increased supply of raw materials; and (iii) as a potential market for industrial products. Government policy regarding the agricultural sector can be said to have an indirect impact on the location of industrial activity, but this is quite an indirect and unintentional role of the State.

One of the objectives of the Industrial Licensing Policy is balanced economic development for different regions of the country so as to reduce disparities in levels of development. Since every investor needs clearance from the licensing authorities, this is expected to have a significant influence on the spatial pattern. On the other hand, the location of public sector plant is entirely the decision of the state—a choice which can play a significant role in altering the spatial pattern of industrialization.

As location of public-sector plants and spatial distribution of licences are the only economic policy variables which explain the direct influence of the state on the spatial spread of industries, available facts pertaining to the location of public-sector projects and region wide-distribution of industrial licences are analysed below to understand whether sustained policies of regional dispersal have been pursued.

If one studies the pattern of location of public-sector steel mills, one can assess the combined impact of deliberate policies of the State to achieve industrial dispersal, materials orientation, and regional pulls and pressures. Bhilai in Madhya Pradesh was one of several potential locations for a steel mill visited by Russian technicians late in 1954. Bhilai was selected because Bhilai not only fulfilled the prerequisite of mill site, it also satisfied the government's ideological and political objectives (Johnson, 1966; 108). The British mission that visited India in 1955 was shown locations in only two states:. West Bengal and Bihar. When Durgapur in West Bengal was selected for the British mill it was announced that the next public-sector plant would be situated in Bihar at Bokaro.

Due to lack of spatial composition of plan targets, the state leaders demanded more and more public-sector projects in meetings of the NDC or through other forms of regional pressures. There are examples of the pressures created by states to demand steel mills. To strengthen their demands, they provided blueprints of the projects in the form of technoeconomic and feasibility studies of the proposed plants, prepared by experts in the field. In spite of the fact that the pricing of steel in India is such that it does not attract any of the industries which have linkages with steel to be located near steel plants, states continued to demand steel mills within their boundaries.[14]

There were demands to locate steel mills at Murmugao in Goa, Billory in Mysore, Bailadilla in Madhya Pradesh and at Vishkapatnam in Andhra

Pradesh. But the feasibility studies of these alternative locations suggested that apart from achieving political ends the locations would be economically unsound. The same applies to the locations of plant at Salem in Tamil Nadu (Johnson, 1966: 294-310).

To put extra pressure on the central government for locating the steel mill in West Bengal, the government of West Bengal proclaimed its own plans to build a mill and then a coke-oven battery at Durgapur. Around that time the centre finalized the location of a British Mill at Durgapur, sufficient popular support having been created in West Bengal for location at Durgapur. These forces must have played an important role in the location of other public-sector plants also.

The lack of spatial distribution of targets in the national plans resulted in increasing importance of the NDC. Hence, in the federal polity the regional pulls and pressures, through the NDC and otherwise, were able to achieve dispersal of public- sector investment.

Since the Planning Commission did not control the spatial distribution of industrial targets, licensing authorities were left to decide the spatial distribution of industrial licences. Industrial licensing policy had the explicit objective of achieving 'Balanced Economic Development' (Hazari, 1867: 1; Estimates Committee, 1972). It was supposed to influence the optimum size of a project, the time phasing and location of the project with the assistance of the Directorate General for Technical Development (DGTD) but due to the absence of any criteria for giving weight to alternative objectives of the policy in the granting and rejection of industrial licences, accompanied by the poor quality of techno-economic studies of the proposed investment conducted by the DGTD, the licensing authorities were left without any direction regarding the spatial distribution of licences. Hence the policy which was expected to influence the spatial pattern of private investment could not achieve its objective.

There have been several inquiry committees to assess the working of the licesing policy. Some of them have inquired into the operational part of the licensing system with special reference to procedures,[15] others have gone into details regarding the impact of licensing policy on the concentration of economic power and monopoly (Hazari, 1967; Dutt, 1969), but none of them has seriously pursued the matter regarding the spatial distribution of licences being issued by the licensing authorities. It is worth mentioning that the terms and references of the Dutt Committee include: 'to study to what extent licences issued are in consonance with the desired objectives of the IPR of 1956, with special reference to regional dispersal' (Dutt, 1969: 4).

While discussing location and licensing, the Dutt Committee noted that 'this aspect came to be increasingly emphasized in considering licence applications. While it was realized that the applicant entrepreneur would himself consider what would be a good location from an economic point of view, the licensing authorities had to keep in mind availability of raw materials and also facilities like power, transport and water.' (Dutt, 1969: 30). This suggests that the choice of location was left to the individual entrepreneur, subject to the availability of infrastructure. The Dutt Committee further admits that the licensing system could not much assist the industrially backward regions, because the number of proposals for locating industrial units in these states was much smaller than for other states, and moreover, 'that among the states which had fewer proposals for licences there was also poor implementation of the approved licences' (Dutt, 1969: 109-110). These points further substantiate the observation that the licensing authorities did not influence the choice of location to any major degree.

The Dutt Committee concluded that licensing, as it had operated during the last 10 years, had not been effective in the attainment of the objective of regional dispersal. The effectiveness of licensing as an instrument for achieving this objective was bound to be limited. The responsibility for preparing an overall plan of development which would take note of the necessity to ensure balanced development of different parts of the country should not be placed on the licensing authority. It was essential that the development plans for the country should include a perspective regarding how different regions would develop over a period of time, and how the spatial distribution of investment, as well as production, would be attempted as part of the plan. Such a detailed scheme is not to be found in the plans (Dutt, 1969; 114).

The importance of public-sector investment in manufacturing began during the Second Five-Year Plan (1956-1961). However, the data regarding the spatial distribution of public sector investment are available only since 1963, in the annual reports of the Bureau of Public Enterprises. The shares of the states in cumulative annual investment in central public undertakings from 1963 to 1983 are reported in *Table 1.* To illustrate how far public-sector investment has been able to reduce inter state industrial inequalities, we have ranked different states in terms of per capita Net State Domestic Product (NSDP) in *Table 2,* and on the Index of Regional Intensity of Industrialization (R_1) in *Table 3.* The Index of Regional Intensity of Industrialization (R_1) is defined here as: the ratio of share of the region in terms of labour employed in the industry and its relative share in terms of population.[16]

Table 1 shows that the combined cumulative percentage share of Bihar (15.9), Madhya Pradesh (23.7) and Orissa (21.4) was 61 per cent in 1963. These three states ranked 14th, 10th and 13th in terms of NSDP (see *Table 2*), and 9th, 13th and 14th in terms of R_1 (see *Table 3*), respectively. Apart from Bihar, the relative shares of these states have declined consistently since 1970, and their combined share declined by 15 per cent over the period 1963-1983.

The combined share of the industrialized and developed states of Maharashtra, Gujarat, West Bengal and Tamil Nadu (see *Table 2 and Table 3* for their ranks in terms of NSDP and R_1 increased by the 2 per cent in 1983.

TABLE I Percentage share of states in cumulative investment (gross block) in central public undertakings (1963-1983).

	Year	A.P.	Assam	Bihar	Guja-rat	Hary-ana	H.P.	Karna-taka	Kerala	M.P.	Maha-rashtra	Orissa	Punjab	Rajas-than	T.N.	U.P.	W.B.
1.	1963	0.8	1.5	15.9	0	-	Neg.	3.6	0.2	23.7	2.0	21.4	2.8*	0.2	0.2	0.2	19.4
2.	1964	1.0	1.8	16.9	0.1	-	Neg.	3.5	1.5	23.8	2.6	19.2	2.6*	0.2	9.1	0.7	17.4
3.	1965	1.5	2.1	17.7	0.1	-	Neg.	3.0	1.8	23.9	2.7	17.4	2.4*	0.2	9.3	1.8	16.2
4.	1966	2.4	1.5	17.0	1.4	-	Neg.	2.9	2.0	23.3	3.1	15.2	2.0*	0.2	9.1	2.6	17.2
5.	1967	3.0	1.4	17.2	1.4	-	Neg.	2.9	2.2	21.8	3.2	14.3	1.8*	0.3	9.3	3.5	17.7
6.	1968	2.9	2.0	17.8	2.9	0.3	Neg.	2.7	2.6	19.6	3.5	14.8	1.2	0.7	8.9	4.8	15.3
7.	1969	2.9	2.1	20.8	2.9	0.2	Neg.	2.7	3.4	18.2	3.4	14.2	1.1	0.9	8.8	4.6	13.8
8.	1970	2.9	2.2	23.0	2.9	0.2	Neg.	2.7	3.4	16.5	3.6	13.4	1.0	1.0	9.2	4.6	13.5
9.	1971	3.0	2.1	24.9	4.2	0.2	Neg.	2.7	3.4	15.5	3.5	12.6	0.9	1.1	8.8	4.3	12.7
10.	1972	3.1	3.1	25.9	4.3	0.2	Neg.	2.8	3.3	14.8	3.6	11.7	0.9	1.4	8.4	4.2	12.6
11.	1973	3.5	3.2	26.9	4.7	0.2	Neg.	2.9	3.2	14.0	3.9	11.0	0.8	1.8	7.5	4.0	12.5
12.	1974	3.8	3.0	27.8	4.7	0.2	Neg.	3.0	3.3	13.8	3.9	10.1	0.8	2.2	6.9	4.1	12.3
13.	1975	4.3	3.2	26.8	4.8	0.3	Neg.	3.0	3.2	13.4	4.9	9.3	1.2	2.6	6.2	4.1	12.6
14.	1976	4.2	3.6	25.1	5.8	0.7	0.2	2.8	3.3	18.2	5.0	8.3	2.2	2.5	6.7	4.1	7.6
15.	1977	4.2	3.4	27.2	5.7	1.5	0.4	2.9	3.0	16.2	6.8	7.0	2.1	2.5	5.1	4.1	8.3
16.	1978	4.4	3.3	25.2	5.6	1.3	0.7	3.7	2.9	16.1	8.1	5.9	2.0	2.5	5.1	4.4	9.5
17.	1979	4.2	3.0	22.6	5.0	1.7	0.8	4.1	3.0	14.5	7.6	5.8	2.7	2.9	4.9	5.2	8.4
18.	1980	5.1	3.4	20.7	5.8	1.6	0.8	4.9	2.9	14.7	8.6	6.2	2.3	2.2	5.0	5.3	10.2
19.	1981	5.5	3.7	19.7	6.0	1.5	0.8	4.7	2.7	14.6	10.0	5.8	2.3	2.0	5.1	5.6	9.7
20.	1982	5.4	5.7	18.0	5.0	1.3	0.7	4.3	2.4	14.2	13.3	5.7	2.0	2.1	4.8	6.0	8.8
21.	1983	7.5	5.5	16.6	3.9	1.1	0.6	3.7	2.2	13.6	14.1	5.4	1.7	2.0	4.7	8.9	8.5

Note: Neg. = negligible; = not available Punjab and Haryana.

Source: Bureau fo Public Enterprises: *Annual report of the Working of Industrial and commercial Undertaking of the Central Government.* Government of India (various issues).

Among the industrialized states, the share of Maharashtra continued to increase throughout the period, and it registered the highest percentage increase in its share from 2 per cent in 1963 to 14.1 per cent in 1983.

In the case of less-developed states, the relative shares of Uttar Pradesh, Andhra Pradesh and Assam have increased continuously. In Andhra Pradesh it increased from 0.8 per cent to 7.5 per cent, in Assam from 1.5 per cent to 5.5 per cent, and in Uttar Pradesh from 0.2 per cent to 8.9 per cent.

It appears that due to emphasis on the development of heavy basic industries during the Second and Third Plans, three steel plants were constructed, one each in these three states, during this period, because all three are blessed with iron ore and coal. This fact can be established if one analyses the relative weight of investment in steel plants in terms of the total investment in these states. In 1981, steel plants accounted for 30 per cent of cumulative investment in Bihar, 73 per cent in Madhya Pradesh, and 85 per cent in Orissa (Bureau of Public Enterprises, 1981).

These facts suggest that after completing important steel plants during the Second and Third Plans, public sector investment got diversified. The relative weight of public sector investment in the steel industry reached about 60 per cent in 1963, and declined to 33 per cent in 1970 (Bureau of Public Enterprises, 1973). Therefore, one can conclude that till 1970, due to the composition of public-sector investment in favour of the steel industry, which uses weight-losing localized raw materials, public investment's spatial distribution favoured materials-rich less-industrialized states.

TABLE 2 – *Ranking of state in terms of Net State Domestic Product (NSDP)(R_1) 1951-1981.*

Sl no.	State	1951	1961	1971	1981
1.	Aandhra Pradesh	6	7	9	6
2.	Assam	3	5	10	10
3.	Bihar	14	14	14	14
4.	Gujarat	2	4	2	3
5.	Karnataka	7	8	5	5
6.	Kerala	8	9	6	8
7.	Madhya Pradesh	12	10	13	12
8.	Maharashtra	5	1	3	2
9.	Orissa	13	13	11	13
10.	Punjab	4	3	1	1
11.	Rajasthan	9	11	7	9
12.	Tamil Nadu	10	6	8	8

Sl no.	State	1951	1961	1971	1981
13.	Uttar Pradesh	11	12	12	11
14.	West Bengal	1	2	4	4

Source: Central Statistical Organization, *National Account Statistical*, Government of India (various years).

TABLE 3 – *Ranking of States in terms of the Index of Regional Intensity of Industrialization (R1)*

Sl no.	State	1951	1961	1971	1981
1.	Andhra Pradesh	8	8	8	6
2.	Assam	12	6	9	10
3.	Bihar	4	9	10	14
4.	Gujarat	-	3	3	3
5.	Karnataka	-	7	7	5
6.	Kerala	11	5	5	8
7.	Madhya Pradesh	7	13	11	12
8.	Maharashtra	2	2	2	2
9.	Orissa	10	14	14	13
10.	Punjab	6	10	12	1
11.	Rajasthan	9	12	6	9
12.	Tamil Nadu	3	4	4	7
13.	Uttar Pradesh	5	11	13	11
14.	West Bengal	1	1	1	1

Public-sector investment in the post-1970 period was more biased towards the chemical and engineering industries. Location of these industries is significantly influenced by market considerations, moreover, as mentioned above, the importance of regional pulls and pressures resulted in the further spatial spread of public-sector investment; however, these forces favoured the developed more than the industrially underdeveloped states, and the position of major cores of the colonial period has been strengthened (Kirk, 1981: 196).

The above analysis provided the shifts in the spatial pattern of public-sector investment. The results were arrived at by using gross-capital block. It is an important fact that the states comprising the Indian federation are not of uniform size; therefore, the analysis based on cumulative annual public-sector investment is not appropriate, unless it is able to account for variations in the size of the states.

To adjust the data of cumulative annual public investment in relation to the size of the states, we have compiled the Index of Regional Intensity of Public

Sector Investment (R_P), which is defined as the ratio of relative share in terms of population.[17] The rankings of different states in terms of this index are reported in *Table 4.* We calculated rank correlation between the Index of Regional Intensity of Public Investment and the Index of Regional Intensity of Industrialization for the years 1961, 1971 and 1981, to assess how much the regional intensity of public sector investment favoured the states with less industrialization. If the spatial spread of public-sector investment is meant to help less-industrialized states, the value of the rank correlation should have a negative sign.

The rank correlation coefficients as obtained in our case are -0.239 for 1961, + 0.096 for 1971, and -0.195 for 1981. These results show that initially public-sector investment helped in the development of less-developed states, but for 1971 it followed the same spatial pattern that is observed in the case of the spatial spread of industrialization. In 1981, we have a weak negative coefficient, which suggests that it is still in favour of developed states.

To analyse the role of licensing policy in the spatial spread of industrialization, we have obtained data reported in a study by the Planning commission (Hazari, 1967) for the period 1959 - 1966, and data reported by the Centre for Monitoring the Indian Economy for the period 1076 - 1980. Different states have been ranked in terms of their percentage share of the total licences issued during the period (see *Table 5*). For the purpose of analysis, we have calculated rank correlation between the states in terms of percentage share in the licences issued and the Index of Regional Intensity of Industrialization for two time periods, 1959-1966 and 1976-1980. For both the time periods the rank correlation coefficient is positive (+ 0.626 for 1959-1966 and + 0.728 for 1976-1980). These coefficients clearly establish that licensing policy has not been able to achieve its objective of balanced regional development.

TABLE 4 – *Ranking of states in terms of index of Regional Intensity of Public Sector Investment (R^P).*

Sl no.	State	1961	1971	1981
1.	Andhra Pradesh	10	11	12
2.	Assam	8	8	4
3.	Bihar	3	2	1
4.	Gujarat	14	7	7
5.	Karnataka	6	9	9
6.	Kerala	11	6	11
7.	Madhya Pradesh	2	3	2
8.	Maharashtra	9	10	6
9.	Orissa	1	1	3
10.	Punjab	7	12	8
11.	Rajasthan	12	14	13
12.	Tamil Nadu	5	5	10
13.	Uttar Pradesh	13	13	14
14.	West Bengal	4	4	5

Sl no.	State	1959-1966[a] Licences ssuied	Share (%)	Rank	1976-1980[b] Licences aIssured	Share (%)	Rank
1.	Andhra Pradesh	121	3.2	9	154	6.7	8
2.	Assam	30	0.7	14	16	0.7	14
3.	Bihar	125	3.3	8	52	2.2	12
4.	Gujarat	292	7.6	4	322	14.	2
5.	Karnataka	143	3.8	7	178	7.8	6
6.	Kerala	85	2.2	11	70	3.0	9
7.	Madhya Pradesh	110	2.9	10	59	2.6	11
8.	Maharashtra	128	32.8	1	612	26.8	1
9.	Orissa	81	2.1	12	25	1.1	13
10.	Punjab	275	7.2	5	187	8.2	3
11.	Rajasthan	63	1.7	13	66	2.8	10
12.	Tamil Nadu	320	8.4	3	184	8.0	5
13.	Uttar Pradesh	206	5.4	6	185	8.1	4
14.	West Bengal	703	18.5	2	171	7.4	3
	Total	3802	100.0		2281	100.0	

Source: ''R.K. Hazari (1967). *Industrial Planning and Licensing Policy, vol. II, Part I, pp. 72-77, Planning Commission, Government of India."* Economic Intelligence Service *(1981). Basic Statistics Relating to the Indian Economy— States,* vo. II, Table 10.28, centre for Monitoring of Indian Economy, Bombay.

IV. Broad Conclusions

Analysis of the nature of the mixed economy and the strategy of regional planning clearly establishes that the core of the planning process in India has continued to be interesectoral in character. To achieve regional equity, government relied on *ad hoc* committees and working groups. These *ad hoc* bodies evolved criteria for the identification of backward regions and also suggested suitable strategies for developing these regions.

Throughout the planning period government policies were based on the assumptions: (a) that the development of infrastructural facilities in different regions will *pari pasu* lead to spatial spread of industrialization; and (b) the development policies of small-scale industries will correct the regional imbalances created by the spatial concentration of large-scale plants.

The anatomy of the changing nature of federalism suggests that the linguistic reorganization of state boundaries resulted in increasing regional pulls and pressures. However, until 1967, due to the continuous hegemony of the Congress Party at the centre and in the states, policy reflected the characteristics of a unitary State. After 1867, regional pressures were quite significant and resulted in the supremacy of the National Development Council over the Planning Commission.

The empirical results regarding the spatial distribution of public-sector investment suggest that, until 1970, public- sector investment favoured materials rich industrially backward states. This bias in public -sector investment can be explained partly with the help of the development strategy of materials oriented industry during this period, and partly by the state's deliberate attempt to achieve regional dispersal. However, after 1970, due to changes in the composition of investment in favour of market oriented industries and the increasing role of regional pressures, public-sector investment assumed a wider spatial spread. But this development was relatively more favourable to already industrialized states.

The licensing policy, which was expected to play a significant role in the spatial spread of industries, could not achieve its objective of regional dispersal. The empirical results clearly show that the spatial distribution of licences followed the same pattern as was observed in the case of the spatial pattern of industrialization.

Acknowledgements

The author is grateful to Professor G.J.R. Linge of the Australian National University, Professor L.S. Bhat of the Indian Statistical Institute, New Delhi, Dr. B.N. Goldar of the Institute of Economic Growth, Delhi, Professor V.S. Minocha of the Delhi School of Economics and to the anonymous referees for comments on an earlier draft of this paper. However, the author alone is responsible for any errors that may remain.

NOTES

1. The federating units in the Indian federal system are known as 'states'. Therefore, we use 'State' for representing the nation- state and 'state' for the federating units.
2. Even in this strategy of promoting small-scale and cottage industries, Gandhi accepted the inevitability of some large-scale firms in crucial sectors which should be owned by the State (Aggarwal, 1940; Prasad, 1957). The ideas of Nehru were available in the report of the National Planning Committee which was set up by the Indian National Congress in 1938 (Shah, 1948). The Federation of Indian Workers felt that for the benefit of the common masses private enterprise with its profit motive would not be suitable (Roy, 1944). The group of industrialists which prepared the Bombay Plan advocated that the State should provide protection from international competition and pleaded for the creation of infrastructure (Krishnamachari, 1962).
3. See Article 31.1, 31.2, and 31.3 of the Constitution of India for the definition of 'Fundamental Right of Property'.
4. See the Constitution of India, Article 39B and 39C.
5. The Planning Commission was established by a cabinet resolution,and consists of a Chairman (The Prime Minister), a Deputy Chairman (nominated by the Prime Mini-

ster) and one or more nominated members who are noted experts in different fields.

6. The Industries (Development and Regulation) Act of 1951 came into force on 8 May 1952 (see Estimates Committee, 1972: 119- 209).
7. The Third Plan was followed by a period of plan holiday (1966-1969) caused by dislocation of aid flows following the Indo-Pak War (1965) and two successive agricultural droughts (1965 and 1966) which led to industrial recession.
8. These 15 indicators represented the demographic features, sectoral growth patterns, literacy, health and level of development of infrastructural facilities in different regions— see *Census of India* (Registrar-General, 1967).
9. Indices which were used by the Pandey Committee for the identification of industrially backward states were: (a) total per capita income, (b) share of Industry and mining in the income generated in the regions, (c) number of persons working in the registered factories, (d) length of roads and railways in relation to (i) population, and (ii) area, and (e) per capita consumption of electricity.
10. Symmetry in a federal state exists if political units comprise equal territories, population, similar economy, etc.; whereas the asymmetric federal states are those where diversities find political expression.
11. The NDC was constituted in August 1952. It consisted of the Chief Ministers of the states, Members of the Planning Commission and the Prime Minister as its Chairman (Krishnamachari, 1962).
12. 'From 1947 to 1967, political control of the Union and states rested with the Congress Party..... The strength of Congress in the Union and in the states gave an impression of political control over the planning process, which may have concealed.. The very real conflicts of interest among particular states' (Lefeber and Chaudhuri, 1971: 52).
13. It is interesting to note that while Indian planning models became fairly sophisticated, there has been no comparable extension of analysis to questions of spatial planning (Bhagwati and Chakravarty, 1971; Bhat, 1972; Prakasa Rao, 1974).
14. Studies to establish the claim of steel mills are by Guha (1957); and by the governments of Madhya Pradesh (1953) and Tamil Nadu (1964).
15. In September 1963, the Swaminathan Committee was constituted, which submitted its reports on 18 march 1964. The committee is known as the Industrial Development and Procedure Committee, and its objective was to trace the efficiency of the procedures being followed for issuing licences.
16. $$R_i = \frac{L_it}{\sum_{i=1}^{n} L_it} / \frac{P_i}{\sum_{i=1}^{n} P_it};$$

Where L_it = industrial workers employed in the ith states at time t;

$L_iT = \sum_{i=1}^{n}$ total industrial workers in the country at time t;

P_it = population at the ith state;

$\sum^{n} P_it$ = total population of the country at time t.

17. The Index of Regional Intensity of Public Sector Investment (R_p) is defined as:

$$R_p = \frac{S_it}{\sum_{i=1}^{n} S_it} \Big/ \frac{P_it}{\sum_{i=1}^{n} P_it}$$

where S_it = the annual cumulative gross public-sector investment in the ith state at time t.

$\sum^{n} S_it$ = the total cumulative annual public-sector investment in the country at time t.

REFERENCES

1. Aggarwal, S.N. (1940). *Gandhian Constitution of India.* Allahabad.
2. Alexander, P.C. (1963). *Industrial Estates in India.* Bombay: Asia Publishing House.
3. Bhagwati, J.N (1962). Economies of Scale, distribution of industries and programming. *Economic Weekly.* 1 September.
4. Bhagwati, J.N. and Desai, P. (1970). *Planning for Industrialization.* New York: Oxford University Press.
5. Bhagwati, J.N. and Chakravarty, S. (1971). *Contribution to Indian Economic Analysis—A survey.* Delhi: Lalwani Publishing House.
6. Bhat, L.S. (1972). *Regional Planning in India.* Calcutta: Statistical Publishing Society.
7. Bureau of Public Enterprises (1973, 1981, and various dates). *Working of Commercial and Departmental Undertakings.* New Delhi: Government of India.
8. Cannon, J.B. (1975). Government impact on industrial location. In *Location Dynamics of Manufacturing* (L. Collins and D. Walker, eds). New York, Wiley.
9. Central Statistical Organisation. *National Accounts Statistics* (of various years). New Delhi: Government of India.
10. Centre for Monitoring Indian Economy (1981). *Basic Statistics Relating to Indian Economy,* Vol. II. Bombay.
11. Chaudhuri, P., Ed. (1971). *Aspects of Indian Economic Development.* London: Allen and Unwin.
12. Crawley, W.F. (1972). *Political direction in India.* In Reforms and Revolution in Asia (G.F. Hudson, ed.) London: Allen and Unwin.
13. Dhar, P.N. and Lydal. H.F. (1961). *Role of Small Scale Industries in India.* Delhi: Asia Publishing House.
14. Dikshit, R.D. (1971). Influence of the government system on location of economic activity. *In Government Influence on Location of Economic Activity* (G.J.R. Linge and P.J. Rimmer, eds.) Canberra: Australian National University.
15. Dikshit, R.D. (1975). (1969). *The Political Geography of Federalism.* New Delhi Macmillan.
16. Dutt, S. (Chairman) (1969). *Report of Industrial Licensing Policy Enquiry Committee.* New Delhi. Government of India.
17. Estimates Committee (1972). *V Lok Sabha, XIX Report.* New Delhi. Lok Sabha Secretariat.
18. Frankel, F.R. (1978). *India's Political Economy 1947- 1977— The Gradual Revolu-*

tion. Oxford: Oxford University Press.

19. Gadgil, D.R. (1972). *Planning and Economic Policy in India.* Poona: Gokhale Institute of Politics and Economics.
20. George, K.K. (1982). The transfer mechanism in India— an appraisal. *In Regional Inequalities in India—Inter-State and Intra-State* (L.S. Bhat *et.al.*, eds). New Delhi: SSRD.
21. Government of India (1968). *Report on the Machinery of Planning.* New Delhi: Administrative Reforms Commission.
22. Government of Madhya Pradesh (1953). *New Iron and Steel Plant.* Nagpur, government of Madhya Pradesh Press.
23. Government of Madras (1964). *Consulting Engineers Report on Neyveli-Salem Plant— A summary.* Madras: Government of Madras Press.
24. Guha, A.K. (1954). *Location of New Steel Plants and Its Economic Aspects.* Alipore: West Bengal Government.
25. Halliwell, J.F. (1968). *Public Policies and Private Investment.* USA: Clarendon Press.
25. Hanson, A.H. (1966). *The Process of Planning— A Study of India's Five Year Plans 1950-54.* Oxford: Oxford University Press.
26. Hanson, A.H. and Douglas, J. (1972) *India's Democracy.* New York: W.W. Norton.
27. Harcourt, G.C. (1968). Investment decision criteria, investment incentives and choice of technique. *Economic Journal 78,* 77-95.
28. Hardgrave, R.L. Jr (1970). *India: Government and Politics in a Developing Nation.* New York: Harcourt, Brace and World.
29. Hazari, R.K. (1967). *Industrial Planning and Licensing Policy.* New Delhi: Planning Commission, Government of India.
30. Hicks, U.K., ed. (1961). *Federalism and Economic Growth in Under-developed Countries— a symposium.* London: Allen and Unwin.
31. Johnson, W.A. (1966). *Steel Industry of India.* USA: Harvard University Press.
32. Kirk, W. (1981). Cores and peripheries—the problems of regional inequality in the development of Southern Asia. *Geography* 66(3), 188-201.
33. Krishanamachari, V.T. (1962). *Fundamentals of Planning in India.* Bombay: Asia Publishing House.
34. Lefeber, L. and Chaudhuri, M.D. (1971). *Regional Development Experiences and Prospects in South and Southeast Asia.* Paris: Mouton.
35. Livingston, W.S. (1952). A note on federalism. *Political Science Quarterly* 67, 81-95.
36. Mahalanobis, P.C. (1963). *An Approach to Operational Research to Planning in India.* Bombay: Asia Publishing House.
37. Marglin, S. (1967). *Public Investment Criteria.* New York: Allen and Unwin.
38. Ministry of Law (1982). *The Constitution of India.* Nasik: Government of India Press.
39. Mohanty, H.B. (1957). *Location of Steel Plant in Orissa.* Bhubaneswar: Industries Department, Government of Orissa.
40. Narang, A.S. (1982). Decentralisation to remove regional disparities. *In Regional Inequalities in India— Inter-State and Intra-State* (L.S. Bhat *et. al., eds).* New Delhi SSRD.
41. National Committee on the Development of Backward Areas (1980). *Report on Industrial Dispersal* (Chairman, B. Sivaraman). New Delhi: Planning Commission.
42. Planning Commission (1951). *First Five-Year Plan 1951- 1956.* New Delhi, Government of India.

43. Planning Commission (1956). *Second Five-Year Plan.* New Delhi: Government of India.

44. Planning Commission (1961). *Third Five-Year Plan.* New Delhi: Government of India.

45. Planning Commission (1969). *Fourth Five-Year Plan (1969- 1974).* New Delhi: Government of India.

46. Planning Commission (1973). *Draft Outline Fifth Five-Year Plan.* New Delhi. Government of India.

47. Planning Commission (1978). *Draft Outline Fifth Five-Year Plan.* New Delhi. Government of India.

48. Planning Commission (1980). *Sixth Five-Year Plan.* New Delhi: Government of India.

49. Prakasa Rao, V.L.S. (1974). *Regional Planning* Calcutta: Dorchester Printing Works.

50. Prasad, P. (1957). *Economic Problem of Public Enterprises in India.* Leiden, Netherlands: H.E. Stenfert Kroese NV.

51. Rajamannar, P.V. (Chairman) (1971). *Report of the Centre— State Relations Enquiry Committee.* Madras: Government of Tamil Nadu.

52. Registrar-General, Census Commissioner of India (1957). *Census of India,* Paper no. I. New Delhi: Government of India.

53. Robinson, K.W. (1961) Sixty years of federalism in Australia. *Geographical Review* 50(1), 10-45.

54. Roy, M.N. (1944). *People's Plan for Economic Development* Delhi: Ajanta Publications (1981 reprint).

55. Rudra, A. (1975). *Indian Plan Models.* Bombay: Allied Publishers.

56. Sachs, I. (1966). *Patterns of public sector in Underdeveloped Countries.* Ph. D. thesis, Delhi School of Economics.

57. Santhanam, k. (1961). *Union-State Relations in India.* Bombay: Asia Publishing House.

58. Seth, V.K. (183). *Spatial Dimensions of the process of industrialisation in India.* Ph. D. thesis, Delhi School of Economics.

59. Shah, K.T. (1948). *National Planning Principles and Administration,* Bombay: Vora and Co.

60. Staley, E. and Morse, R. (1965). *Modern Small scale Industry for Developed Countries.* Englewood Cliffs, NJ. McGraw Hill.

61. Wadhwa, C.D. (19770. *Some Problems of Indian Economic Policy.* Delhi: Tata McGraw Hill.

Section III | India in Regional and Global Context

With the particular weaknesses and strengths of its society, economy and physical geography, India has attained a particular status among the countries of the world. Judged from whatever scales, it is behind several countries in terms of its economic strength and possession of coercive power. A recent estimate of the International Monetary Fund based on purchasing powers of national currencies places India behind the U.S., Japan, China, Germany and France. But she has not achieved that clout in the world economy because of her low share in global trade. Whereas China is being viewed as the next superpower because of its growing economy and possession of instruments of coercion, India is now being looked at as a regional power in Asia. A status of a country is not fixed for all times to come. Some of the European countries which were considered as world powers have slid down the wealth and power scales and others have moved upwards on them. The location of each country among the constellation of countries keeps changing according to its size, speed and direction, which are attained by social cohesion, economic growth, level of science and technology, and armed strength, and the nature of politics within the territorial extent of the country. Section II explores and illustrates how with her internal weaknesses and strengths the state of India is conducting her external relations with her neighbours and others in the world from a particular position of power and influence, and how geopolitical theories of earlier years have been rendered irrelevant in the post-Cold War world at the close of twentieth century.

XIV India and Her Neighbours

C.S. VENKATACHAR

The Indian Ocean and the high Himalayas have effectively isolated India and barred easy movements from without. This has made the Indian mind largely isolationist. Indian isolationism has been accentuated by three other characteristics— wholly India. Indian society has always been continental; until recent times there had been little mobility of movement even from region to region.India's vast spaces have the effect of imprisoning the minds of her people; little interest is evinced beyond the vast internal horizons. The outer geographical framework of India was permanently fixed by Nature; at times the expansionist ambitions of an outsider may have resulted in the nibbling of small fringes of territory in the northwest or the northeast regions of India. Such temporary shiftings of frontier in the remotest parts were inconsequential events in India's long history. The energy of everyone—indigenous or the foreigner— was fully extended in reaching up to the limits set by geography and there was no need to look at India's neighbours with greed or envy or covet any part of a neighbour's territory. India's pacific intentions were mainly dictated by this, apart from any philosophic detachment. Thirdly, Hinduism as religion is exclusively confined to its own territorial limits and has no extra-territorial affiliations unlike Christianity and Islam. Hinduism has been responsible for political incapacity and pragmatic indifference to the existence of neighbours. When the Chinese pilgrim Huen-Tsang decided to return home after his sojourn in India it is reported: "The Monks of Nalanda, when they heard of it begged him to remain, saying, 'India is the land of Buddha's birth, and though he has left the world, there are still many traces of him. What greater happiness could there be than to visit them in turn, to adore him and chant his praises? Why then do you wish to leave, having come so far? Moreover, China is a country of 'Mlecchas', of unimportant barbarians, who despised the religious and the Faith. That is why Buddha was not born there. The mind of the people is narrow and their coarseness profound, hence neither saints nor sages go there. The climate is cold and the country rugged.'

The master of the law replied, "Buddha established his doctrine so that it

Reprinted from C.S. Venkatachar, "Geographical Realities of India", 1955, *The Eastern Economist* Pamphlet No. 34, pp. 40-53.

might be diffused to all lands. Who would wish to enjoy it alone and to forget those who are not yet enlightened? Besides, in my country the magistrates are clothed with dignity and the laws are everywhere respected; humanity and justice are highly esteemed and old men and sages are held in honour.....How can you say that Buddha did not go to my country because of its insignificance?"

Yet it would not be wholly correct to say that India's mind was completely impervious to all external contacts. There have been phases in India's history when lively interest was evinced in her neighbours—near and far. There were episodes rather than a continuous process. It is not easy to explain the waxing and waning of this national tendency. Possibly it may be due to disturbed conditions in Indian society resulting in too much preoccupations with internal consolidation accompanied by the withdrawal of the gaze from outwards, and partly to the innate conservatism of a populous, self-sufficient country of peasant culture with no urge for trade and commerce with distant countries. Though Hinduism and Buddhism spread far and wide particularly in Southeast Asia it was not the state authority. Politics, commerce and religion played an insignificant part in actitivizing the passive attitude towards India's neighbours. India has meant for her neighbours than the neighbours have meant for India. She has been a lodestar for the ambitions of people with predatory instincts, and from the peaceful ones she has earned respect and regard. But whether peaceful or warlike, they have failed to impress her mind and rouse her interest.

While considering who were the neighbours who in the past mattered, we may take note of the fact that the attitude towards neighbours was conditioned by the differing characteristics of the North and South Indian mind. Northerners imprisoned in the vast spaces in the plain and shut off from the Himalayas developed a 'closed' mind. By contrast the peninsular people on India's coastal areas had their gaze seawards and had a long tradition of migration across the seas to the countries adjacent to India.

In the pre-maritime age India, rather than the plains of the north of India, had contact with the nomads of Central Asia through the Hindukush 'divide' and that contact was regulated by a traffic roundabout in the Oxus-Jaxartes Basin. What was India's response to the constant movement of traffic into India from this area? Indian attitude was one of complacency. India was content to tackle the intruders as they came and after the entrance to India had been pierced and they had set foot on Indian territory. No serious thought was given to the problem of the distant mountainous frontiers. Physical barriers may be static but the working of the minds of men behind them is not; it is the men and ideas behind the barriers which matter. Physical geography generally acts as a challenge; it had a contrary effect on India and did not create a dynamic thought on the politico- military problems of the Hindukush.

The regulatory function of the traffic roundabout was such that at any given time it would permit only a small element of migratory stream to pierce the Hindukush. There is the six-hundred mile stretch of the Hindukush from the borders of Chitral to Herat. All the accessible passes from the northern slopes of the mountain to the Kabul Valley cross the main ridge along a line of about 150 miles, of which the centre is about 60 miles due north of Kabul and the main routes from all of them converge on to the Kabul plateau. From Balkh, the vicinity of our roundabout — the earlier routes crossed the Hindukush at Nil Pass or the Ak Robat Pass. Both routes converge on the Bamiyan Valley from where a principal route in Buddhist times descended into the Ghorband Valley. Huen-Tsang followed this route in the seventh century. This was the highway for the peaceful Buddhist pilgrims, caravan traffic and cultural missionaries. The great conquerors who came to the Punjab followed shorter and more difficult routes. Alexander (328 B.C.) and Timur (A.D. 1398) crossed by the Khawak pass (11,640 feet) and Babar (A.D. 1504) by the Qipchak pass (13,900 feet). All these descended into the Valley of the Kabul river and followed a more direct route to Kabul itself.

The geopolitical problem on either side of the Hindukush was to 'abolish' this divide. This abolition depended upon quiescence on the part of the nomads in the open corridor and the existence of a strong northern Indian empire. There have been periods in Indian history when, from the Indian end, the Hindukush was effectively abolished such as by Asoka and by the Moghuls, till the later Moghuls lost by stages their hold on Trans-Oxiana. At other times, even in the absence of a strong central government in the North, Indian occupation extended up to the Kabul Valley so long as there were no counter-threats from across north of the Hindukush. This state of affairs prevailed in the ninth and tenth century of the North Indian history before the Islamic onslaught began. On other occasions the divide was abolished by the movement of nomads, but their safety lay in securing lodgement either on the southern slopes of the Hindukush or in the Peshawar Valley, if not into the plains of Delhi. The capital of Kanishka's Central Asian empire was in Peshawar. Earlier, the Bactrian Greeks who had occupied the Kabul Valley ruled the north-western parts of India. Generally, a movement to the north of the Hindukush caused no seismic tremor in the plains of the Punjab; when the southern slope of the Hindukush was occupied then the security of the northern Indian plains was definitely threatened. This principle of Indian geo-politics has often been illustrated. The Mongols who swept across the corridor and scourged the greater part of Asia and Europe in the thirteenth century did not cross over the Hindukush and swoop down upon the north western regions of India. India remained unmolested by the Mongol world-conqueror. When Babar was fighting for his patrimony in Farghana his activity was not felt in India. As soon as he crossed Valley his mastery over India came within practicable range.

India's neighbours across the mountain barriers were not civilized states and kingdoms but nomads who were brought into relationship with her over the Hindukush mainly as a result of the conflict between the Siberian wind-swept Steppe and the Chinese Sown. Long before this conflict started, of which we have some recorded history, we have an earlier glimpse of the Eurasian Steppe nomads— some hordes of them had, like locusts, settled in the Punjab plains. Historians think there was a nomadic eruption from the Eurasian Steppe in the eighth and seventh centuries B.C. The universal state of Darius had to deal with these troublesome neighbours in the area in which we are concerned — the Oxus-Jaxartes basin. The Persians knew one branch of the Saka nomads who were their allies, another their nominal subjects. The Greek Herodotous was aware that Darius had annexed a small fringe of territory bordering on the Indus— possibly up to Multan and the lower Indus up to the sea-board. Beyond, in the interior was a great and populous country, for Herodotous speaks of India as by far 'the largest nation in the world'. The Sanskrit term Bahlika must be presumed to be a generic term to cover the Eurasian nomad peoples, including not only the Bactrians proper (Bhallas), but also, Malavas, Kshudrankas, Madda, Madraka, Kamboja and what not— names preserved in the later Indian historical traditions. These swarms were in the Punjab and beyond, waiting to be legitimized by Hinduism according to Vedic orthodoxy.

There followed a peaceful interlude till a Chinese Caesar arose to deal with the nomads who were menacing North China. The Chinese took vigorous steps for outflanking the powers of the Huns. The Han Emperor Wu-Ti ordered an investigation into the state of the open corridor by his envoy, Chang-Chien. This envoy discovered that the Huns had driven a horde called Yuechi who had moved towards Ferghana and in doing so, they had pushed a people known to Indian history as Sakas settled in the corridor somewhere between Jaxarites and Lake Issyk-Kul. The Yuechis were not interested in getting involved with the more ferocious Huns. They moved nearer than the traffic roundabout and displaced the Bactrian Greeks who were ruling in the area north of the Hindukush. The displaced Greeks moved south to the Kabul Valley, soon to succumb to the still southward moving Yuechis. The Yuechi hordes figure in Indian history as the Indo-Scythians whose great ruler was Kanishka. Having successfully abolished the Hindukush divide, Kanishka held sway over a great part of the north Indian plains. The more docile Sakas had to get round the Hindukush at Herat; they proceeded to Baluchistan from whence they drifted to Western India and Gujerat. The Hindukush tapers out and ceases to be a barrier at Herat; here the Hindukush can be outflanked by an easy movement. It is for this reason that the nineteenth century rulers of the British Indian Empire were frightened by the name Herat and by its threatened outflanking by the Iranians or the Russians.

The Han rulers of China had a firm grip over the Huns who had for a time become the tractable auxiliaries of the Chinese. Some of the Hun hordes driven away by the Chinese moved on to Oxus- Jaxaries basin. There were at least four branches of the Huns whose exact affiliation is obscure. The fourth century A.D. witnessed the eruption of the Huns over Asia and Europe. Attila's men went to attack Europe on the Scythian Broadway. The white Huns harassed the Persians from 420 to 557. A branch of the Huns passed through the traffic roundabout and attacked the Gupta Empire in India about A.D. 455. The collapse of Persian resistance enabled the hordes to swarm into India. In the middle of the sixth century the Huns on the Oxus were overthrown by the Turks and the Hun marauder Mihiragula was overthrown by a local Indian ruler.

Tai-Tsung, the great ruler of the Tang dynasty, successfully broke up the hegemony of the Turks from Mongolia to the Aral Sea. Grousset quotes from a Turkish inscription: "The sons of the Turkish nobility became slaves of the Chinese people. They submitted to the Chinese Khagan and surrendered their empire and their institutions. This was one of the period quiescence in the corridor. The traffic roundabout continued to divert small streams of nomads—not all fiercely warlike as the Huns, possibly more peaceful—the hordes of Gurjaras, Maitrakas and the rest; these, according to Vincent Smith, had been converted into Rajput clans and into the Hindu castes of Gujars, Ahirs, Jats and others.

Indian civilization dealt with two types of nomadic movements. By far the more frequent was the folk migration of peaceful pastorals who were absorbed into the Hindu fold; the other established its political rule in the southern slopes of the Hindukush and often became a menace to the Gangetic Sown. These nomads were either overthrown or were thoroughly Indianized. Whether they were peaceful or were such that in either case the stream that entered India cut itself away from the main traffic of civilization in Central Asia. Even when the marauders established principalities and became local rulers they were Indian rulers with no external affiliation. They stood or fell depending on their own resources and fortune within India. India captured them and then "closed" their mind.

India dealt with the nomads with a profound passivity as a local problem and not as an external problem. These troublesome neighbours who harassed India off and on for one thousand five hundred years were disregarded with masterly inactivity; they were considered at best a nuisance and at worst a temporary menace. India was once and for all rid of their menace when the Russian and the Chinese empires advanced into the open corridor.

II

If Indian attitude was static across the mountains, it was more dynamic—only for a time —across the seas. Here Indian physical geography was, unlike the formidable Himalayan barriers, of assistance. The roof on the top of the

Indian world could never be the highway. The natural exit out of India is by sea. Nomadism and north Indian history created an inverted view of Indian history and India's contact with her neighbours. The shutters and windows for India's exterior were located in peninsular India: only India chose, for some time to close them and turn her gaze inwards.

On the western sea-board of India three points of contact with the Fertile Crescent and the Mediterranean world manifested themselves from the first century B.C. The northernmost point was located somewhere near the present day Karachi. The middle point was in Western India, the principal port being the present-day Broach. The third focal point was the Malabar Coast. The northernmost point named by ancient geographers, Barbaricon, has vanished into the mists of the past. Ships coming to this port from Egypt simply coasted round the southern coast of Arabia and Baluchistan. This port came into use for trade between India and Rome via Egypt before the discovery of monsoon winds or when the silk route was disturbed by the movement of the nomads. The discovery of the monsoon winds made it unnecessary to hug on to the coast, and a great frequency of voyages across the Arabian Sea as well as to the Bay of Bengal and further on to the Indies became possible. If Barbaricon was active as the port for the northwest regions of India, the ancient port of Barygaza (Broach) was the outlet from the midlands of the north and the central parts of India to the Roman world. On the Malabar Coast was developed the great entrepot trade between the far-eastern world and the Mediterranean, and the traffic on this maritime road remained uninterrupted for nearly one thousand five hundred years. It was the maritime counterpart of the Silk Road on land and bridged across the seas, two continents. The land route was famous for silk; the maritime for spice. The continuity of the maritime route was maintained by the early discovery of the monsoon winds, by the development of hardy sea-going and sea-worthy vessels, by the development of the science of navigation and by the energy, adventure, and intrepidity of the mariners of the Indian Ocean Basin. There have been no more daring mariners than the pre-Islamic people of southern Arabia, and they were the most successful middle-men in the Greco-Roman world handling the maritime traffic across the Indian Ocean. India's commercial and political interests were limited towards the West. The Arabian bridge-head fell under the influence of Persians and the Mediterranean world. India had lively contact with this area and exchanged goods and commodities, but there was little active colonization.

It is on her eastern sea-board that India developed mere intimate contact with the societies in farther India. Here the Andaman and the Nicobar Islands in the Bay of Bengal and the Islands of the Indies made voyages more easy and frequent. Colonization was developed, followed by a strong drift of a prolific variety of Indian cultural forms. Indian migration was in volume very small; there was no alteration in the basic population structure of the indigenous societies. Indian culture was not intrusive. India made a precious gift of her discovery of alphabets to these area; if China had consented to accept this philological equipment its written character would have lost its terror. India's services to the south eastern Asian people included also the conferring on them of a strong protective cultural armour, later enabling them to resist the inroads of the westerners in the Maritime Age. These societies, small in population and confined to deltaic areas, did not go under western impact as did the civilization of Mexico and Peru. Lying between India and China the independence of the southeast Asians was not threatened by any expansionist ambitions of their two powerful neighbours. India and China never attempted to have a common frontier at the expense of the people of farther India.

The voyages across the seas were the reverse of the activities of the nomads in the land. Indian population was not transplanted; there was no forcible coalescence of cultures, no breaking up of local societies and no conflict. Higher elements of culture were integrated with the pre-existing culture forms by the people themselves. There was no desert nor the fervour of a militant religion like Islam to smother the life of those societies. India withdrew her contact with them somewhat abruptly, but they had reached a stage when they could develop on their own lines even though they lost all contact with a country which had exercised no form of imperialism over them.

For strange reasons India chose to forget the existence of these neighbours. The political vacuum created by this forgetfulness— a somnolent period of Indian history—was filled in by a peaceful Arabian interlude. Mackinder wrote of the Saracen sea-power:

"But the Saracens were not content with a dominion based only on the means of mobility proper to their Steppes and deserts; like their predecessors, the Phenicians and Shebans, they took to the Sea.....eastward from Yemen, at the mouth of the Red Sea and from Oman, at the mouth of the Persian Gulf, they sailed on the summer monsoon to the Malabar Coast of India and even to the far off Malay Islands and returned home on the winter monsoon. Thus the Arab shows sketched out a sea-empire, extending from the Straits of Gibralter to the Straits of Malacca, from the Atlantic Gate to the Pacific Gate."

He considered this vast Saracen design suffered from one fatal fact; it lacked manpower in its Arabian base to make it good and cautioned against losing sight of this warning given by history.

Would it not however be more correct to say that the nomads had no sea sense, they only used the waterways to get across? The idea of developing power on the seas was a legitimate discovery of the west Europeans. The Saracens and the Turks while fighting the Christians on land made ineffective naval thrusts in the eastern Mediterranean, with the object of capturing a few islands, the strong-hold of the Christians. There is little evidence of sea-power as a weapon in the hands of Islamic rulers. Two instances appear to attest this view. From the decline of the Hindu influence to the entry of eastern powers into the Indian Ocean the Arabs were peaceful traders Islam spread to Indonesia in a perfectly peaceful manner. The Ottoman Turks no doubt gloried in their title of the 'Khakans of the two Sea'; they were really more formidable at Constantinopole as the Ghazi-i-Rum. Had the Turks any recognition of the importance of sea-power, they would have turned their attention to the Indian Ocean when the Portuguese outflanked their land power, took them by the rear and intrigued with the Safavid Rulers of Perisa and obtained a foothold at the head of the Persian Gulf.

III

On the eve of the Maritime Age India was in a state of isolation in relation to her neighbours. She had lost temporarily all knowledge of the exterior world; only at the extreme tip of the peninsular south the maritime traffic continued.

A new power pattern was developed by the western powers who broke into the Indian Ocean basin. As life in the land corridor was hushed into silence, the Indian Ocean was transformed into an arena of conflict. Vasco da Gama was a great geo-politician of the Indian Ocean basin. He secured all the keys to the entrances which were later seized by the British. The British organized India as a land power and turned it into a hub of a new system of security and defence. This organization was from the sea and not from land— a fact which often escaped the notice of the Anglo-Indian administrators of the nineteenth century. The new system of power pattern was grouped around India. There were three concentric zones, the first of which were 'soft' attached areas such as Bhutan, Sikkim, Nepal and possibly Afghanistan. Then came the ring of states such as Iran, Arabia, Tibet and for sometime even parts of Sinkiang. The third and the outermost circle represented the line of India's 'interest' now fixed on one side of the maritime route from eastern Mediterranean through the Middle-East to the Indian Ocean and on the other side, on Indo-China and Indonesia. In such a strategic conception the term 'neighbours of India' acquired an entirely new meaning. The historical neighbouring areas of India remained the same as in the pre-maritime age. The area contiguous to India became potential threats to the safety of India for reasons which lay altogether outside Asia. The neighbourhood of these areas to India attained a new significance because of the rivalry of the European powers in Europe. "Except for the sake of Indian security", asks an English writer, "what interests would Great Britain have had in the Persian Gulf, Tibet or Sinkiang?' British power had to be extended far beyond the borders of India so as to meet the threat from Europe into the Indian Ocean. To protect the maritime route to the Indian Ocean, India's neighbours had to be defended. The empire had no doubt to be defended no on Indian but upon Britain's needs in the West. This changed pattern had no correspondence to the one time activities of the nomads in the neighbourhood of India.

Let us look again at the state of India's neighbours with the decline of the western dominance in Asia and in the light of changed and changing contexts of power pattern. The stateless nomads have disappeared; no longer will they harass from the backyards of Central Asia, the Islamic lands, the Hindus of the north Indian plains and the Chinese in the Yellow River Basin. The areas tenanted by the nomads have been occupied by two mighty land powers with contiguous frontier and common political ideologies.

The European conflict during the nineteenth century which extended to the non-European parts of the world assumed the form of rivalry of empires; in the mid-twentieth century the European conflict is ideological. The fight today is not much over the marginal lands: it is over the minds of men. What over-hangs the high Himalayas is not the threat of the advance of arms but the march of ideas. When the nomads overran the Gangetic plains, India had not to defend the Empire of her mind, she had to admit some socially undesirables into the folds of her civilization. The rulers of India in the nineteenth century had to defend a far away land empire based on sea-power against the advances of Cossack horsemen following in the wake of the rail road on the Central Asian Steppe. Today the empire of her mind rests on the strength of her own ideology, her social and political structure and on the consolidation of her basic fundamental unity.

But unlike the past, India cannot afford to have a closed mind or an indifferent attitude to the happenings across the high Himalayas; she has to be watchful. Likewise her neighbours across the seas are no mere curiosities or play-things of intermittent interests. Their friendship, their well-being and their independence are matters of paramount importance. In the colonial phase of the contact of the southeast Asian societies significant upsets came about in their placid social and economic life. The working of the labour and economic policies of the western colonial powers converted South Asian territories into offshoots of India and China. The fruit of these past policies is seen in the existence of plural societies, in the conflict between the indigene and the migrant people where the acquisitive instinct of the later operates to the detriment of the interests and welfare of the former and in the creation of a new class of unassimilable population with attendant racial complications. China and India have survived both the barbarian attacks and the western dominance in the maritime age. The southeast Asian countries, thanks to their cultural armour, have also survived and some have already become independent; others are in the process of shaking off their colonial masters.

The renewal of India's contact in the Age of Freedom will bring into prominence three main tendencies. In the context of the present-day political world India's contact with the southeast Asian countries may have to take the form of close economic and maritime ties. Secondly, the national societies of the Burmese, the Thais, the Cambodians, the Laotians, the Anamites, the peninsular Malays and the Indonesians should be assured of their survival with unfettered opportunities for full growth and stature, unimpeded in any manner by the intrusive presence in their midst of migrant population. Thirdly, that exçellent principle of the past that India should never attempt to have a common frontier with China at the sacrifice of the independence of the people of southeast Asia should be a fixed and unalterable policy.

"He would be a short-sighted Commander who merely manned his ramparts in India and did not look beyond"; so said Lord Curzon when India was conceived as a fortress in an Imperial Defence System. Today India is a fortress for democracy and peace. It is no less essential for her to man her ramparts and constantly look beyond. This she has to do in response to her geography and strategic position. One lesson of her past history she has to unlearn and that is to be indifferent to her neighbours.

XV The Sea

K.M. PANIKKAR

Next to the Himalayas, what has affected Indian history more than any other geographical factor is the Indian Ocean. The following bare facts will show how greatly India is dependent on the ocean that washes the three sides of her coast. India has no ports except on the Indian Ocean. Over 85 per cent of the sub-continent's trade is carried over the waters. The raw materials necessary for her industries come almost exclusively by the sea. Normal communications with other countries, apart from the small percentage which uses air transport is still mainly across the seas. A great portion of her internal trade even today is carried by country craft from port to port along the entire peninsular coastline. The history of Peninsular India from the earliest days has been intimately connected with the Indian Ocean.

Of the great oceans of the world the Indian Ocean is alone an embayed ocean. For the most part its area is walled off on three sides by land, with the southern side of Asia forming a roof over it. The continent of Africa constitutes the western wall, while Burma, Malaya and the insular continuations protect the eastern side. The vital feature which differentiates the Indian Ocean from the Pacific and the Atlantic is not the two sides but the sub-continent of India which juts out far into the sea for thousand miles to its tapering end at Cape Comorin. It is the geographical position of India that changes the character of the Indian Ocean.

Compared with the other oceans, this feature stands out most clearly. The Arctic and the Antarctic, circling the poles have but little connection with inhabited land. The Pacific and the Atlantic on the other hand lie from north to south like gigantic highways. They have no land roof, no vast land area jutting out into their expanse. The Pacific roof is at the end of the two continents, in the icebound zones of East Siberia or Alaska. Considered geographically the Indian Ocean in spite of the vastness of its surface and the oceanic character of its currents and winds, is more like a landlocked sea.

Another important feature of this Ocean is the distribution of islands and archipelagoes near its continental shores or in its vastness. Apart from Ceylon which is attached geographically to India, and Madagascar which is near to the African mainland, the oceanic spaces do not possess the same distribution

Reprinted with permission of the publisher from K.M. Panikkar, *Geographical Factors in Indian History*, Bhartiya Vidya Bhavan, Bombay, 1955, pp. 56-71.

of islands which is so marked a feature of the Pacific. Even in the case of the Atlantic, Iceland, the Azores and the West Indian Archipelago, not to speak of Great Britain and Ireland, constitute a geographical feature of great importance.

The Indian Ocean washes the entire African east coast up to Somaliland, the south coast of Arabia, the southern shores of Iran and Baluchistan, the peninsular India, and Sumtra. Its eastern and western entrances are guarded by two narrow straits, the Bab el Mandeb and the Straits of Malacca both of which can be easily controlled. From the Bab el Mandeb, the entrance is into the Red Sea, which being an inland sea is controlled by the lands on either side. The Straits of Malacca lead to the vast expanse of the Pacific, but here again the lands on either side become so narrow as to be easily capable of effective control of egress and ingress.

The main islands in the ocean are Ceylon, which is so close to India as to lose its insular character, and Madagascar which by its size and position provides an ideal cover to the South-East coast of Africa. Ceylon has at least two fine harbours, Colombo and Trincomalee, whose importance has been recognized from time immemorial and Madagascar has unique facilities in Diego Suares which the Third Republic vainly converted into a base second only to Singapore in the Indian Ocean. The other islands of importance whose geographical situation has to be noticed are Socotra on the Arabian coast., Zanzibar and Seychelles on the East African side, Mauritius and Reunion on the tropic of Capricorn. The Laccadives and the Maladives in the Arabian Sea near the Indian coast, the Bahrein group near the Persian Gulf, the Andamans and the Nicobars in the Bay of Bengal and Penang are the other island features which have a great bearing on maritime history.

The distances have also to be kept in mind. The Andamans are over 700 miles from the nearest Indian port and over 400 miles from the Burmese side. Mauritius lies 2,094 miles away from Colombo, while Socotra is no less than 1,000 miles. Penang is situated 1,278 miles away from Ceylon. the Laccadives and the Maldives, small islets dotted over a great length to the South-west of India provide no suitable harbour. In fact except Daman and Diu the Indian coast has no suitable islands near enough to afford cover.

The bays and bights of the Indian Ocean require special consideration. To the north-west corner is the Persian Gulf, a protected landlocked sea with Ormuz commanding the entry. With the valley of the Euphrates and the Tigris and the historic land of Mesopotamia as its hinterland this sea has in the past had a dominant role in shaping the history of oceanic navigation and may in the future have even a greater influence than in the past.

The Arabian Sea, a vast expanse separating the two peninsulas of India and Arabia and bounded on the north by the barren coast line of Persia is one of the vital seas of the world. As a result of the seasonal monsoon it has been for at least 3,000 years a great highway of commerce and intercourse. The

Indians, the Phoenicians, the Arabs— in fact all the seafaring nations of the East—have considered this to be the chief area of navigation.

The Gulf of Aden is the bottleneck which is created by the projection of Africa and the south-eastern tip of Arabia, the two shores running more or less parallel here and ending at the straits of Bab el Mandeb. The entry to this gulf from the east side is controlled by the island of Socotra. For ages it had been the home waters of the Arab corsairs who had their base in Aden, a covered and almost hidden post on the coast of the tip of Arabia.

The Bay of Bengal, lying to the east of India and separating Burma and Malaya from the peninsula of Hindustan is also governed by the monsoon. Further, it is said to be one of the four sea areas in the world subject to the phenomenon of revolving storms. From the ports on the east coast of India argosies have sailed this sea from the dawn of history and the colonization of the Pacific islands by the Hindus shows the extent to which this sea had been explored and navigated at least 2,000 years ago.

The Gulf of Malacca is like the mouth of a crocodile, the Peninsula of Malaya being the upper and the jutting end of Sumatra being the lower jaw. The entry to the Gulf can be controlled by the Nicobars and the narrow end is dominated by the island of Malacca.

The wind currents of the area are specially important and are perhaps unique in the world in their effects. The monsoon whose regular action was discovered for the West by Hippalus in A.D. 45 dominated the navigation of the Arabian Sea and the Bay of Bengal till the invention of steamships in the nineteenth century. The south-west monsoon which blows across the Arabian Sea determined for many thousand years the trade route which the ships followed. Following this wind they sailed from the Arabian and Red Sea ports to the coast of India. Equally its flow into the Bay of Bengal and its return by the same route after a period of calm governed the sailing in that area also. Thus for a period of well over five months the direction and route of sailing were determined by these winds and the navigators of the Indian Ocean who had closely studied the action of these phenomena were able to make full use of them.

Allusion may also be made to the revolving storms of the Bay of Bengal, an ever present danger to shipping in that area. Another factor of importance relating to the Indian Ocean is its essentially tropical character. Its northern end does not go much beyond the Tropic of Cancer. Its landbounded area is therefore free from the effects of icebergs and other oceanic obstructions arising from the frozen polar regions. Heavy mists, fog and other difficulties which prevented early navigation are also absent in the Indian Ocean. The rigours of climate are also greatly modified by the geographic situation.

The lack of well-defined "streams" like the Gulf Stream and its counterpart in the Pacific may also be noted. The importance of these streams has been more climatic than navigational and as such their absence has not materially affected the development of sea-going traffic in the India Ocean.

Partly perhaps as a result of the monsoons, and partly as a result of the earlier growth of civilization, the Indian Ocean was undoubtedly the first centre of oceanic activity. The first naval and oceanic tradition in fact grew up in the lands washed by the Arabian Sea. European writers, ignorant for the most part of even the main facts of Asiatic history, have assumed it as an axiom that navigational tradition first developed around the Aegean. Mackinder for example states: "Modern research has made it plain that the leading sea-faring race of antiquity came at all times from that square of water between Europe and Asia which is known alternatively as the Aegean Sea and the Archipelago, the chief sea of the Greeks. Sailors from this sea would appear to have taught the Phoenicians their trade.......". Perhaps he was thinking in terms of the development of sea-faring traditions in Europe, but in terms of world history this statement is obviously inaccurate. Long before sea-faring developed in the limited Aegean waters, oceanic navigation had become common with the coastal people of Peninsular India.

A recent writer has noted that the South-East Trade Wind of the Indian Ocean seems designed for the use of big sailing ships. "So are the other winds of that great ocean—the wild westerlies of the far south, the dependable monsoons which made possible the sailing commerce of the ancient East and still bring fleets of Arab, Indian and Persian shows upon their voyages between the Persian Gulf, India and Zanzibar. The west winds could blow a square rigged ship from Good Hope to Australia in three weeks and less though the distance is 6,000 miles.

It is not the Vindhyas and the tableland alone that separate Aryavarta from the Deccan. The main differentiating characteristic is the attitude towards the sea. To the people of North India, with their Central Asian traditions, the sea meant very little. Their conception of politics is well expressed by Kalidasa, who describes the ideal kings of the Raghu Dynasty as *Asamudra Kshiticsanam*—rulers whose territory extended to the shores of the sea. The idea of ruling the seas, far less of ruling lands across the seas, never entered the minds of the monarchs of the north while it was normal conception with Kalingas, the Cholas and the Pandyas. This is not to say that the Aryans after their arrival in India did not become familiar with the sea or with maritime navigation, but to them population movements, political power, etc., were bounded by the sea. In South India on the other hand population movements and cultural developments would seem to have been determined by the sea. The prehistoric finds at Maski in Mysore and other centres establish maritime relationships with the West, the Arabian Sea coast and the Mediterranean basin, while the prevalence of Austric languages even today and the continuance of kinship systems among the Dravidian populations similar to those of Polynesians indicate population and culture movement from the

Pacific. The prevalence of the forms of Austric language in the Nicobars and its spread not only to different parts of India but to Madagascar, where it was powerful enough to establish a permanent dominance, are evidence of large-scale population and culture movements from the Pacific. The Polynesian origin of the dominant racial groups in Madagascar is generally accepted now and a movement on so vast a scale could not have left South India untouched. In fact Mrs. Irawati Karve in her recent work which deals exhaustively with kinship organization among Aryan and Dravidian races, establishes the fact that the kinship system of the Dravidian peoples, and even of the Marathas who have been Sanskritized in their language, bear striking similarity to the system prevalent in Polynesia. This, together with such other evidences as the introduction of the outrigged canoe, and the coconut, from the Pacific islands would seem to establish the fact that the racial and cultural relationships of South India were with the oceanic people even in prehistoric times. Mrs. Karve seems to hold the view that the dominant peoples of South India who originated in the Pacific were moving towards the north and occupying the tableland when the Aryans were settling down in the Punjab and expanding into the Gangetic Valley and the two peoples met and mixed only in Gujarat where she has been able to find surviving traces of the Polynesian social culture. The momentum of both these parallel movements having broken down long before they reached the Deccan, the two people seem never to have met racially, though in historic times, Hindu religion and Sanskrit culture attained a pre-eminence which gave a top layer of Aryanism to these peoples.

Sylavain Levi has pointed out that the sea routes to the East from the ports of South India had come into common use many centuries before the Christian era. When we come to the historic times, the maritime character of India's peninsular civilization becomes even clearer. Early Tamil literature is replete with allusions to maritime trade. In a Tamil classic of the first century A.D. *Kaveripattanam*, the great port at the mouth of the river Cauveri is described thus: "The sun shone over the open terraces, over the warehouses near the harbour and over the turrets with windows like the eyes of deer. In different places of Puhar the onlooker's attention was arrested by the sight of the abode of Yavana whose property never waned. At the harbour were to be seen sailors from distant lands". That this was not merely poetic imagery has now been proved by the excavations at Arikamedu. An extensive trading centre with large warehouses for foreign merchants has been brought to light where apart from other interesting things, dated Roman pottery has been discovered. According to the *Cambridge Ancient History* Italy imported more from India than from any other country except Spain. Pliny computes Rome's annual payment to India at 550 million sesterces, and the yearly loss to Rome in the balance of trade at 100 millions.

It is also necessary to emphasize that unlike the northern kings who were

Asamudra Kshitisa, kings whose territories extended to the sea, the southern rulers always proclaimed maritime dignities. Thus the Andhra kings assumed the role of *Tri Samudradhipati* while even up to modern times the Zamorins of Calicut described themselves as the sovereigns of the mountains and the sea. From the ports of the East Coast a regular colonization, as we know, was undertaken of the islands of Indonesia and of the territories of South Asia. Periplus mentions three ports in the Tamil country of which Kaveripattanam was the centre, as the places from which great ships which he calls *Colandia* sailed to the islands of the Pacific. Ptolemy also notes the ports to the north of Chicacol as centres of navigation to Malaya and the Archipelago. We have also the authority of the Chinese pilgrim, Itsing, for the statement that there were direct sailings from Negapatam to Kedah in Malaya. While there is no doubt that other areas also shared in this work of colonisation, the very fact that Chola, Pandya, Pallava and Malayala are names for tribes inhabiting certain areas of Sumatra clearly establishes the direct relationship between the South Indian kingdoms and the colonists of Indonesia.

That the political history of the southern kingdoms in historical times has been connected more with Ceylon and Farther India than with any kingdoms north of the Vindhyas is also a well- established fact. In fact both the Pandaya and Cholakings were engaged in intermittent warfare with the Sri Vijaya monarchs for over a century and a half.The Chola emperor Rajendra in A.D.1007 led in person an expedition against Sri Vijaya, captured Kedah and established Chola power on the Malaya Peninsula.Over 120 years later the Sri Vijaya king, Chandra Bahu, in turn, attacked Ceylon in alliance with the Pandaya ruler. These facts may indicate the extent to which the southern kingdoms were connected politically with the states across the seas.

How much South India with its overseas commerce was dependent on the sea was made clear when the naval power of the Portuguese, represented by ships carrying heavy guns, appeared before Calicut in A.D. 1498. The mastery of the Indian Ocean passed to the Portuguese and along with it the trade of India within a period of five years. It is true that the rulers of the coast especially the Zamorins of Calicut fought hard for over a hundred years to keep their coast clear of the Portuguese, but India's authority over the waters of the Indian Ocean vanished from A.D. 1503, when at the battle of Cochin, the Portuguese admiral gained what in effect became a decisive victory —a victory more important in its historical consequences than the battle of Trafalgar. From that time to the withdrawal of British navy from its bases in India, the Indian Ocean was no more than preserve of European nations who held Indian trade and economic life at their mercy.

The control of the Indian Ocean and with it, of the trade of India, Indonesia and Burma was exercised by a very simple process, which the genius of Albuquerque design and elaborated. It was by a control of the approaches to the Ocean itself that Albuquerque ensured the mastery of the Indian seas for the King of Portugal. He sealed the entry from the Pacific by the control of Malacca, from the Red Sea by the control of Socotra, from the Atlantic by his establishments on the African Coast, and finally he provided a base by the conquest and colonization of Goa. As long as the Portuguese were able to maintain this system they held the mastery of the Indian Ocean. When by the conquest of Malacca, the Dutch made the first breach into Albuquerques's system Portuguese authority in the Indian Ocean vanished and gave place to the rivalries of the great maritime nations of Europe—England, France and Holland. When Finally with the departure of Suffren from the East, Britain became the sole mistress of the Indian Ocean, it is to the Albuquerque system that the British reverted. After the Napoleonic wars when the picture becomes complete Britain is firmly established as Aden, controlling the entrance into and exit from the Red Sea, at Singapore, guarding the narrow straits connecting the Pacific with the Indian Ocean, in the Cape Colony, able to intercept any European power venturing to threaten her authority. The strategy of Albuquerque found its culmination in the policy of the British Empire in the 19th century.

When it broke down, it was again because a power operating from the Pacific was able to break the defences of Singapore. The Japanese by conquering Malaya and Singapore entered the Indian Ocean for a short period, demonstrating thereby that the period of the domination of the Indian Ocean from the Atlantic was over.

It is, however, well to remember that the land powers of Europe sought more than once to enter the Indian Ocean by other routes. The first attempt in this direction was by imperial Germany under William II. William evolved the grandiose scheme of a direct entry from the side of land into the Persian Gulf. His project was to connect Berlin with Baghdad by a direct rail route. The Berlin-Baghdad Railway was thus a historic reply to Vasco da Gama's achievement, the attempt of the land power to out-flank the sea. It short-circuited both the Atlantic and the Mediterranean and gave entry into the Indian Ocean in a manner which could set at nought the carefully planned controls of Albuquerque's strategy, inherited and perfected by the British.But the Scheme was never fulfilled because the British, alive to the danger of their power in India, were able to checkmate it in time.

A second and equally grandiose attempt was by Mussolini, the successor to the pretensions of the Venetians, to destroy whose monopoly the Portuguese, inheriting the transition from Genoa, had sailed round the Cape. With a large land empire in Abyssinia, Mussolini thought he could meet the challenge of sea power and reach the Indian Ocean direct. If the Italian Somaliland and Abyssinia could be joined up Italy would have entered the Indian Ocean in force. But the war came too soon and Mussolini's dream turned finally into a nightmare.

From the point of view of India the control of the Arabian Sea from across the African coast remains a possibility. The European powers who have colonies on the coast may in time become independent powers and develop considerable naval strength. A new European dominated state in Kenya, the Central African federation and other political growths of which we see the beginnings today, apart from South Africa itself, may introduce new factors in Indian Ocean policies. The eclipse of old powers and the weakening of old concepts should not lead us to the facile assumption that the Indian Ocean will remain isolated in future from the main currents of world politics.

The short period of Japanese supremacy in the Indian Seas following her capture of Singapore and the occupation of the Andamans demonstrated that the threat to the ocean from the side of the Pacific is a real one. Till the end of the 19th century the Pacific was not one of the vital centres of sea power. The end of the first war marked a definite shift of interest. The Washington Conference on the Limitation of Naval Armament in the Pacific was the visible expression of this change. Today the Pacific is a potential area of naval rivalry, though the Japanese challenge has been for the time being eliminated. But we shall be making a great mistake if we think that the Pacific has been abandoned by the other Asian powers to the U.S.A. Few people remember that the Chinese Navy under Cheng Ho in the 15th century controlled the Pacific Mediterranean and his ships even sailed the Indian Ocean on commercial expeditions. The development of combined Sino-Soviet navy based on Port Arthur may not immediately involve a challenge to American supremacy, but once the industrial potential of the Chinese Republic is fully harnessed, there is nothing to prevent her from claiming her place in the Pacific sun. India's area of interest in the past (under the British) extended only to Aden in the West and Singapore in the East. With the Pacific in storm the whole issue requires careful reconsideration.

It is outside my present purpose to discuss in any detail the effect of 450 years of European control of our seas. Those who are interested in the subject may read it in my *Asia and Western Dominance* which tries to give a connected picture of the political, economic and cultural effects of that control. For our purpose I shall state my point of view in a few propositions which have special relevance to the present subject.

(i) The control of the sea by a foreign power changes the economic structure of India, no doubt slowly but effectively. India's trade, unlike that of China, was at all times predominantly maritime. In China it followed the caravan route, the route of the camels. Chinese economy was therefore continental. India on the other hand, her land mass closed on three sides by mountains, had her trade mainly along sea routes. When these sea routes came to be controlled by the European powers, the economy of India began to be dependent to a large extent on the foreign merchants. When in the 18th and 19th centuries commerce developed on a world scale, the emphasis in India shifted from the interior to the coast and the great cities of Calcutta., Bombay and Madras began to take the place of Delhi., Agra and Allahabad. The development of her internal communications in the 19th century, mainly with a view to bring the produce of the Gangetic Valley to the coast, the neglect of the vast area of the interior untouched by those rail-roade, the maintenance of a feudal system of indirect government over large areas mainly in the tracts away from the coast, the discouragement of industrial development, the creation of an indigenous subsidiary financial class operating over the whole of India from their headquarters at the port cities and acting as agents and distributors of foreign commerce, the ruin of the native industry of India—these were the results of the control of the sea which Britain came finally to exercise in the 18th and 19th centuries.

(ii) While naval power was helpless against organized states, the history of India from A.D. 1500 to the time of the British conquest clearly demonstrated the fact that the most organized land power was unable to protect its interest on the sea against maritime powers. Akbar at the height of his power could not afford protection to the Mecca pilgrims. Even the imperial ships could not proceed to the holy city unless provided with Portuguese passes. Whenever the lords of the sea, whether the Portuguese, the Dutch or the British, tried to defy the empire on land, not only the Moghuls, but even the small rulers of the coastal states were able to teach them a lesson, as in the case of Aphonso Albuquerque at Calicut or the British factories in Bengal when they defied Shah Jehan, or the Dutch in their final fight with Travancore. But where the land power became weak or was torn by dissensions, naval power was able to consolidate and expand finally even to conquer the land masses.

(iii) Sea power is ubiquitous. Mahan the prophet of indivisible sea took as his text the quotation from Genesis: "And God said, let waters be gathered in one place". The oneness of the sea is an obvious geographical fact and its importance to an area like India, exposed on three sides to pressure from the sea, need not be emphasized. In respect of land power proximity is a major consideration. The Roman Empire was the most powerful military state in its day, but it presented no threat to India. Neither the Hans nor the Tangs in the days of their greatest expansion thought of a conquest of India. Distance and geographical barriers effectively control land empires. Distance, however, is no bar to the expansion of naval power. Portugal with hardly three million people at the time was able, from her distant base, to control the Indian Ocean and monopolize the Asian trade. Britain with hardly twenty million people was able to undertake a conquest of India, from her bases in the North Sea half way across the world.

(iv) Conquest of a country by a sea power differs from conquest by a land power. It is a significant fact that though large portions of North India were conquered many times by foreigners, India was never in the past ruled by a monarch who did not have his capital in India—except when it came to the naval power of the British. The Kushans, the Huns, the Turks and the Moghuls conquered many portions of North India, but their empire was in India. Kanishka ruled from Peshawar and his successor in the third generation became Vasudeva. The Huns, Mihiragula land Toramana, came from outside, but even when they oppressed the country and pillaged the people, they became Indians with no interest outside the country. The Khiljis and the Tughlaks have been absorbed in the population of India. The Moghuls, once they settled in Agra, lost their contacts with Central India and became a national monarchy. In her long history, the British are the only people who ruled India from a distant capital and treated India as a subordinate part of a larger whole. In fact the only foreigners who ruled India were the British. If this position is examined, it will be seen, that the indivisibility of the sea and the ubiquitousness of naval power were responsible for the long domination of India from London. This simple fact has lessons for us.

XVI India In Asia: Walking On A Tightrope

WALTER K. ANDERSEN

The political turmoil that has swept over the southern tier of Asia during the past eighteen months has forced India to adopt a more cautious approach in the conduct of its foreign affairs. New political actors playing different roles have complicated its efforts to advance key foreign policy goals. The insurgency against the Marxist regime in Afghanistan has generated tensions in Pak-Afghan relations that threaten to upset the structure of regional cooperation India has been attempting to build. Further, the collapse of the Pahlavi Dynasty in Iran removed a government that since 1974 had sponsored programmes to draw the South Asian states closer together. Continuing domestic unrest in Iran and the potential for instability in other oil producing states of the Middle East endangers India's supply of crude. Conflicts in Indochina have stalled the Sino-Indian normalization process, as well as blocked New Delhi's efforts to serve as a bridge between the ASEAN states and Vietnam. Inter-Arab disputes have made it difficult for India to preserve good ties with the various sets of antagonists. The instability of Asian states to develop mechanisms for mediating differences raises the spectre of increased superpower involvement in Asian affairs.

In contest, the era of relative stability between 1974 and 1978 created opportunities that made it possible for the managers of India's foreign affairs to pursue policies that enhanced the country's regional and international status. At the regional level, New Delhi sought to reduce the areas of conflict that had inhibited India from playing a more dynamic international role and had provided incentives to its South Asian neighbours to seek outside support in their bilateral disputes with India. Simultaneously, India moved to improve relations with China in order to open the way to the conduct of a more flexible foreign policy. The Desai government's policy of "genuine nonalignment." envisaging a balance of forces that denies either superpower a predominant

Reprinted from *Asian Survey*, Vol. 19, No. 12, 1979, pp. 693-710, by permisssion of the Regents of the University of California.

(The opinions and views in the article are of the author and not necessarily of the U.S. Department of State.)

position in Asia, provided the theoretical underpinning of India's efforts to improve relations with the U.S. and China, while preserving good Indo-Soviet ties.

The collapse of the Desai government in July 1979 will probably not produce any major revisions in India's foreign policy. The caretaker government of Prime Minister Charan Singh has pledged its commitment to the broad foreign policy objectives of its precedessors. None of the three major contenders for the prime ministership made foreign policy an issue in the campaign. Nevertheless, the destabilizing trends in Asia may force the new government to devise different strategies to preserve the gains achieved over the past several years. This article analyzes India's responses to the shifting balance of forces in Asia.

Regional Considerations

The 1971 Indo-Pakistani war left India unquestionably the preeminent power in South Asia, but it also limited New Delhi's diplomatic maneuverability by further arousing Beijing's suspicions of India as an instrument of Soviet interests in South Asia. The reduction of China's antagonism called for policies that would prompt its reassessment of India's role in South Asia. A first step was to convince the Chinese that India was not a threat to its regional neighbours, most notably Pakistan, with which China had developed close ties. The 1972 Simla Accord, setting down the guidelines for the bilateral resolution of Indo-Pakistani problems, laid the foundations for a good neighbour policy. This approach was pursued even more vigorously after the 1977 electoral victory of the Janata Party.

The government of Morarji Desai attempted to establish a more cooperative framework of relations with India's South Asian neighbours through a series of confidence-building measures. Despite strong domestic pressures to speak out on the trial and execution of Zulifikar Ali Bhutto, the government issued no statements. It supported Pakistan's application for membership in the nonaligned movement (after the government of President Zia-ul-Haq had withdrawn from CENTO). It remained diplomatically neutral on Pak-Afghan differences. Desai publicly accepted Zia's claim that Pakistan's nuclear programme had peaceful ends. Similarly, a five-year water-sharing arrangement acceptable to Bangladesh was negotiated, and Bangladeshi insurgents were denied the use of Indian sanctuaries. The stalled trade and transit negotiations with Nepal were concluded. Diplomatic relations with Bhutan were upgraded to the ambassadorial level.

Political turmoil around the western fringes of the subcontinent, however, poses a direct threat to regional stability. The emergence of a Marxist government in Afghanistan backed by the Soviet Union and the collapse of a supportive regime in Iran exacerbated Pakistan's already live sense of insecurity. The new Islamic government in Tehran remains friendly, but, wracked by internal dissension, it is not in a position to exercise much influence on behalf of Pakistan. The situation assumed a more threatening character when Kabul began to accuse Pakistan (and Iran) of collusion in the Afghan civil war, and warned that Afghanistan could retaliate by supporting the "liberation struggle" of Pakistan's Paktoon and Baluch-speaking population, a thinly-veiled Afghan rejection of the Durand Line frontier. Pakistan's fears were further aroused when the Soviet media questioned Pakistan's disclaimers of interference in Afghan affairs.[1]

India clearly hopes to insulate South Asia from the unrest in Afghanistan. The regional destabilizing potentialities of the situation there are considerable. A victory of the Islamic insurgents could add momentum to the further Islamization of Pakistani society and politics, a development with possible anti-India overtones. Should the Marxist government crush the rebels it might then be tempted to incite Pakistan's tribal population—the Paktoons and the Baluchis. The continuation of the civil war would drive still more refugees, now estimated between 175,000 and 225,000 into Pakistan. An escalation of the fighting raises the spectre of increased Soviet assistance to the beleaguered Amin government and, simultaneously, a worsening of Pak-Soviet relations. If, at this juncture, friendly Islamic states and/or the West provide Pakistan with weapons in an effort to contain the spread of Soviet influence, India's fears of superpower confrontation in its neighbourhood would be realized.

India has adopted three general approaches to limit the effects of the Afghan instability on South Asia. It has remained strictly neutral on the Pak-Afghan situation to assure to Islamabad that India is not part of any move to isolate Pakistan. Pakistan reciprocated by drawing closer to India. Foreign Secretary Shah Nawaz's March 1979 visit to India, initiated by Pakistan, underscored the new cordiality. Both sides pledged themselves to quicken the pace of rapprochement within the framework of the Simla Accord. More importantly, they agreed to meet on a periodic basis to discuss bilateral issues.[2]

Secondly, India has used diplomacy to encourage Afghan and Soviet moderation. Foreign Minister Vajpayee, during his September 18-19, 1978, visit to Kabul, advised his hosts to remain nonaligned and to promote trust in the region.[3] Prime Minister Desai in his June 1979 trip to the USSR reportedly suggested to the Soviets that they encourage the Afghan regime to broaden its base of support and to stop blaming others for the civil war.[4] Perhaps as a signal of Soviet intentions, Yevgeniy Primakov, head of the Oriental Institute in Moscow, told the delegates at the International Association of Political Science that "Marxist-Leninists" are opposed to the export of revolution."[5] Finally, India has retained close ties with Kabul to provide it a diplomatic option other than the Soviet Union. However, India has neither the economic resources nor the political influence to prevent the situation from deteriorating.

Uncertainties over India's security environment have buttressed the case of those who want India to acquire state of the art weapons, which can only be accomplished by substantial purchases from foreign sources. India signed an agreement last fall to acquire 40 British-built Jaguar deep penetration strike aircraft, with another 110 to be manufactured in India under a licensing arrangement. It has ordered 70 Soviet T-72 tanks,[6] and, according to a May 1979 statement of former Defence Minister Jagjivan Ram, India is searching for additional tanks with the most advanced weaponry.[7] There are reports that India is seeking 20 submarines to augment and ultimately replace the existing 8 Soviet-built Foxtrots.[8] It has placed an order for vertical take-off Harriers for the I.N.S. Vikrant aircraft carrier.[9]

India, on the other hand, remains unsympathetic to Pakistan's sense of military vulnerability, now exacerbated by the cut-off of U.S. military assistance and by the threat on its northern borders. Some Indians argue that Pakistan's recent purchase of 32 Mirage-5 fighters is a misappropriation of scarce funds on grounds that the real threat to Pakistan's integrity is economic stagnation. Analysts maintain that such sophisticated weapons as the Mirage encourage Pakistani aggressiveness against India.[10] The 1977-78 *Report* of the Ministry of Defence asserts that Pakistan's search for "more sophisticated arms...is not in consonance with the process of normalization [between India and Pakistan]."[11]

Pakistan's perception of vulnerability, however, is likely to drive it on to prepare for a worst case scenario that could damage Indo-Pakistani relations. Its refusal to abandon the development of a centrifuge enrichment facility capable of producing weapons grade uranium has the potential of spoiling the cordiality that now characterizes Indo-Pakistani relations. An exchange of letters between President Zia-ul-Haq and Prime Minister Desai in early 1979 reveals how difficult it will be to overcome the problem. Desai restated his government's unilateral rejection of a nuclear weapon and asked for a similar guarantee from Pakistan. Zia responded by proposing that the two sides agree either to bilateral or multilateral inspections of their nuclear facilities, or to a South Asian nuclear free zone.[12] The Indians predictably rejected both schemes. Foreign Minister Vajpayee informed Parliament that Indian participation in a nuclear free zone arrangement made no sense if China were not included.[13] Prime Minister Desai told a group of foreign correspondents that India would agree to inspection only if all other countries, including the five nuclear weapons states, also did so.[14] At the May 23-July 1, 1979, UN Special Session on Disarmament, India rejected the concept of geographically limited nuclear free zones on grounds that the proliferation issue called for a global solution.

Yet, Desai tried to defuse the disruptive potential of Pakistan's nuclear programme on bilateral relations by publicly accepting Zia's disclaimers. He further argued that India's security position would not be significantly affected even if Pakistan acquired a nuclear weapon,[15] a stand not widely accepted either in his own ruling Janata Party or among opposition groups. Desai, who admitted that India's 1974 nuclear explosion had aroused the distrust of neighbours, apparently hoped that the progressive improvement of Indo-Pakistani relations would substantially reduce Pakistan's sense of insecurity and thus induce it to impose self-restraints on its nuclear programme.

Prime Minister Charan Singh (Desai's successor) assumed a harder public line on the Pakistan nuclear programme soon after taking office. In his August 15 Independence Day address, Singh, in a departure from his prepared text, stated that "we have information that they are trying to make the bomb" and that it could only be directed against India. While careful to note that India "up to the present" has renounced "a bomb," he warned that "I and my colleagues will have to perhaps rethink this issue...if Pakistan continues with its intention."[16] This statement met a barrage of criticism in India and was not subsequently pursued by Singh, or the other leading candidates, in the parliamentary campaign. President Zia for his part two weeks later again affirmed the peaceful nature of Pakistan's nuclear programme, though he also asserted that "whatever the attitude of others, Pakistan, God willing, will acquire a reprocessing plant under any circumstances."[17] Pakistan's de facto decision to maintain a nuclear option probably will not in itself result in an Indian reassessment of the ground rules that govern bilateral relations. A testing programme or Indian information that such a programme was imminent, however, almost surely would.

While the instability in Nepal, Bangladesh and Burma is worrisome, it does not now pose a serious security problem. The recent up-surge of cross-border raids by dissident tribals in the mountains of the northeast can be met by the strong military forces stationed there. The reappearance of such groups again demonstrates the weak control maintained by Burma and Bangladesh over their borders. The political turmoil in Nepal is not likely to be exploited by the Chinese as long as it seeks to lure India away from the Soviets. Indeed, both countries want stable political conditions along their shared Himalayan frontier.

Balancing China and the USSR

The process of Sino-Indian normalization took shape in 1976 with the exchange of ambassadors after a hiatus of 15 years, but unresolved bilateral issues, most notably their disputed borders, were formidable obstacles to further movement forward. The Desai government's policies of good neighborliness and genuine nonalignment apparently convinced the new Chinese leadership that the time was ripe to break through the logjam. If successful, China would relieve the security threat against its southern frontiers as well as score a diplomatic coup against the USSR. Wang Bingnan, President of the Chinese People's Association, during his March 7-23, 1978 goodwill tour of

India, invited Foreign Minister Vajpayee to China for high-level political talks. The Indians accepted, though Vajpayee, describing his trip as a "probing mission," cautioned that not too much should be expected from this initial contact. To calm Soviet misgivings, he also maintained that improved Sino-Indian ties would not be at the expense of Indo-Soviet relations.

Vajpayee's talks in Beijing went well. However, China's military incursion into Vietnam in February 1979 effectively stalled the rapprochement process. Latent Indian fears of Chinese aggressiveness were aroused. Vajpayee's presence in China at the time of the attack[18] embarrassed the Desai government domestically. China's comparison of the "lesson" it was administering Vietnam with the "lesson" it taught India in 1962 reopened painful wounds. These images tended to overshadow the progress that had been made.

Vajpayee had extensive talks in China with Foreign Minister Huang Hua, Vice Premier Deng Xiaoping, and Premier Hua Guofeng, and the discussions covered the full range of Sino-Indian relations. The Chinese decision to permit a consideration of the border represented an important concession to India. Chinese officials earlier had argued that it should be taken up after the two sides had settled less contentious problems. Vajpayee informed Parliament that his host had proposed expanding the volume of trade as well as establishing scientific, cultural and educational exchange programmes. He was assured that Chinese assistance to India's tribal dissidents "was a thing of the past."[19]

The next round of talks are to take place in New Delhi. So far, no date has been set and no initiatives on this matter are likely until after India's parliamentary elections in January 1980. A resumption of fighting in Indochina at that time would surely further delay the dialogue.

The stalling of the normalization process provided the Soviets an opportunity to push for a restoration of the special political relationship that had been inscribed into the 1971 Indo-Soviet Treaty of Peace, Friendship and Cooperation. Premier Kosygin, in his fortuitously timed visit to India from March 9-15, 1979, attempted to project the USSR as a stable reliable ally ready to back India against an "aggressive" China. He hammered on the theme that the interests of the two countries coincided and that they should therefore develop joint approaches to the "threat" from China and to American "destabilizing" activities in West Asia.[20]

Desai adroitly avoided identifying India with the strident anti- PRC line adopted by Kosygin. Moreover, he resisted Soviet pressures to recognize the Vietnamese-backed Heng Samrin Regime in Kampuchea. He informed Parliament that India would defer recognition until a government there is in effective control of the country. Vajpayee stated on a television news programme soon after the talks had concluded that the joint communique had omitted any reference to Kampuchea because of differences on the subject. He also noted that Kosygin's public attacks on China while he was in New Delhi could cause problems for India in the conduct of its relations with Beijing.[21]

While only one major substantive agreement was signed, Kosygin carefully laid the groundwork for future accords. He employed the diplomatic instruments of economic and military assistance to draw India and the USSR closer together. Kosygin sancioned the delivery of 600,000 tons of oil in exchange for Indian rice to help alleviate the oil shortages created by the disruption of Iranian crude deliveries. A long-term economic agreement designed to protect the Soviet stake in the future development of India's heavy industry was concluded. Press reports note that Kosygin also informed the Indians that the Soviets were willing to sell them a wide range of sophisticated military equipment[22] and to construct a "large" nuclear power facility in India.[23] Following up on the Kosygin offer, an Indian delegation left for Moscow in September to evaluate an array of military items, including frigates, tanks, missiles, and aircraft according to a wire service report.[24] The Soviets expressed their willingness to be forthcoming. At the conclusion of the meetings, on October 5, a Soviet spokesman said the USSR would allow no obstacles including prices, to impair its assistance to India's defence needs. Soviet military offers are clearly intended to deny the West a significant role in the modernization of India's armed forces. The nuclear proposal was meant to provide New Delhi an alternative to the "unreliable" West.[25]

However, these offers pose some problems for the Indians. Indian industry has reached the stage where Western technical inputs are generally more desirable than what is available from the Soviet Union. The unutilized 250 million rouble aid offer extended during Foreign Minister Gromyko's April 1977 visit underscores India's difficulties in meshing projects with Soviet assistance. Efforts to broaden the source of military supplies, demonstrated by the selection of the British Jaguar, will be one significant consideration in a decision on military imports. Finally, India may want to avoid more of the problems it has experienced in trying to deal with foreign nuclear collaborators.

The Kosygin visit probably left the Indians confident that they occupy an important place in Soviet strategy and that New Delhi can disagree with Moscow without sending bilateral relations into a nose dive. The results suggest that New Delhi considers itself sufficiently secure to reject Soviet blandishments that restrict India's diplomatic flexibility. At the same time, the visit reassured the Indians that the Soviets are prepared to back India if it has problems with China and the U.S.

While the Indians have demonstrated their commitment to an independent foreign policy, the continuing instability in Southeast Asia limits their options vis-a-vis China. The expulsion of the pro-Chinese Pol Pot regime in Kampuchea by the Vietnamese set in motion a polarization of forces in the region. Vietnam's subsequent incursion into Kampuchea hastened the process. Indochina has once again become an arena of proxy wars, this time between

China and the Soviet Union. Under these circumstances, any significant forward movement in Sino-Soviet relations would give Beijing a diplomatic advantage at a time when China and the USSR confront each other in a conflict situation. New Delhi has put off making major decisions on the conduct of its relation with China and will probably continue to do so until the crisis in South-east Asia is defused. Nevertheless, India's decision to downplay reports of Chinese (i.e., Tibetan) graziers on Bhutanese land demonstrates a desire to keep relations on an even keel.

At the same time, New Delhi has adopted an evenhanded policy on the controversy in Southeast Asia that identifies India with neither side. It condemned the invasions of both China and Vietnam and it called on both to pull back their troops. At the various nonaligned meetings during the past year, it evolved a policy that would deny seating either Kampuchean delegation. Consistent with this stand, India, at the beginning of the 34th UN General Assembly, voted against seating the Pol Pot delegation without supporting its replacement with one representing the Heng Samrin government. India's response to this issue has complicated its relations with the ASEAN states, which support seating the Pol Pot regime as the legal representative of Kampuchea. Dinesh Singh, a former foreign minister, was dispatched by Desai to Southeast Asia in May 1979 to discuss the Indochinese situation with government leaders in the ASEAN states and in Vietnam. There are reports that the ASEAN officials were irritated over what they interpreted as an effort on his part to drum up support for the Heng Samrin regime.[26]

The Chinese for their part have given no sign that they are losing patience and appear prepared to wait until India's domestic and international situation becomes more conducive for the negotiation of specific agreements. In the interim, they have contacted the Dalai Lama's "government-in-exile" to negotiate an arrangement under which he and eighty thousand Tibetan refugees in India could return to Tibet. If successful, one irritant in Sino-Indian relations would be resolved.

India's cautious treatment of the Sino-Indian rapprochement process reflects the continuing value of its Soviet connection. Moscow, the only country that can provide military hardware on a timely basis, remains India's largest and most reliable source. It is India's second largest trading partner (after the U.S.) and a major source of crude oil. Moscow's diplomatic support in the UN and elsewhere may again be needed if relations with Pakistan and China deteriorate. Strategically, the Soviet's role as a counterweight to China is still an important consideration. Nevertheless, as long as India's security is not directly threatened, Moscow cannot expect a revival of the close political relationship of the 1971-1972 period. India's drive for an influential balancing role in the Asian scale of power calls for an independent and flexible foreign policy.

Indian Ocean Concerns

India has two broad goals in its Indian Ocean policy: removing superpower naval forces and bases and building up its navy to give India an independent capacity to defend national interests. To achieve the first, India has advocated the idea of a "zone of peace," which it first raised at the 1970 Lusaka nonaligned meeting. However, it has stopped short of backing demilitarization schemes that place limits on regional powers. At the July 2-5, 1979, UN General Assembly debate on the Indian Ocean, the Indian delegation rejected Pakistani proposals calling on littoral states to prepare a "code of conduct" and to establish a "reasonable ratio" of forces. It also opposes nuclear free zone proposals as backdoor attempts to impose on India discriminatory nuclear nonproliferation restraints not demanded of the nuclear weapons powers.

India initially expressed more concern about the U.S., which possessed a naval facility at Diego Garcia, than the Soviets. Indeed, the surge of Soviet naval strength in late 1971, reached three weeks after a U.S. task force was dispatched to the Bay of Bengal during the Indo-Pakistani war, was considered a supportive move. However, the stationing virtually permanent naval forces by both sides and the potential use of such forces to support friendly littoral states in conflict situations and to exert pressure in defence of their own national interests have given rise to a more evenhanded call for their withdrawal. At a press conference immediately after taking office, Desai stated that "the United States should realize that they should not have their activities here. We should also see that the Soviet Union should not have its activities here."[27] India applauded the two sides for opening naval limitation talks in 1977, but was dismayed that the now-stalled negotiations were considering only the "freezing" of force levels.

The Indian navy, prior to 1971 the stepchild of the services, has come to play an increasingly important place in defence planning. The 1971 Indo-Pakistani war, demonstrating the potential of missile-armed fast patrol boats in future regional conflicts, gave the navy a mission.The need to protect India's growing external trade, its 200 mile wide economic zone and its offshore exploitation of resources have given additional weight to arguments favoring the expansion of the navy. The failure of bilateral and multilateral efforts to remove the naval forces of the superpowers, the rising tide of instability in West and Southeast the superpowers, the rising tide of instability in West and Southeast Asia, and the uncertainty over future Chinese naval activities have further spurred on the defence planners to build up India's naval power to prepare it for any worst case scenario.

To carry out its mission, the Navy has pushed the development and acquisition of technologically advanced missile-armed naval craft capable of patrolling beyond coastal waters. In the last two years, it has acquired light new Osa-class missile-armed fast patrol boats, and as many as six Nanuchka class corvettes armed with SS-N-II missiles. In addition, the acquisition of three

Kachin class destroyers and more support craft will permit the navy to operate in a blue water mode. Former naval chief-of-staff A.K. Chatterji argues that the long- range objective of India's naval programme is a "force equal in size and competence to the naval forces of any one of the superpowers now formally operating in the area."

But, the final decision on the further acceleration of the naval programme lies with civilian bureaucrats and politicians who, as Raju G.C. Thomas points out,"are unsympathetic o to sizeable demands on budgetary resources based on vague concepts of seapower and long range economies. They will be reluctant to reorient the budgetary allocations of the three services, whose proportionate shares of the defense budget have remained roughly the same since 1973-1974, unless a seaborne threat were to become visible and credible. Recent developments have strengthened the case for the naval proponents, but not sufficiently to convince the decision makers to alter significantly the budgetary pattern established over the past several years.

West Asia

Oil and Pakistan are the two considerations that shape India's policy towards the states of western Asia. Close ties are cultivated with them to ensure India a steady flow of oil imports. Good relations are also important as a counterbalance to Pakistan's Islamic ties.

Two-thirds of the oil consumed by India during 1979 was imported, most of it from Iraq, Saudi Arabia, and the United Arab Emirates. The effect of the revolution in Iran demonstrates India's vulnerability to the instability sweeping through the region. Oil imports from Iran, the largest single source in 1978, plummeted from four-and one-half million metric tons in that year to about one million metric tons in 1979, causing a serious crude shortage during the first quarter of the year. The strain on India's economy was compounded by three OPEC oil price increases, driving the direct sale prices from about $13 dollar/barrel at the start of 1979 to $23-$24/barrel at the end. Spot markets had oil priced at over $30/barrel. As Iranian crude production plunged, Indian negotiators fanned out to arrange for increased deliveries from Iraq and the Soviet Union as well to establish new supply links with Libya, Algeria, and Saudi Arabia. It simultaneously moved to establish good relations with the Islamic government in Iran by immediately recognizing the Barzargan government and by dispatching a delegation to visit Ayatollah Khomeini.

The oil weapon wielded by the oil-producing states is a powerful one since a reduction in supplies would disrupt India's economy. Libya has already threatened to cut off deliveries for reasons that are still unclear. Alternative sources could quickly be found if it did so since a relatively small proportion of India's imports comes from Libya. However, should a country like Iraq, which supplied over 40% of India's imported crude in 1979, turn off the tap or reduce sales, the impact would be catastrophic. Consequently, India has been extremely careful to avoid irritating the Arab states. In an effort to maintain good relations, India has tried to avoid entangling itself in inter-Arab squabbles, and to follow the Arab consensus on the Egypt-Israeli peace process.

These efforts have been complicated by close Indo-Egyptian relations. The problems came to the surface at the various nonaligned meetings during the year as the hardline Arab states pushed vigorously for Egypt's expulsion. Shortly before the June preparatory meeting of the Nonaligned Coordination Bureau in Colombo, Foreign Minister Vajpayee and Samarendra Kundu, minister of State for Foreign Affairs, visited seven Arab states to sound out their views on the subject and to persuade them that such action would jeopardize the unity of the nonaligned movement. Shortly after their return, Prime Minister Desai told a group of foreign corespondents that "there is no provision for removal of anybody as long as the country fulfills the commitments of the nonaligned movement, and Egypt fulfills it." While standing by Egypt on this question, India has consistently opposed the Egypt-Israeli Treaty. On his return from the Arab states in late May, Vajpayee repeated the Indian position by stating that the treaty "falls short of a comprehensive solution of the problem. Unless there is a total vacation of all Arab territories occupied by Israel and the inalienable rights of the Palestinians are restored, there can be no lasting peace in the region..

This balancing of interests is a tricky, and possibly dangerous, business as it could touch off a backlash from such hardline states as Iraq. Not only must India worry about the possibilily of an oil boycott, but also renewed Arab diplomatic support for Pakistan on such bilateral issues as the disputed state of Jammu/Kashmir. Former Defence Minister Jagjivan Ram, now head of the Janata Party, underscored the problem when he warned Desai in mid-July that his "impromptu reactions of belated responses have threatened to alienate friends, such as those of the Arab World." The new government that emerges after the parliamentary elections will have to decide how to fine tune the Egyptian issue in a way that does not threaten vital national interests.

Conclusion

The fluid political situation in Asia, particularly along its volatile southern tier, has radically altered the stage on which India conducts its foreign policy. The rising tide of instability in the area provides the superpowers with opportunities to shape events in ways that undermine the political influence of competitors. The most direct threat of outside involvement is now the projection of Soviet power in Afghanistan, in Southeast Asia, and in the northwestern Indian Ocean littoral states. Countering moves by the West and by China, either directly or through states that feel endangered by the actions of Moscow and its allies, heighten the potential for proxy wars in these areas. India's major foreign policy concern has been, and is likely to remain, insulating South Asia from these developments. The good neighbour policy serves this objective and it will remain a central plank in India's foreign policy. Nonalignment is a complementary strategy since it reduces the chances of South Asia becoming an arena for such disputes. In Moscow's calculations of the Afghanistan situation, for example, the USSR will have to consider whether destabilization in Pakistan and Iran will induce New Delhi and Beijing to resolve more quickly those bilateral differences that have kept them apart. Consequently, there will be a strong compulsion for Indian leaders to devise strategies that prompt both Beijing and Moscow to act in ways that do not undermine India's national interests.

However, India's efforts to insulate South Asia are complicated by unstable political conditions among regional states. Ethnic, religious, political and economic problems jut across national boundaries. When tensions break to the surface in one state the impact is often felt in the others, giving rise to bilateral problems. India's neighbors have a record of seeking outside supporters under these circumstances. Since India is the most powerful country in the subcontinent, a conciliatory attitude on the part of leaders is key to the maintenance of regional cooperation.

Equally complicated will be India's efforts to limit the damage caused by conflicts among the states of West Asia and Southeast Asia. While the new government will probably try not to antagonize either side, neutrality will be difficult to sustain in the face of incompatible Indian interests. India can also be expected to use international forums to prevent the superpowers from intervening in these conflicts, either by advocating the regional resolution of conflicts. If such efforts fail, pressures for accelerating the modernization of India's military programme will mount to enable the country better to deter potential threats to its security.

REFERENCES

1. A five part series of articles under the pseudonym "Petrou" that appeared in Pravda between March and July 1979, hammered away on this theme.
2. See a discussion of the meeting in *Times of India* (Bombay). June 7, 1979.
3. FBIS, Delhi Domestic Service, June 20, 1979.
4. *New York Times,* June, 15, 1979. The Karachi Domestic Service, June 14, 1979 reported Desai saying that the Afghans should stop blaming Pakistan.
5. TASS, August 17, 1979.
6. *World Armaments and Disarmament* SIPRI Year book: 1979 (London: Taylor and Francis ed.), p. 214.
7. FBIS, Delhi General Overseas Service, May 7, 1979.
8. *Washington Star,* July 9, 1979.
9. SIPRI, pp. 214-215.
10. For example, see P.N. Haksar, "Appearance and Reality," *Seminar,* No. 225 (May

1979), pp. 39-41.

11. Annual *Report, 1977-78,* New Delhi: Ministry of Defence, Government of India. p. 2.
12. For accounts of this correspondence, see FBIS, Karachi Domestic Service, March 21, 1979; *Hindu,* March 19,1979: *Eastern Economist,* May 7, 1979.
13. FBIS, Delhi Domestic Service, May 5, 1979.
14. FBIS, Delhi ISI, June 2, 1979.
15. Ibid.
16. FBIS, Delhi Domestic service, August 15, 1979.
17. FBIS, Karachi Domestic Service, August 30, 1979.
18. His visit, scheduled to conclude on February 18, was broken off a day early because of the fighting in Vietnam.
19. FBIS, Delhi ISI, February 21, 1979.
20. See account of visit in *Far Eastern Economic Review,* March 23, 1979.
21. FBIS, Hong Kong AFP, March 16, 1979. While the Chinese did deliver a protest note, they handied the issue in a low-key fashion.
22. *Times of India* (Bombay), April 17, 1979.
23. *Economic Times,* May 2, 1979.
24. Reuter, New Delhi, *September* 21, 1979.
25. The Soviets had already signalled their desire to collaborate in the development of India's nuclear power industry when they sold it 200 tons of heavy water. A trilateral agreement with the International Atomic Energy Agency was signed on Nobember 17, 1977, to place it under safeguards. FBIS, Delhi Domestic Service, July 27, 1978.
26. *For Eastern Economic Review,* August 10, 1979.
27. FBIS, Delhi ISI, March 24, 1977. Richard Haass notes that the Soviet Union maintain 17-22 vessels in the Indian Ocean. The US, besides the three ships of the Middle East Force in the Persian Gulf, rotates a task force from the Seventh (Pacific) Fleet, according to Haass, "Naval Arms Limitation in the Indian Ocean." (London International Institute of Strategic Studies, March - April, 1978), pp. 56-57.
29. Quoted in SIPRI, p. 377.
30. Raju G.C. Thomas, "The Indian Navy in the Seventies," *Pacific Affairs,* 48:4 (Winter 1975-76) p. 515.
31. The army, navy and air force shares of the defence budget have generally been around 70%, 20% and 10% respectively. The naval share of capital expenditures has remained between 40-50% since 1973-74. *Defence Services Estimates,* 1973-74 through 1978-79.
32. Estimates for oil statistics drawn from *Financial Express,* January 4, 1979: January 21, 1979; March 19; April 1970; as well as from the *Overseas Hindustan Times,* May 31, 1979.
33. There are conflicting reports on Libya's reason, some state that Libya was angered by the low level of nuclear assistance provided by India. Others that Libya made the threat because India sold Libyan crude on the spot market. See *Times of India* (Bombay), August 28, 1979; and *Hindu,* August 28, 1979.
34. FBIS, Delhi ISI, June 2, 1979.
35. FBIS, Delhi Domestic Service, May 29, 1979.
36. *Hindu,* July 17, 1979.

XVII South Asian Regional Security: Political and Socio-Economic Dimensions

K.P. MISRA

> Self-interest itself should drive every nation to the wider cooperation in order to escape disaster in the future and build its own free life on the basis of others' freedom.
>
> —Jawaharlal Nehru

The problems of security at various levels—individual, sub- national, regional and international—are inextricably interwoven.[1] In fact they are different points along a continuum: Their integral nature is evidenced by the fact that if security at one point is endangered, the other points cannot remain entirely unaffected. All the same we find that studies on security generally limit their scope to national security. The reason for this is not hard to seek. Notwithstanding the influence of some schools of thought which emphasized the pivotal position of the individual in any durable security arrangement, and in spite of the considerable growth of internationalism, the nation-state continues to be the most decisive social-political organization in the world today. Our allegiance to it clearly surpasses our loyalty to all else. But it is not being said that this is where our allegiance ought to lie. Hence national security has a primacy over the security of all other units and organizations.

We recognize the primacy, though not exclusiveness, of security at the national level. The situation being what it is, there is hardly any such thing as a security of South Asia as such or as a whole. This is not to suggest that the South Asian countries cannot meaningfully coordinate their efforts in dealing with the problems of national security. By doing so they logically promote security interests of South Asia also.

While identifying the components of the concept of security, some key problems confront us. One of these is the bewildering complexity of contemporary technological sophistication. Even apparently simple things have now been discovered to possess dimensions which were hardly suspected a few years ago. This development has serious implications not only for security

Reprinted with permission of the Editor from *Punjab Journal of Politics*, Vol. 9, 1985, pp. 99-110.

matters but also for wider social situations. It has immensely contributed to political complexity. The ruling elite, which is already exhausted, does not have the capacity to absorb the intricacies so created. Many a time it finds itself at the mercy of the technologists. It is no wonder, therefore, that the average citizen should find it almost beyond his competence to understand problems of security. This is dangerous for political systems which are on the whole democratic in character. The fact is that today few problems are more important for a citizen than those relating to the security of his nation. Briefly, today's technological and political complexity and the inability of the citizens in general, and the busy decision-makers in particular, to comprehend it mark the contemporary security situation.

Till a few decades ago, the concept of security was confined to narrow limits; it was almost a habit of mind to think of a nation's or a region's security problems almost exclusively in terms of military hardware. This kind of thinking promoted wars and armed conflicts and placed the security of large parts of the world in jeopardy. Two world wars were fought within the span of a single generation and numerous local conflicts erupted from time to time. The South Asian countries have known it at their own cost. Not only South Asia, but the other two subregions in Asia, Southeast Asia and West Asia, have also witnessed bitter conflicts. The issues which generated these conflicts are still alive and may lead to conflict again.

What factors have been responsible for these conflicts which have endangered the national security of countries in South Asia. The pivotal factor has been that the nations of South Asia are engaged in a desperate endeavour to build viable political systems, often out of unpromising and divided elements. Their efforts at socioeconomic reconstruction are generating tensions. Most of these areas are weak and poorly defended. Taking advantage of these sources of insecurity, powers external to the region have moved in and have, through their manoeuvres and machinations, through supply of arms and the sinews of war, and through extension of support to unpopular and undemocratic regimes, added fuel to the fire. The irony of the situation is that these powers have done what they have done ostensibly with a view to strengthening their own security. They appear to be oblivious of the basic point that they "cannot find their equilibrium in themselves." As Rabindranath Tagore once said: "Power has to be made secure not only against power, but also against weakness, for there lies the peril of its loosing balance. The weak are a great danger for the strong as quicksand for an elephant. They do not assist progress because they do not resist..."[2]

There is now a fairly wide agreement about the components of security. The tragedy is that the powerful countries are unable to pay heed to them in practice. They ignore the lessons of history and even now conceive security substantially in terms of military hardware.

Since the traditional hardware approach to security has been found wanting in many respects, many enlightened people are pleading for a wider approach to the concept of security. If all relevant factors and forces are kept in view, in general terms one may say that the security of a nation is a condition of being protected or of not being exposed to danger. It means well- founded confidence on its part. It implies freedom from danger and from a fear of danger. Lastly, a nation's security consists in the stability that it enjoys. Thus viewed, the frame-work of any valid concept of security must be multidimensional and rest on socio-economic, politico-cultural, ethnic, and ideological considerations, in addition to military aspect.

During the last four decades many new nations have come into being. Some of their leaders, most notable among them being Jawaharlal Nehru, enunciated a set of principles which later came to be known as nonalignment. Though negative in appearance it had considerable element of positive content and above dimensions of security were inherent in it. Nonalignment envisaged a foreign policy orientation which was different from usual traditional model. Its basic motivation was to create a situation conducive to maximum socio-economic development of the newly-born states. The assumption was that this alone would subserve the real needs and requirements of security. But this sort of thinking has been shared by some leading westerners also. For instance, Robert McNamara points out:

> In a modernizing society security means development. Security is not military hardware; though it may include it; security is not military force though it may involve it; security is not traditional military activity though it may encompass it. *Security is development and without development there can be no security.* A developing nation that does not, in fact, develop simply cannot remain secure for the intractable reason that its own citizenry cannot shed its human nature.[3]

This discussion of the general, theoretic aspects of security is relevant to the South Asian situation. Attempt may now be made to discuss the situation in the region concretely and more specifically.

The subregion or subsystem of south Asia should or should not consist of Burma and Afghanistan has been a subject of difference of opinion. But what is generally agreed to is that it comprises India, Pakistan, Nepal, Bhutan, Sri Lanka and Maldives. These seven states have nearly 20 per cent of the world population which lives on only 3.3 per cent land area of the world. Demographically, thus, this region has one of the densest populations of the world. The rate of population growth is also very high in the region, about 2.2 per cent. India stands as a collosus among its neighbours; its territory extends between the Himalayas to the Indian Ocean. It has the second largest population and the seventh largest territory in the world. Two other countries—Pakistan and Bangladesh—are also not small; they are tenth and eighth in world population rating. Nepal and Bhutan are landlocked Himalayan states. On the other hand Sri Lanka and Maldives are island states. India is the only country which has common frontier with all other countries in South Asia, in some cases it is land frontier and in others sea.

With regard to national resources, the situation in the region is far from satisfactory; it has less than one per cent of the world's brown coal and lignite, crude petroleum, natural gas, copper, gold, silver etc. These meagre resources are unevenly distributed among the countries in South Asia; most of them are confined to India. This paucity can be considerably compensated by substantial resources of the Himalayas and of the sea provided concerted efforts are made by the countries of the region. The South Asian countries are amongst the least developed whose annual per capita income ranges from US$ 20 to $ 200. Their economies are diverse, from island economies, to landlocked to continental. Apart from area and population India's preponderance should also be evident from the fact that its gross domestic product is 77 per cent of the region.[4]

From the view point of military expenditure so far the region should not cause much concern because 20 per cent population of the globe spends only one per cent of the total. However the capability of India and Pakistan to produce nuclear weapons is recognized to be considerable. India's capability was demonstrated by the 1974 explosion conducted in Rajasthan but it set a unique example by declaring that explosion as peaceful and that it had no intention of becoming a nuclear-weapon power. In the meantime there are conflicting reports about the nuclear capability which Pakistan has developed during the last decade or so. Its capability is based on enriched uranium which need not be tested through explosions for making weapons as is the case with regard to plutonium. The Western news-media has published unconfirmed reports that it has already done clandestine testing on the Chinese territory. Whatever the details, there is no doubt that both India and Pakistan have reached a threshold stage. Thus the nuclear factor has added a new dimension to the security situation in South Asia. Compared to the West Asia and Southeast Asia, South Asia is relatively less important for great powers but events in Afghanistan since December 1979 have somewhat changed the situation which is indicated by a $ 3.2 billion military and economic aid sanctioned to Pakistan by the United States in 1981. Thus on account of happenings outside the region there is a real danger of greater arms race between India and Pakistan.

Another noteworthy aspect of South Asia is diversity in its political systems. While countries like India and Sri Lanka are functioning democracies, based on secular ideals, Pakistan and Bangladesh and to an extent Nepal are authoritarian systems. India's two next door neighbhours in the subcontinent are being ruled by military dictators who are desperately attempting to acquire political and constitutional legitimacy through carefully calculated and controlled political exercises. This difference has had serious implications for security. In a recent significant writing this point has been ably highlighted.

> It is not just India's size that creates insecurity among neighbours; it is also the paranoia of their rulers, fearing that the lure of democracy and friendship might prove too tempting for their people, particularly when amalgamated to the truth of a common culture. That is why these regimes need barriers. Democratic wafting across the borders could fuel movements to urges overthrow the vested interests who have seized power both in Pakistan and Bangladesh. If India, with its immensely greater problems, can afford democracy, why cannot Pakistan or Bangladesh? This is the question which terrifies the mullah-military combination both to the east and the west of India. Similarly, if India destroyed feudalism, why should Nepal continue to live in the nineteenth century.[5]

The threat perceptions to national security in some countries of the region have sometimes been interpreted as individual or regime security without any reference to the needs and requirements of the people. In one situation this led to the disintegration of the country.[6] A popularly responsive and politically responsible government's perceptions are not very often so off the mark from reality as they are in autocratic systems. In South Asia this feature in some countries has encouraged tension and contention. Domestic conflicts and lack of stability accounts for some of the difficulties in the region. The linkage between domestic politics and foreign policy cannot be ignored anywhere but in South Asia this has been of considerable significance.

This becomes obvious when foreign policy perspectives of South Asian countries are taken into account. Ever since the closing years of the 1940s the growth of their foreign policies has not been in the same direction. Due to security perceptions of the ruling elites of Pakistan, the country chose to become a part of the alliance systems sponsored by the West. Rightly or wrongly, and history proves wrongly, Pakistan thought that the main source of insecurity came from India and this could be taken care of with the military and political help of the West. This situation has handsomely contributed to Indo-Pakistani wars in 1949, 1965 and 1971. However, the country now has formally walked out of the alliances and has sought and secured membership of the nonaligned movement but the remnants of the previous foreign policy are still evident. Pakistan has become once again a part of the strategic consensus of the West which is a principal feature of the new or the second cold war. As against this, all other South Asian countries, including of course India, have been nonaligned. Through this policy they have attempted to reduce the role of outside powers in the region, an effort which has suffered reverses on account of Pakistani counter policy. This aspect has vital bearing on the security of South Asia.

Connected with this is the fact that three of the six nuclear- weapon powers are around South Asia. China is the largest Asian power and large parts of the Soviet territory are Asian,; both have close physical proximity with major countries of the region. Their presence in the north as well as the south of the region (in the Indian Ocean) is significant. The United States has naval bases in the Indian Ocean. These three powers have played quite a role in South Asia. The situation has undergone a qualitative change since the Soviet action in Afghanistan in December 1979. Pakistan now is not only informally a part of the "strategic consensus" of the United States but is also treated as a kind of frontline state for "containing the Soviet power."Thus the renewed cold war since the closing years of the 1970s has not left the South Asian region untouched.

The evolving relationship of the United States with Pakistan in the 1980s has witnessed a situation in which the former has forgotten its human rights and nuclear non-proliferation professions. Its consequence is the transfer of sophisticated weapons to Pakistan which India perceives as a threat to its security. It may be remembered that since it is only through Pakistan that the US can supply arms and other military assistance to the rebels of Afghanistan, the military rulers of Pakistan found it easier to secure arms supply. On the others hand, India's position in relation to the Soviet Union also improved. The Soviets agreed to help India in the latter's defence modernization programme. India also secured more petroleum and petroleum products against rupee payment. The Soviet Union, it is interesting to note, initiated trade relations with the private sector in India.

An important dimension which is bedewelling relations between South Asian countries is the problem of ethnicity or ethnic minorities. All the countries in the region are plural societies and their demographic structure is diverse. Many of these problems and tensions, to begin with, appear domestic, between a minority and a dominant ruling class. Later on in some cases, they assumed regional character. In India ethnic conflicts could be contained in some cases easily while in other the difficulties were great. The problem of Sikhs is an example of a difficult situation. It has assumed international dimension. India's relations with Pakistan went sour and it has been pleading with countries like the US, the UK and Canada that the Sikh activities hostile to India, in these countries should be curtailed, or if possible, stopped. In Sri Lanka the conflict between Sinhala majority and Tamil minority has assumed alarming proportions. Indo-Sri Lankan relations have touched a new low and many powers external to the region are gradually getting involved. The level of violence is increasing, leading to enormous loss of life and property. There are severe strains on Sri Lanka's political unity and territorial integrity. At the moment there seems to be no way out. India's stand is that Sri Lanka's unity and stability must be preserved and at the same time a reasonable autonomy should be granted to the Tamils. Pakistan has been the worst sufferer. In 1971 the country was dismembered, a process in which India had to play a role. But what is unfortunate is that even after the events of 1971, Pakistan is still facing problems of ethnic movements and minority communities. Similarly, Nepal and Bhutan are not free from problems and one does not know as to when the subterranean currents become strong, causing concern to the country concerned and the region.

It may be added that the ethnic tensions have increased in the countries of South Asia, many a time on account of discriminatory political, security, socio-cultural and economic policies of the dominant classes. Also, because practically all ethnic groups have their kinsmen in some neighbouring country of the other, a purely domestic problem after a certain stage becomes an important foreign policy matter. Finally, since India is geographically in the centre of South Asia, ethnic problems practically in all countries of South Asia affect it. Sometimes India becomes an aggrieved party and sometimes its neighbour is placed in this position.

These problems and differences notwithstanding, there are important common points among South Asian countries. Firstly, their historical experience is the same. Most or the South Asian countries were under the British rule or domination. This experience is important because it subsequently shaped the thinking of public opinion leaders in many respects. Secondly, there is close cultural affinity between the peoples of these countries. Cultural and civilization emanating from Hinduism, Islam, Buddhism and even Christianity intermingled in this part of the world and produced a kind of composite outlook which is rather unique. Related with this is a kind of linguistic nearness on account of which a vast majority of people in the region can communicate with each other through English or a mix of Hindi and Urdu or some variant of these. This establishes a kind of instantaneous rapport among the peoples as soon as they get an opportunity to meet with each other within or outside the region. Thirdly, and perhaps most importantly, the problems of poverty and of social evils, though not identical, are similar. In spite of the fact that India has better scientific and technological infrastructural resources compared to other countries in the region, there are certain areas of economic activity which call for mutual beneficial cooperative relationship in the region.

It is in this background that efforts at achieving greater regional interdependence, collective self-reliance, and peace and stability in South Asia in the coming decades may be analysed.

Amongst the freedom-fighters it is understandable to have a degree of idealism in their approach. The outlook of the independence movements in South was no exception to this general rule. Hence, before and soon after 1947-48, political leaders of India, and also of other countries, talked of closer and friendly relationship for future good among themselves. They even talked of establishing an Asian federation to cope with the new and emerging challenges. They thought that through closer relationship common policies in the areas of trade, defence and culture could be evolved to the advantage of all countries in the region. The organization of the Asian relations conference in Delhi in March 1947, and a couple of years later, a conference on Indonesia, were in fact manifestations of this desire. In a less enthusiastic form this thinking lingered on till early 1955 when the Afro-Asian Conference was organized at Bandung (Indonesia).

In spite of these activities, a degree of disenchantment grew among South Asian countries from the early 1950s onwards and a variety of events—domestic as well as international—contributed to this. The strong winds of cold war did not spare the region. Intra-regional differences grew and the domestic difficulties became evident and even actue in some cases. There was no doubt some linkage between developments at the domestic, regional and international levels. The situation grew so conflictual, particularly between India and Pakistan, that for about three decades, 1950-80, no significant initiative was thought of in favour of regional cooperation though such efforts were made in many other regions of the world. On both sides of South Asia, significant endeavours had been made in South-east and West Asia. It was only around 1980s that some thinking in this direction was made and soundings began.

Perhaps there is no other region in the world where initiatives for regional cooperation were taken too belatedly and so hesitatingly. It seems that there are two basic reasons contributing to this situation. Even at the risk of some repetition they may be identified as follows:

(i) Wide disparity in power (military, economic, population, etc.) among the countries in the region.

(ii) A variety of disputes/differences between the countries of the region. Mostly they are bilateral and include political, boundary, ethnic and economic fields.

Needless to add that in the context of these points India's position is central, both literally as well as metaphorically. This fact has a variety of implications but its meaning for security is perhaps the most important. As Sri Lankan scholar has pointed out.

> .....it could be said that one of the biggest dilemmas of South Asian politics that India conceives of her neighbouring countries as lying within the Indian defence perimeter and being integral to the security interests of India, while India's neighbours themselves regard India itself as the source of their own insecurity against whom it is necessary to organize their own security interests, sometimes even on an extra-regional basis.[7]

Notwithstanding India's preponderance in the region, as we have pointed out earlier in this paper, a move for regional cooperation was made by Bangladesh and after some consultations was accepted by the countries of South Asia. The inaugural meeting was held in April 1981. So far seven meetings have taken place at the official levels—Colombo, April 1981; Kathmandu, November 1981; Islamabad, August 1982; Dhaka, March 1983 New Delhi, July 1983; Male, February 1985; and Thimpu, May 1985. In addition to these, Foreign Ministers of South Asia have met in New Delhi, 1983; and in Male 1984. Preparations are also being made to organize a summit meeting in Dhaka towards the end of the year 1985.

The Delhi meeting of Foreign Ministers laid down the following objectives of the South Asian Regional Cooperation (SARC):

(i) to promote the welfare of the people of South Asia and to improve their quality of life;
(ii) to accelerate economic growth, social progress and cultural development in the region and to provide all individuals the opportunity to live in dignity and to realise their full potential;
(iii) to promote and strengthen collective self reliance among countries of South Asia;
(iv) to contribute to mutual trust, understanding and appreciation of one another's problems;
(v) to promote active collaboration and mortal assistance in the economic, social, cultural, technical and scientific fields;
(vi) to strengthen cooperation with other developing countries.
(vii) to strengthen cooperation among themselves in international forums on matters of common interest; and
(viii) to cooperate with international and regional organizations with similar aims and purposes.

There is a clear understanding among the countries in South Asia that cooperation under the SARC will be based on the principles of sovereignty, territorial integrity, political independence and non-interference in internal affairs of each other. It is also agreed that these efforts will complement other forms of bilateral and multilateral cooperation. Also, with a view to achieving harmonious functioning of the SARC, it is understood that bilateral and contentious issues will be avoided.

Keeping these general principles in view, a few areas have been chosen for cooperation. They are: agriculture, rural development, telecommunications, meteorology, heath and population, transport, postal services and scientific and technical cooperation.

For South Asia where regional cooperation was non-existent, it is indeed welcome that during the last 4-5 years some efforts have been made. There is no doubt that different countries of South Asia do not have the same approach towards regional cooperation; their enthusiasm for cooperation is also not the same. There are, in fact, serious political and security problems among them and these problems have primacy over other issues. Whatever has been done no doubt constitutes small beginning but perhaps an irreversible movement has begun which in due course may develop into more mutually beneficial activity. But for this, greater political will is needed which, unfortunately has not been forthcoming in adequate degree. An encouraging feature in South Asia is that several segments of public opinion are supporting the idea. A Pakistani, for instance, pleaded:

> The South Asians will not be able to carry on with their individualistic and egotistic policies much longer. The sooner they accept one another's importance and instead of constantly trying to indirect one another start cooperating among themselves, the easier it would be for them to face the future which has been pronounced as bleak by all the wise men of the world.[8]

The best insurance against insecurity of South Asian countries individually as well collectively is to enter into meaningful cooperative endeavour at the bilateral, regional and international levels which are interlinked. The cooperation can be in multiple areas but the key word must be self-reliance and good of the people in general. Keeping long term perspective in view, countries in our region have to keep the external forces out and work for organizing their domestic societies on the basis of social justice.

This is easier said than done but the lessons of history do not provide for any escape routes to our rulers. They must generate necessary political will and give a push to efforts at cooperation and development which alone can ensure security. Borrowed strength, whether it is of arms or of any other kind, is not going to bail them out.

REFERENCES

1. Here the author draws upon his earlier publication "Framework of Security for Asia," *Pacific Community* (Tokyo), Vol. 7(1976), pp. 506-18.
2. Rabindranath Tagore, *Creative Unity* (New York: 1922), p. 122.
3. Robert S. McNamara, *The Essence of Security* (London: 1968), p. 149. Italics added. It is ironical that in practice McNamara's own country ignores this approach.
4. For a fuller description of the resources, see M.L. Qureshi, *Survey of Economy, Resources and Prospects of South* (Colombo: 1981).
5. M.J. Akbar, *India: The Siege Within* (Penguin Books, 1985) p.99.
6. A case Study of Pakistan is found in K.P. Misra, "Intra-State Imperialism: The case Study of Pakistan," *Journal of Peace Research,*(Oslo), 1, (1972), pp. 27-39.

7. S.U. Kodikara, "Regional Roles and Behaviour in South Asia: A Theoretical Framework of Regional Cooperation," (Unpublished) pp. 9-10.
8. Ibnul Hasan, "Regional Approach Inescapable in South Asia, "*The Muslim* (Karachi) (27 September 1982), p. 6.

XVIII The Indian Subcontinent: Contemporary Structure of Power and the Development of Power Relations

ASHOK KAPUR

The history of war and diplomacy in South Asia (excluding Afghanistan) since 1947 reveals an incremental development of an Indocentric power structure. (A "power structure" implies the existence of continuity and regularity in relations among states.[1]) South Asian international relations have passed through several stages of development— from conditions of near anarchy in 1947 to orderliness in regional diplomatic and military affairs. This essay outlines the pattern of development (from "near anarchy" to "order" and from "insecurity" to "relative security") from the late 1940s to the 1980s and provides a sketch of the "international relations" of "India" since the days of the East India Company. Each stage in the subcontinent's diplomatic and military history has distinct characteristics: (1) key players; (2) salient objectives; (3) select methods of interstate action; (4) particular opportunities for diplomatic and/or military action; and (5) particular set(s) of mental (psychological), domestic, and external compulsions that motivate leadership activities. By utilizing the historical approach we can build an assessment of the pattern of development (the line of change) of the region's power relations/power structure as well as its diplomatic, military, cultural, and economic boundaries (limits and constraints). No attempt is made here to speculate about the future of the regional security system.

The subcontinent's structure in 1988 rests on an interplay among three types of South Asian states. My contention is that subcontinental power relations are not simply the result of the "Indo-Pakistan polarity" as is widely and mistakenly assumed in Western academic studies and policy analyses.[2]

Reprinted from *Asian Survey*, Vol. 18, No. 7, 1988, pp. 693-710, by permission of the Regents of the University of California.

The attitudes and policies of the three party cluster of South Asian states (India, Pakistan and Sri Lanka, Nepal, Bangladesh, Bhutan and Maldives) on regional affairs depend on the internal political developments in the states, their economic and military strength, their relations with India, and the cultural sensitivity to, and the political utility of Indian and Pakistani conceptions about regional security for the elites of the other South Asian states. Only by examining the diplomatic and military history of the subcontinent in terms of attitudes, policies, and interactions among the three types of the state actors can we begin to understand the nature, scope, and direction of the subcontinental power structure and the regional security system and, incidentally, to assess the validity of the conventional scholarship.

The emergence of an Indocentric power structure in the subcontinent is an important development for several reasons. First, it took shape in the short span of 40 years. This is a significant achievement because the geographical and strategic unity of the Indian subcontinent was broken after the retreat of British power in 1947, with the new South Asian states lacking practical experience in diplomatic and military affairs.

Second, the present architect of regional order, India, lost time in organizing subcontinental power relations because India's mentor, Nehru, offered poor leadership in regional diplomatic and military affairs from 1947 to 1964. Still, despite Nehru's arrogance and indifference to the region, the intellectual and policy burden of the past which Nehru's successors inherited, and the growth of internal and external compulsions during the Nehru years.[3], post Nehru Indian political and bureaucratic elites formed conceptions, political will, and methods of action that incrementally succeeded in organizing regional power relations after 1971.

Third, the regional power structure was conceived and formed in the framework of India's isolation in Asia and in the context of sets of "positive" and "negative" external alignments. These alignments, which crystallized in the mid-1950s and were active throughout the 1947-88 period, were mostly negative from India's point of view. They motivated India to organize an Indocentric power structure as a substitute for Nehruistic globalism. That is, the power structure changed because of and despite the presence of hostile foreign policy alignments. As a result of the incremental growth of the Indocentric power structure in the region, the impact of the negative foreign alignments presently stands diminished. But the negative alignments have been slow to change in response to the India-orchestrated power structure.

And last but not least, the Indocentric power structure faces a problem of external (within Pakistan and other South Asian states) consent. On the one hand a pattern of asymmetrical distribution of economic and military power (capability, strength) has emerged within the region. This is the basis of interstate bilateral negotiations— e.e., between India and Pakistan and between India and Sri Lanka. But on the other hand, because of the mental, domestic, and external compulsions of South Asian states, formal public interstate agreements are not easily gained. The negotiation "framework" and "process" is manifest at two levels of interstate action—bilateral (India-Pakistan, India-Sri Lanka) and multilateral (the South Asian Association of Regional Cooperation or SAARC). The negotiation "outcomes" (agreements/disagreements) are governed by considerations of power and prestige of the state actors/leaders; the disutility of general war in the region or between hostile pairs of states (especially India-Pakistan); the fear of an uncontrolled arms race that would divert valuable resources from the economic, social, and political development of South Asian countries; and competing conceptions of "regional security" that correspond to the threefold classification (India, Pakistan, and the other South Asian states) in the subcontinental system.

Considerable progress is to be found in the emergence of a negotiation framework, revealed by the asymmetrical development of military and economic power in the region and by the development of salient principles and precedents in South Asian international relations. These may be summarized as follows:

A big state (India) shall not attack smaller neighbours except in self-defence (e.g. 1965 war).

In a situation of social and/or military conflict, the negotiation process must not be lost—i.e. "keep talking especially in a crisis" (e.g. 1971 Bangladesh war and the Sri Lankan crisis from 1983 to the present).

Help minorities in neighbouring states when the minorities are mistreated (East Pakistan Bengalis in 1971 and Tamils in Sri Lanka), and when military action against the minorities creates political and military threats for a neighbour (India in 1971 and 1983 on).

Avoid an unstable military balance or a two-party military symmetry and political/psychological bipolarity between a bigger state (in population, territory, economy, and military fronts/threats compared to its regional neighbours) and a smaller but territorially and diplomatically dissatisfied neighbours). That is the practitioners in India seek to shape the negotiating framework by a policy of continual development of a stable asymmetrical military imbalance. In the Indo-Pakistan context this means that there must be at all times a big gap in military armament and in nuclear capability if the negotiation process is to be guided toward a search for results that rests on the traditional principles of diplomacy—i.e., reciprocity, compensation, compromise, and intervention.

Extend Indian political and military authority in the border areas (Ladakh,

Sikkim, Arunachal Pradesh, Andamans) and secure frontiers in select cases (Sri Lanka) by military and diplomatic means— i.e,, rely on good fences rather than buffers in dealing with neighbours in boundary areas.[4]

My contention is that the development of a negotiation framework, and its recognition by smaller states in the subcontinent and by the great powers and multilateral fora,[5] indicate that "system boundaries" have emerged. Two in particular deserve mention. First, the geographical scope of the regional security system concerns the diplomatic and military activities of the seven SAARC states.[6] This geographical delimitation applies to policy issues connected with conventional military armament and economic and diplomatic-political relations among the seven states. Excluded from the negotiation framework is the nuclear issue, inasmuch as missile threats to the subcontinent emanate from the presence of nuclear forces that are deployed by the USSR, the U.S., and China in the Asian mainland-Indian Ocean area.[7] Burma and Tibet are outside the regional security system. Tibetan security is connected with the Chinese security machinery, and Burma was disconnected from the Indian subcontinent by the British India government in the 1930s and has remained so. Afghanistan has been connected with the Soviet security system. Political and military developments in Afghanistan impact on Pakistani security, on U.S.-Pakistan relations, and on Indian thinking about Soviet attitudes and policies concerning the region. However, Afghan developments have not altered the motivation for, and the line of development of an Indocentric regional security system that was set in motion before 1979. Thus, in my calculation, Afghan developments since 1979 have not been a catalyst in the formation of the subcontinental security regime, particularly since the order of battle of the Pakistan defence forces before and after 1979 reveals a continuous preoccupation with India, and not Afghanistan. In this sense the geographical boundary of the subcontinental security system is still the Durand Line— Pakistan's border with Afghanistan.

Secondly, particular normative principles govern subcontinental power relations. They represent the diplomatic reference points and red lines or boundaries of the regional system. In addition to the principles and precedents mentioned above, post-1971 subcontinental international relations utilize the principle of bilateralism to address contentious conventional military and diplomatic issues and regional multilateralism to address functional issues like terrorism and the drug traffic. These principles have important implications. They exclude great power — i.e., extraregional— intervention in regional affairs. They also exclude intraregional third party interference in bilateral military-political affairs. They do not exclude war as the ultimate method to settle controversy, but these principles nonetheless seek to limit to the key bilateral contenders the geographical scope and the decision-making process in war and peace issues.

As such, bilateralism is the opposite of internationalism, as well as of regionalism. In the latter two approaches the decision-making process is meant to be broadened to include extraregional great powers rather than being delimited to those states whose interests are directly involved. Bilateralism is a negation of the superpower doctrine that the superpowers have a right to be involved in regional disputes because they know no frontiers and have worldwide interests. The message of bilateralism is that the superpowers should mind their own business and discharge their international responsibilities in the one area where their policies and attitudes are the most threatening to world peace—i.e., in the area of nuclear weapons and nuclear disarmament. Implicit here are the notions that superpowers tend to view regional security affairs in East-West terms, and that their globalist attitudes are not directly relevant to regional disputes, consequently, wrongheaded superpower intervention tends to confuse the regional issues, to increase the intransigence of one or the other regional state, to delay conflict resolution, and to freeze local conflicts. Bilateralism, therefore, is an approach that seeks to reduce and, whenever possible, eliminate the harmful influences of superpowers in regional affairs. In this way, bilateralism is meant to choke foreign interference by cutting off the external supply of oxygen in the form of foreign support for the locally dissatisfied state. Bilateralism is constructive because it is meant to induce bilateral negotiations between public enemies. This is done by addressing causes of conflict and only by managing their symptoms. Here the negotiation framework is not geared to crisis management or peacekeeping as these are understood in Western international relations theory and practice. Instead, the aim is ambitious, to move peoples and governments of hostile pairs of countries in the direction of conflict resolution based on realities as well as reciprocal compromise.

Bilateralism is the opposite of Western and Soviet attitudes and policies on intervention. These justify intervention on grounds of military and economic necessity (for US/USSR interests), efficiency (it works), and ethics (it promotes desirable values—e.g, world and regional order and humanitarianism). Bilateralism is an opposite to the extent that it seeks to put brakes on superpower diplomatic and/or military intervention in regional affairs by raising (militarily) the costs of such intervention and by inducing (diplomatically) superpower respect for territorial boundaries. It is different in yet another sense if we compare the objectives and consequences of intervention. In theory and in practice the Indocentric conception of regional security does not exclude military and diplomatic intervention, as when Indian military aircraft violated Sri Lankan territorial sovereignty by dropping air supplies over Jaffna (1987), and when Indian covert and overt military support was given to Bangladesh "freedom fighters" (1971). But here the superficial similarity ends. The typical objective of superpower intervention in Third World conflicts in the post-1945 era is either to try to brutally suppress a Third

World state (U.S. in Vietnam, USSR in Afghanistan), to change a regime (U.S. in Chile), or to maintain a situation of manageable and continuous regional instability by policies that favour conflict management and avoid conflict resolution (U.S. and USSR in the Israeli/Arab dispute, the Iran/Iraq war, Southern African conflicts, and South Asia up to 1965).

The novelty of India-induced bilateralism and diplomatic-military intervention as it has evolved since 1965, lies in the search for connections in the attitudes and policies of policy communities of hostile pairs of states (India and Pakistan, India and Sri Lanka), between a "quick war" (1965, 1971), when war becomes necessary, and a quick search for normalcy. This is done by means of extended bilateral negotiations to settle, and not simply manage crises. In this sense bilateralism is ambitious and goes beyond the stated aims of superpower—orchestrated crisis management, which is to prevent conflict from going beyond a particular level of discomfort for the superpowers. Bilateralism is tedious because it endeavours to tap into the cultural or psychological underpinnings of bilateral conflicts so as to understand and isolate cause and find cultural and diplomatic solutions of lasting value. Bilateralism seeks, by a process of extended diplomatic (government-to-government, societal (people-to-people) and academic (nongovernmental expert in one country to counterparts on the "other side") dialogue, to change the mental outlooks and the political atmospheres of countries and peoples locked into a hostile mode. Inasmuch as India and Pakistan, other South Asian states, and the great powers recognize bilateralism as a desirable norm in subcontinental international relations—as a reference point to guide complex interstate negotiations and as a constraint against those who would seek military or international solutions to bilateral and regional issues—this norm must be viewed as the cultural and the diplomatic boundary of the region.

Stages of Development of Power Relations

The emergence and importance of these system boundaries, principles, and precedents in the subcontinental system of states can be studied by comparing different states/periods in the area's diplomatic and military history. I sketch below the broad outlines of the comparative picture. My contention is that subcontinental international relations have moved (a) from feudalism to order/managed disorder (1600s-1947s); (b) from near-anarchy to order (1947-1971); and (c) from insecurity to security (1971- 1987).

The First Stage, 1900-1947: Organizing Power Relations Through Controlled Political Disorder. This period is the immediate context in which post-1947 subcontinental international relations evolved. British India, not Mughal "India" or ancient "India," was the crucible in which modern subcontinental power relations were generated. The story begins with an assessment of the imperatives, methods, and objectives of the East India Company and the British India government. In the 1600s there was no political entity that could be called India. Borders, as we understand term today, were no man's land; "Indian" kingdoms (e.g., Asoka, the Mauryas) reflected the limits of elephant and horse power and the personality of the rulers. Under the Mughals the sword of Islam went to territories unknown to the invaders, spreading first to the lush Gangetic plains and then south wards. The pattern of "government"— compared to the meaning of the term today— was to conquer the land and install an overlord; control by the king or the emperor henceforth depended on his imperial lieutenant's personal loyalty. There *was* no conception of centralized or decentralized systems of government or power politics. There was a conception of revenue administration, which simply meant that the overlord extracted revenue from individuals. The notion of a continuous and reciprocal relationship between ruler and ruled did not exist.

This was the setting of disorder when the Europeans stepped onto the subcontinent. Portuguese, Dutch, French, and English came from the sea, unlike the Greeks, Huns, and Mughals. Europeans at the time did not have a concept of India. To them it was a region where one sought to recognize the immediate princely neighbour, assess his strength and weakness, and acquire an understanding of his external and domestic enmities. The East India Company acted on the basis of its commercial interest and well defined strategic conceptions. It possessed, from the beginning of its contact with "India" an awareness of British sea power and its importance for the Company's activities. It recognized the importance of ports and forts as commercial and political bases for purposes of political and military aggrandizement against neighbours. The method was to create political disorder by nursing neighbourly enmities, and to join the weaker side. The strategic aim of the East India Company was not to create an all-India political order. Rather, it followed a concrete set of political-military principles. First, it continually sought East India Company involvement in princely affairs with a view to setting up a prey. Aggrandizement was the dominant motive. Second, except as was expedient, the Company did not treat the princes as sovereigns, and it did not give them peace. Instead the Company sought to develop a variety of rights—to collect revenue, to maintain law and order— which created opportunities for direct interference or intervention. Third, the Company always had relationships with some of the feudal princes, but these were usually temporary. Occasionally princes became allies (e.g. the Nizam of Hyderabad). The maxim that in power politics there are no permanent friends or permanent enemies, but there are always permanent interests is relevant here.

In sum, the attitude of the East India Company was that it should continually expand its power and its power relationships with the princes. In the international relations of the Company in "India," power relations and political order were two different concepts. A belief in the first did not necessarily imply a belief in the second. Rather, the linkage was between power relations and political disorder.

After the Mutiny of 1857 the British government took over the Indian possessions of the scandal-ridden East India Company and set out to create two kinds of power relations in the area. The first consisted of bringing the Company's territorial conquests under direct British relations with over 500 Indian princely states. The princely states were managed by the British Crown according to the strategic principles of the East India Company. This princely order/managed disorder system continued until Sardar V. Patel, independent India's first home minister, unified the princely states into the Indian Union in 1947 through administrative action.

The Second State, 1920s-1940s: Origins and the Early Framework of Indian Diplomatic Thinking. World War I was a watershed in modern Indian history, politics, and external relations. India's political master, England, faced Germany, a landpower. In spite of its naval strength and strategic calculations, England lacked military strength to challenge Germany on the ground. A debate followed between the England- based Secretary of State for India, and the Indian Viceroy and Commander-in-Chief of the Indian Army. The former wanted Indian troops to fight Germany, the latter resisted. The viceroy's concerns were that Indian soldiers lacked political consciousness, they had been used mainly to fight internal wars and external wars against the brown people (in Afghanistan, Tibet, China). If they were used against German whites, they could easily turn against the English whites. The viceroy was overruled and Indian troops were used in the war. Having defeated white soldiers, Indian troops developed a political consciousness. This experience marked the beginning of "Indian" nationalism, before Gandhi and Nehru shaped the Indian political scene.

The Bolshevik Revolution was the second major event that affected the growth of Indian political consciousness. Mahatma Gandhi and Nehru enlarged this consciousness but they did not create it. The exposure of Indians to Bolshevik Russia reinforced the idea of revolt against British rule. The growth of British racialism against Indians after World War I and the hatred of Indian intellectuals for British racism worked in favour of the Bolsheviks.

Thus, anticolonialism became a focal point in Indian thinking during the period between the wars as a result of the effects of World War I, British racialism, and the Bolshevik Revolution. Indian Congress Party leaders embraced the popular resentment against racialism; by giving it an anticolonial focus, they used it to enlarge Indian public consciousness. This buildup rested on two sets of ideas and external circumstances. First, the Bolshevik Revolution was projected by Moscow, and accepted by Indian intellectuals, as a genuine sympathy between two oppressed societies. The

1971 Revolution was projected as the model by which to bring down a rich and powerful class. The ability of Indian soldiers in the first World War showed Indians that white power could be defeated and the Bolshevik model reinforced the necessity for revolt. M.N. Roy and Nehru played an important role in building this uncritical but psychologically comforting and politically relevant link between Indians and Russians. Second, Indian anticolonialism was turned against the British Conservatives because colonialism was associated with the Conservatives and anticolonialism with the British Labour Party and the Fabians. Hence, Indian intellectuals intuitively turned against the British Conservatives and toward Moscow and the British left. This reaction reflected emotion, not assessment. This was the bad effect of anticolonial thinking in India at the time, one that has endured to this day.

The interplay between the three factors (WWI, racialism, and revolution) created a bias in mainstream Indian intellectual opinion against British racialism and colonialism, and in favour of Soviet socialism. The negative (anti-British and anti-West) and the positive (pro-USSR and pronationalism) perceptions provided mutual reinforcement. This set of biases had its immediate policy use; it aided the Congress Party drive for Indian independence. But it also hindered long-term theory development and policy analysis and development on the Indian side. It shielded Indian intellectuals from a critical examination of the nature of Russian (pre-1917) expansion and the geostrategic principles that pointed to a continuity between Russian and Soviet expansion from the Urals to the Pacific and into Central Asia. It also blinded Indian foreign policy elites, especially Nehru, from assessing and publicizing the British Indian principles of strategy and their continuing relevance to independent India, i.e., the need to watch neighbours and to organize their power relations. In hind-sight, Indian thinking at the time was governed by utopianism (false expectations) about the Bolshevik Revolution and the Stalinist USSR) and high ideas (on racial and colonial issues). Real politic considerations of military affairs, especially about subcontinental regional security issues, were downgraded or neglected. The disorientation in Indian strategic thinking reflected the outlook of the Congress Party elites, particularly Nehru who was the party's main spokesman on foreign policy from the 1920s up to 1947. The mental compulsions were self-inflicted. This disorientation in Indian thinking had crystallized before 1947, and the Nehru years should be assessed in this context.

This period overlaps with the first one, but it merits separate treatment because of the second military activities in the Nehru era reflected the orientation of the second stage. In this sense the origins of post -1947 Indian foreign policy and regional power relations lie in developments and attitudes that were formed earlier.

The Third Stage, 1947-1964: From Near-Anarchy to India-Pakistan and

India-China Polarity and Cold War. The main regional players in this period were India and Pakistan. This period is important because a fundamental shift occurred from a condition of anarchy or near-anarchy in the context of the partition of British India in 1947 to a condition by the mid-1950s of loose bipolarity and cold war in Indo-Pakistani relations represented by competitiveness and hostility at diplomatic, cultural, military, and economic levels between the two political systems. But overall throughout this period, the relationship between the two states was one of neither total war nor peace, nor was there continuous war preparation. The Kashmir dispute was the most divisive issue between India and Pakistan, but it was not the central one for other members of the subcontinental system or even for East Prakistan. In other words, Indo-Pakistani polarity was not then the central determinant in the foreign affairs of most countries in the subcontinent. The period is important also because Nehru's reference of the Kashmir issue to the United Nations put it on the international agenda and opened the region to external diplomatic intervention; and the U.S-Pakistan security relationship opened the region to external military intervention, which was followed by India-USSR ties. However, as subsequent history demonstrates, the external interventions failed to establish an "intrusive system"[8] In this period there was no regional power, no bipolarized regional power, no bipolarized regional power structure, on permanent intrusive crisis manager, and no conception of an operative Indocentric (India-managed) regional power structure.[9] But there was an Indo-Pakistani cold war.

This period does not reveal an overarching Indian concept of regional security, nor did India possess the power to enforce regional peace. Nehru had multiple agenda. (1) globalism and utopianism—expressed by his concern with global disarmament, anticolonialism, and world peace, (2) national security by military means—rejected by Nehru; (3) realism by diplomatic means-revealed by Nehru's systematic effort to build a Soviet link from the 1920s and especially during the Korean War. (4) subcontinental regionalism (multilateralism) and geopolitics—neglected by Nehru, as an approach, although it is true that he did not attack neighbours and he sought bilateral negotiations on contentious issues; (5) Asianism—revealed in Nehru's vision of united Asia. These central elements in Nehru's mental outlook dominated Indian diplomacy because India's foreign policy establishment at the time was small and Nehru practically *was* the Foreign Office. India did not produce regional power structure because the regional approach was at best peripheral in Nehru's thinking, and he did not even have defined conception of Indian power. Nehru had a conception of India as a model of peaceful and democratic development and a conception of global economic and social security through peaceful change. Nehru was an Asian but not truly a South Asian in his mental outlook.

The subcontinental implications of this stage may be sketched in the following terms. Up to the late 1930s, the British Government of India had a clear policy of spheres of influence, buffers, "political order through disorder vis-a-vis princely states" in India. It took independent decisions that reflected the primacy of strategic principles and regional geopolitics. At the time, the regional presence of the Indian viceroys extended to Burma and the Gulf. Russia was the perceived threat and the world debate between Russian and British power continued to guide British diplomacy and strategy in the subcontinent. The Government of India and China had a vague relationship in the form of British interest in forming and maintaining a buffer between India and China in the Tibet/Central Asia region. The British Indian government avoided seeking an explicit India-China border; by 1954 in Nehru's India, the border and the border dispute had become explicit. The British Government of India had a regional Indocentric strategic conception and policy, whereas independent India under Nehru acquired an attitude and a vision of Asianness, which was formed in the preindependence days and lasted until 1962.

Nehru's all-Asia vision did not include an emphasis on managing Indian or regional security. Nehru patronized small South-Asian and South-east-Asian States. Before 1963 these states had urged regional cooperation. Their ideas then were (1) to develop a regional community so that ethnic conflicts could be absorbed by the larger community; (2) to help each other; and (3) above all to develop an economic Bandung, a Sri Lankan idea. However, the brief of Indian officials then was to break up the idea of an economic Bandung and press instead for bilateral links. The small states disliked this as it was likely to hurt their interests. As recently as 1982, India's approach was to sabotage regional cooperation but to do it cleverly— i.e. to appear to be cooperating and to blame Pakistan for problems. This approach of the Nehru days showed a lack of Indian vision about regionalism at the subcontinental and Southeast Asian levels, in contrast with Nehru's vision of all Asian unity. Still the events of this period—the conflicts between India an Pakistan, and India and China— revealed the shift from near-anarchy to polarity and cold war within the region (India-Pakistan) and extraregional polarity and cold war—except for the brief fighting in October 1962—between India and China (1954 and on).

The Fourth Stage. 1964-1984: The Indira Gandhi Years[10]. This period saw a major change in India's approach to foreign and military affairs, and this affected subcontinental power politics. Several circumstances contributed to the change. First, Nehru's vision of Asian unity was found to be irrelevant to India's foreign policy after 1962. Nehru's globalism—stressing peace diplomacy rather than military security—was in shambles. In the subcontinent, hostile foreign policy alignments had crystallized: the U.S. Pakistan and Pakistan-China combination were aligned against the India-USSR combination, and the leaders of other South Asian states were distrustful of Indian intentions even though India under Nehru had not attacked its neighbours and had not meddled in their politics in a big way. That is the external setting had changed and India's external compulsions had formed by the end of the Nehru era and the beginning of the Indira Gandhi era.

Second, the domestic compulsions of the Nehru and the Indira Gandhi eras also differed. Nehru had betrayed the massive trust of the Indian people. They had believed in his speeches and had repeatedly noted for him. In return he gave them a China problem, a loss of self-confidence and prestige, and no real solution to the problem of poverty. Indira Gandhi was aware of this, and she recognized the need to restore public trust in the Nehru family. She did so by engaging in high political drama at home (via socialist talk, antipoverty electoral slogans, bank nationalization and termination of princely privileges) and military-diplomatic action on the regional and world state—e.g., the 1971 war.

Third, the mental outlooks differed, Indira Gandhi lacked Nehru's vision or grand strategy. Her personal compulsions were to shine on the stage, to divide her domestic and international enmities (divide and rule), and to take advantage of domestic and external opportunities on an ad hoc basis. Her approach was circumscribed compared to Nehruistic global and Asian vision, but it was more purposeful in one sense and destructive in another. In the policy arena, she allowed her bureaucracy—which was responding to Indian external and domestic pressures and circumstances—to develop state intervention to gain economic security (as to food self-sufficiency) and military security. India's move in this direction was not the result of Indira Gandhi's design. However, it happened in her time because of the revised orientation of India's nationalistic bureaucracy and as a reaction to Nehru's wrong military and diplomatic policies. At the same time, there was recognition of the utility of his broad approach that advocated a mixed economy and diplomatic bridge-building vis-a-vis the great powers irrespective of their ideological nature. Yet the Indira Gandhi era was also destructive. She damaged Indian political institutions, as Indian cabinets and parliaments became talking machines and paper factories that had no impact on the prime minister-bureaucracy- business coalition that actually possessed the economic and electoral power. In this sense, the driving force of Indira Gandhi's approach to domestic and foreign/military affairs was Indira-centric rather than regional or global.

Nonetheless, the first steps to lay the foundations of an Indocentric regional power structure took place in her time, though more by accident and circumstance than by design. External opportunities, or rather mistakes by her foreign enemies, helped her fashion them to her own and India's advantage. The Bangladesh diplomatic-military operation in 1971 provided a line of action that had implications for the development of an Indocentric regional power structure as well as for Indian relations with the great power. Out of the

1971 diplomatic action came the principle of bilateralism (e.g. the Simla Agreement). Out of the 1971 military action and India's agricultural and industrial performance also came the recognition of India as a middle as well as a regional power. While Nehru had a vision of India as a great power, Indira Gandhi by the early 1980s had acquired a vision of India as middle power and a view that middle powers such as France, Canada, and Australia could temper the international environment and the superpowers view of their role and importance. This approach derived from a study of U.S. political- military ties with Pakistani military governments that facilitated American intervention in subcontinental affairs, and Soviet intervention in the Tashkent Agreement. It created a belief that the superpowers were harmful influences in subcontinental security affairs and needed to be curbed.

The Fifth Stage, 1984- Present: Rajiv Gandhi's Subcontinent. This stage may be assessed from two different angles. First, Indira Gandhi's legacy to her son, to India's foreign policy machinery, India's political system, and to South Asian states was bewildering. She made Indian politics, Indian diplomacy, and Indian security decision making in crisis situations into an Indira-centric enterprise. She damaged domestic political institutions; the Indian National Congress Party, which had been the nursery for developing Indian political leaders, became a breeding ground for mediocrity, sycophancy, and political corruption. She injured the incentive and the capacity of the Indian Foreign Ministry to anticipate and to manage external crises. In regional affairs she was snobbish and interventionist toward India's smaller neighbours. Unlike Nehru, who neglected them, she interfered in their domestic political affairs. Thus, a part (but not necessarily the most important part) of the Sri Lankan crisis was of Indira Gandhi's making. The training of Tamils for military action was her idea, and this broke the norm of a big country not harassing a small country by sub rosa means.

The Indira Gandhi legacy was one of short term gains for her and long term problems for India and its domestic and foreign policy institutions. Rajiv Gandhi inherited the problems, but even his mother, had she lived, would have had a tough time coping with them as the evolution of the Punjab conflict since 1983 has demonstrated.

Comparing the compulsions of Indian decision makers in the Nehru, Indira Gandhi, and the Rajiv Gandhi eras, we find that in the Nehru days the subcontinental power relationships were unimportant and Delhi's ties with the great powers were of central importance. Secondly, the interface between domestic problems (threats to the political system and threats to territorial integrity) and external problems was not a direct one in the India-China border clash or when Indira Gandhi decided to intervene in the Bangladesh crisis (1971). This interface surfaced in the late 1970s and the early 1980s. It is revealed in big way in the Indian Pujnab/Pakistan and Tamil/Sri Lankan situations. As a result of the interface, the approach of the Indira and Rajiv Gandhi Government to regional foreign affairs became a manifestation of the approach to the internal political and cultural situations, and it revealed a fundamental shift in India's approach from globalism to regionalism in foreign affairs.

To an extent, the problem of Mrs. Gandhi's legacy to her son and to India was compounded by Rajiv Gandhi's lack of perspective and experience in dealing with pressing practical issues within India and in the region. Rajiv Gandhi talked about India in terms of the 21st century and high technology, and this served the well-being and ambition of Indian urban elites. But it failed to connect with issues of Indian poverty, communalism in state and national politics and in Sri Lanka feudalism and corruption in the Congress Party, and fundamentalism of different kinds that affects secularism and the fabric of Indian society. The issues here are not mainly about the imminence of a Sikh state; or the scale of terrorism; or the linkages between drug, liquor, and arms trading; or fundamentalist causes in the Aghanistan-Pakistan-India nexus. These are symptoms that can be managed. Rather it is the cultural side of fundamentalism that must be addressed by the development of grass-roots political participation. The ailments of democracy are best handled by more grass-roots democracy rather than by manipulative democracy from the centre. India is not a corporation to be managed by executive authority. Peoples and institutions with their respective cultures, histories, and psychologies have to be mobilized, cajoled, persuaded, or coerced by political action, and by sentiment and vision. The corporate model of efficiency and chain of command activity is unworkable in the large, diverse, individualistic, and egoistical society that is India. Here we find that the Rajiv government's approach is problematic; it fails to secure the consent of its constituents, and a similar problem lies in its approach to region.

Note here a fundamental contrast between the approach to politics and society by Mahatma Gandhi and to an extent by Jawaharlal Nehru on the one hand, and Indira Gandhi and Rajiv Gandhi on the other. Mahatma Gandhi, even more than Nehru, sought to build people, build leaders with grass-roots experience who had risen through the ranks by constantly rubbing shoulders with the folks and who, by the time they reached centre stage, possessed valuable experience in dealing with practical issues. Under Indira and Rajiv Gandhi, no leaders of stature and practical experience have been allowed to gain independent political authority. Instead of the Gandhian emphasis on individual and social change, the legacy of the Nehrus has been to stress government intervention and overcentralization of power at the top of the political and policy process. Instead of developing a cadre of experienced political leaders, the Nehru's political culture has sought to develop state institutions around the notion of dynastic rule, which invariably encourages sycophancy and careerism. During 40 years of Indian independence, generational change has brought in managers, not leaders. This approach not only has destroyed leaders at the centre, but also in Congress-dominated states. Today, state level leaders exist only in non-Congress governments, and they have emerged despite the Nehrus.

But there is a second perspective from which to assess the present stage. Despite the major defects in the thinking and approach of the Nehrus toward Indian politics and India's neighbours, and despite the domestic and external compulsions that have grown up against India in the region's diplomatic and military history, India's military strength and its diplomatic will to intervene— ableit on an ad hoc and clumsy basis — has also increased, it has thus far organized regional power relations. Because media headlines revel only the difficulties, it is prudent to indicate that the benchmark of India's success is its ability to survive domestic, regional, and international strains on a prolonged basis and still be able to establish a forceful presence by a mix of diplomatic and military action. By present signs, India's military and administrative presence in Sri Lanka is likely to be of long duration, as the history of the psychological gulf between Sri Lankan Tamils and the Sinhalese population goes back to 1949, and the logic of the country's domestic politics requires a strong foreign third party to pacify and democratize the island state. India's air, naval, and military development and Indian diplomacy reveal a southward and a power projection orientation. With Pakistan the situation is likely to be one of long term and noisy cultural conflict, a controlled conventional arms race, peace offensives, and ambiguous nuclear diplomacy. There are no signs that India is willing to provide Pakistan with the concessions it seeks (i.e., on Kashmir or a South Asia nuclear weapon-free zone), and there is no middle ground at present on which to develop formal, public reciprocal obligations other than tacit and historical rules of the game. In other words, unless India breaks up, the utility of the Indocentric conception of regional security management (but not leadership with consent) has gained ground.

The issue of consent is pertinent vis-a-vis the small South Asian states and particularly with Pakistan. At present, Indo- Pakistani differences are at a manageable level even as the mistrust is profound. As long as Pakistan fails to outrace India in conventional armament and in nuclear weapons capability, and it fails to achieve Khalistan (a Sikh state), it will remain, in Indian thinking, at best a dangerous nuisance and not the second equal pole of power and attraction in the subcontinent.

The future of Indo-Pakistani power relations and the issue of consent depend in part on the evolution of Pakistani thinking. Pakistanis have to make up their minds about the geographical direction of their destiny. Assuming that the past glory of the Mughals in the subcontinent cannot be restored and that any such ambitions only excite Hindu nationalism, does the Pakistani elite and public want Pakistan to be left alone by the Indians as a Muslim state whose future lies with West Asia? If Pakistan sees itself as a moderate and modern Muslim state in the Arab/Persian worlds, it makes sense to promote the notion that Pakistan must be a strong state whose future orientation must be to attend to its cultural, economic and strategic interests westward. Or do Pakistanis see themselves as South Asians where the pressing need is to form an Indo-Pakistani rapprochement through the development of reciprocal obligations at different levels of bilateral discourse—economic, cultural, military— and not simply through the old pattern of competitive arms racing and cultural insularity? At present, thinking about such questions is underway in both countries but if the later route is taken, the task entails cultural reengineering on a bilateral basis and it is early to expect a breakthrough.

Summing up and Research Implications

This essay points to an evolution of power relations in the subcontinent from the days of the East India Company to the present. A historical approach has been employed to trace the shifts. The purpose of the paper is to shape the research agenda. No claim to finality is made, as the topic is conceptually and empirically vast and complex, and a lengthy study is required to examine the details and the implications. The bias here is to examine from an Indian perspective the evolution of the subcontinental power structure—the underlying approaches and settings, the stages of development, and the competitive conceptions. A thorough study, however, must evaluate the region's diplomatic and military history at two levels (1) actor-specific analysis that is country-specific and leadership motivation/perception-specific as per our list of attitudes/policies set out in the beginning of the article; and (2) regional crisis-specific outcomes and processes. The later perspective has been emphasized here, but the evolution of attitudes and policies of the region's seven states—a divided into three types of subcontinental actors—is of such importance as to deserve a detailed analysis as well.

REFERENCES

1. Buzan maintains that India and Pakistan are the primary actors in South Asia, and "internal changes......have not altered the basic pattern of relations. Neither have external ones, which have so far tended more to reinforce, rather than to undermine, the existing bipolarity."
2. B. Buzan and G. Rizvi, ed., *South Asian Insecurity and Great Powers* (London: Macmillan, 1986), p. 22.
3. For a preliminary analysis of the "burden of the past," see Ashok Kapur, "Indian Security and Defence Policies Under Indira Gandhi." *Journal of Asian and African Studies,* 22: 3 and 4 (July, October 1987), pp. 175-93.
4. Frontier refers to areas beyond a country's territorial boundary. The difference is sketched in A. Lamb, *Asian Frontiers* (New York: Praeger, 1968), pp. 4-9.

5. Thus, Pakistan (Bhutto and Zia governments) accepted the Simla Agreement (1972) as the basis of Indo-Pakistan normalization. The Indo-Sri Lanka Accord (1987) has been endorsed by SAARC, the U.N., and the great powers.
6. There are suggestions to include Afghanistan and Burma in SAARC. If this happens, our analysis will require reassessment.
7. The premise is that neither India nor Pakistan currently possesses or deploys nuclear forces in the region.
8. L.J. Cantori and S.L. Spiegel, in the *The International Politics of Regions* (Prentice-Hall, 1970), speak of an "intrusive system." This refers to the active participation of the great powers in the regional balance of power. My contention is that the U.S., China, and the USSR are great powers. They have interests in South Asia but, singly or collectively, they did not form or control the subcontinental power balance in the 1947-87 period.
9. Leo Rose asserts that the "basic principle underlying India's regional policy since 1947 has been its undeclared claim to hegemony in south Asia," in J.N. Rosenau, et. al., eds. *World Politics* (New York: Free Press, 1976), p. 214. My analysis indicates that Nehru had no coherent regional policy, and there is no evidence to validate the notion of an "undeclared claim to hegemony."
10. The Shastri era was brief but eventful, but is not discussed in this paper.

XIX Economic Interests of Big Powers in the Indian Ocean Region

S.S.BHATTACHARYA

Introduction

Since the mid-sixties the big powers have been accelerating their naval activities in and around the Indian Ocean.[1] The militarization of Indian Ocean has become the subject of intense international debate. Most of the literature on this topic deals with the military and strategic aspects of the big power presence in the Indian Ocean and hardly any attention is paid to their economic interests in the littoral countries. Economic calculations play an important role in the politics and defence strategy of the big powers. It was economic interests that lured the maritime powers of the West to enter the Indian Ocean which eventually led to the establishment of colonial rule in Asia and Africa. Though decolonization has led to an attenuation of these interests, the economic links of big powers with the littoral areas are still strong and these considerations play a major part in the Indian Ocean politics.

The Soviet Union and China are the new competitors for the economic exploitation of the Indian Ocean though unlike the Western maritime powers, they did not establish any colony in the region. Neither are their industries dependent on the raw materials of the countries of the Indian Ocean region. However, they are trying to develop economic and technical cooperation with countries around the Indian Ocean. They consider increased naval activity in the Indian Ocean by other big powers as a threat to their strategic interests in the region.

From the very beginning the European nations were interested in the Indian Ocean as a maritime passage to the resource rich countries of the East in which they developed trade and political interests. These interests seem to have been sustained over the centuries. Even today the big powers are critically dependent on the minerals of the Indian Ocean region, particularly oil.

This article is an attempt to make a quantitative assessment of the big powers' economic interests in the region. The data presented, however, may

Reprinted with permission of the Asstt. Director from the Journal of the Institute for Defence Studies and Analyses, New Delhi, Vol. 10, No. 3, 1978, pp. 261-287.

not be quite uptodate or fully adequate. The idea is not to attempt a precise statistical analysis in terms of the big powers' trade and investment stakes, but to have an overall picture of their economic interests in the region. That will give us an idea of what is really behind the so called strategic interests, which the big powers arrogate to themselves being the guardians of the present international system.

Raw Material Base The countries around the Indian Ocean possess raw materials which are vital for the growth and development of the major manufacturing industries of the big powers. Malaysia and Indonesia, for instance, are the leading producers of natural rubber in the world. In 1973, the Indian Ocean region, as a whole, produced 3,120,500 metric tons of natural rubber which is 89.41% of world production (Table 1).

TABLE 1 – *Natural Rubber Production (Thousand metric tons)*

Region	1970	1971	1972	1973
Burma	7.1	9.4	7.5	7.5
India	89.9	98.9	1209.1	123.2
Indonesia	815.2	819.3	818.7	885.8
Malaysia	1,269.2	1,318.5	1,324.9	1,556.4
Singapore	1.0	1.1	0.9	0.9
Sri Lanka	159.2	141.4	140.4	154.7
Thailand	287.2	316.3	336.9	382.0
Total	2,628.8	2,704.9	2,743.4	3,120.5
World	3,100	3,075	3,110	3490
World percentage	84.8	87.96	88.21	98.41

Source: U.N Statistical Year Book, 1974

During 1970-73 the average production of natural rubber in the Indian Ocean region was 87.41% of world production. About 75% of this was exported to the West European and American markets.

In the countries on the western fringe of the Indian Ocean, plantations had been developed by White settlers with Western Capital. More than half the cash crops in these countries are in the hands of foreign settlers. Though the Indian Ocean countries are major producers of items such as tea, coffee, sugar cane, cotton, jute, and wool, they export a major part of these to the developed countries. Most of the littoral countries are deficient in food. Only Australia, Thailand and Burma have a sizeable surplus of foodgrains for export; others depend mainly upon the U.S.A. India, for example, imported one billion dollars of foodgrains from America in 1976. This imbalance in agricultural production in the countries around the Indian Ocean is partly a legacy of their colonial past.

Southeast Asia, especially the equatorial belt, is an important timber producing centre of the World. Japan, France, West Germany and a number of other countries depend upon this region for supply of timber and logs. South Kalimantan in Indonesia exported timber worth $ 43.1 million to Japan in 1976.[2] Similarly, during 1976 timber exports from Maluku and Riau regions of Indonesia reached $ 24.8 million and $ 4.5 million, respectively. Indonesia's timber exports in 1976 totalled 18,649,000 cubic meters worth $ 78.1 million.[3] After Japan, other bulk importers of timber from Indonesia are France, West Germany and Italy.

In Malaysia, Sabah's log exports for the first nine months of 1976 were M$ 750 million, and Sarawak's timber exports for the first eight months of 1976 were M$209.1 million.[4] Major importers were Japan and the East European countries. Japan imports about 40% of its logs from Sabah. In 1976, Thailand's exports of wood and its products are expected to be 1,100 million baht.[5]

India's lumber exports in 1975 totalled about $ 16.1 million of which about 40% were destined for Japan. In 1977, India had imposed a total ban on export of wood materials in move to build its domestic plywood industry. This action is however, expected to deal a heavy blow to Japan's furniture industry which totally depends on imports of Indian rosewood,[6] which alone accounted for 90% of Indian wood exports to Japan.

Besides the agricultural products, Annexure-I shows that the countries around the Indian Ocean contain huge deposits of mineral wealth. In 1973, the Indian Ocean region produced about 70% of tin ore of the world. Malaysia is the largest tin producing country in the world. It produces about 40% of the world's total, followed by Indonesia, 16% and Thailand, 13%[7]. World's 79% of gold is produced in the Indian Ocean region. South Africa and Australia are the major gold producing countries. About 28% of world's manganese comes from the Indian Ocean area. Australia and South Africa contain 37% of world's uranium resources. In 1973 their reserves were 161,000 metric tons and 202,000 metric tons, respectively, when world's total deposits were placed at 861,8000 metric tons. The Indian Ocean region in 1973 also produced about 32% of world's chromium, 16% of iron ore, 12.47% of lead ore, 11.32% of tungsten, 11.02% of nickel ore, 10.29% of zinc and 30.26% of antimony.

An essential radioactive metal, thorium, is useful for nuclear reactor fuel and fuel cells such as those used in various NASA programmes. Almost no thorium is produced in the United States. About 60% of the thorium used by the United States is imported from Australia and the balance from Malaysia.[8] The United States depends on several Australian raw materials for its industrial development. Chief among these are zinc, tungsten, titanium, thorium, manganese, lead and iron ore. Australia now supplies some 65% of titanium which is vital for U.S. space industries, missile construction, supersonic aircraft and so on.[9] South Africa is the major supplier of strategic minerals such as corundum, lithium and amosite asbestos to the U.S.A.

Although the security of the United States depends in large part on the continued availability of more than thirty crucial minerals available only or primarily form foreign sources, it also depends upon shipping to transport these raw materials. According to a report on minerals published by the American Mining Congress in June 1973,[10] the United States is still a major producer of minerals and metals. But as its mining industry has not kept pace with the expansion of the economy it will have to depend upon imports of raw materials. Moreover, as the E.E.C. and Japan have concluded long-term contracts to procure the bulk of the growing production of minerals in African, Caribbean and Southeast Asian countries, the competition of the U.S.A. to get access to bulk quantities of raw materials form these regions may be quite stiff in future.

Oil: Oil is the most vital mineral in the Indian Ocean region which plays a crucial role in the development of modern industries. Crude oil production in the Middle East in 1975 was 1,104 million metric tons or 36.37% of the total world production.

Major industrial countries, U.S.A., Japan, U.K., France, West Germany and Italy, imported 37.6%, 75.0%, 70.8%, 78.3%, 68.0% and 81.6%, respectively, of their total crude imports from the Middle East in 1976,[11] (Annexure-II). Saudi Arabia is the single largest crude oil producer and exporter in the Middle East, Japan being the largest single importer of Saudi Arabian crude oil during January-November 1976.[12]

TABLE 2 – *World crude oil production (Thousand metric tons)*

Region	1973	1974	1975
World	2,850,806	872,604	2,702,140
Middle East	1,065,191	1,104,986	982,700
World percentage	37.36	38.47	36.37

Source: *Petroleum Economist* (London), January 1976, Vol. XLIII, NO. 1. p. 7.

During this period Japan imported 508.73 million barrels, amounting to 19.06% of Saudi Arabia's total production of 2 billion barrels. France ranked second with 12.33% followed by Italy 6.88%, Holland 5.29%, England 5.15%, the U.S.A. 4.68% and Spain 4.27%. By area, Western Europe was the largest importer of Saudi crude oil during that period accounting for 43.56% of the total, followed by Asia and Far East which imported 29.0%.

In 1974, Saudi Arabian crude oil reserves (excluding those in Saudi-Kuwaiti Neutral Zone) were estimated by Aramco at about 103.5 billion barrels. Proved and probable reserves together were estimated at 172.5 billion barrels. Saudi oil reserves are greater than those of any other country, greater than those of the U.S.A. and the Soviet Union combined and represent nearly one-fifth of the world reserves.[13]

Indonesia had outstripped Canada as the third largest oil exporter to the United States in the first quarter of 1976.[14] Indonesian share of the American oil imports which totalled about 4.9 million barrels daily was ten per cent, i.e., 497,000 barrels daily, while Canada exported to U.S. 424,000 barrels daily.[15] Known oil reserves in Indonesia in mid - 1976 amounted to between 10,000 and 15,000 million barrels.[16]

U.S. dependence on Arab oil rose sharply in the first six months of 1976. American imports of crude oil from Arab countries (about 2,100,000 barrels a day) represented in that period about 12.4% of U.S. consumption of petroleum products, against only 6% in the first half of 1975.[17]

The oil embargo exercised by the Arabs in 1973 showed how vulnerable the Western military system is to such embargoes. It had a serious impact on America's maritime activities. Steaming time for naval ships was reduced by 20% and military flying time by 18%. In order to save fuel the speed of ships returning to the United States was reduced; a 50 mile an hour speed limit was declared for 158,000 vehicles in the inventory; and a temperature of 65 to 68 degrees was set for military buildings.[18] The U.S. military, which consumes 670,000 barrels a day, had been buying nearly half its petroleum in foreign countries, most of it from the Arabs. The Navy consumes about 220,000 barrels of oil and jet fuel each day. For a major mobilisation, it is estimated the navy would have to nearly triple that amount, or about 600,00 barrels.

Though U.S. oil reserves total 39 billion barrels,[19] and during the next ten to fifteen years, there would be enough crude oil in the world to meet a three per cent annual increase in demand, the U.S. plans to pump 1 billion barrels of oil into underground salt domes by the end of 1982 at a cost of eight billion dollars[20]. This would give the United States the ability to tide over an oil embargo lasting at least six months. The energy experts have calculated that any new embargo would stop only 50% of the nation's flow of oil from abroad.[21] Still, unless the developed countries of the world discover some new sources of energy as a substitute for oil they will have to depend upon countries around the Indian Ocean for their oil supply.

The oil price increase had profound adverse impact on the economies of developed countries. During 1974 and 1975, the U.S. economy was hit by severe recession and high inflation. In May 1975, unemployment reached 10%, the highest in 30 years. Substantial personal and corperate tax cuts were introduced, despite Congressional opposition, to help stimulate the economy. Japan had to drastically revise its economic and social development programme (1970-71) which envisaged an average annual growth rate of 10.6%. Japanese GNP declined by 1.8% (in real terms) in 1974. In France, inflation increased from about 7% in 1973 to nearly 15% in 1974, while a trade surplus of 6,360 million francs in 1973 was succeeded by a deficit of over 16,000 million francs in 1974. There were a number of bankruptcies and unemployment rose to over half a million.

In Britain also the economic situation worsened in 1974 as prices rose by 19.1% and the deficit in the balance of payments rose to £ 3,668 million. Statutory wage controls had to be abandoned. Inflation reached 26.9%, unemployment was at its highest level since the second World War, industrial production fell to the level of the late 1960s with very low investment.[22]

The oil embargo brought about a change in the West's political attitude towards the Arabs in their confrontation against Israel. On 16 October 1973 the Arabs launched their oil war by cutting back production by 25% and announced that they would step up the cut by 5% every month. A total embargo was imposed on supplies to USA and Holland and a differentiated cut to various countries depending upon their relative support to the Arab cause. The nine Western European countries, excluding Holland, issued a political declaration early in November 1973 calling for the implementation of the Security Council Resolution-242 of 1967 which directed Israel to vacate Arab occupied territories. Japan also, in November 1973, extended categorical support to the Arab cause against Israel by supporting Resolution-242.

In fact the threat to the economic interests of Western countries caused by the oil embargo was so serious that the U.S. Vice-President, Gerald Ford, warned that he would retaliate by stopping shipments of food from America to Middle East countries.[23] The U.S. Secretary of Defence, James Schlesinger, threatened that the Arab nations would run the risk of having force used against them if they utilized the oil embargo to cripple the largest mass of the industrialized world.[24] The United States was reported to have drawn up a contingency plan to invade Saudi Arabia in the event of another West Asian war, involving oil embargo. In turn, Saudi Arabia threatened that it would blow up its oil wells if the United States tried to seize them. However, good sense prevailed and nothing happened.

Apart from the oil embargo, Western powers are afraid of any cut in oil production or any rise in oil prices. The 10% hike in oil prices in January 1977 had affected a 0.33% decline in the real growth of the European Community.[15] It is estimated that this increase would cost the United States 2.7 billion dollars and Japan 1.7 billion dollars in extra expenditure a year. The Japanese case clearly demonstrates the adverse chain reaction of oil price increases on the industrialized economies. According to the Japanese Economic Planning Agency, an oil price increase by 10% will reduce Japan's real economic growth rate by 0.6%. Besides, it will result in an increase of about 1% in Japan's wholesale prices. Moreover, the electricity-gobbling aluminium industry will be hit hardest by the price hike as it would add about 10,000 yen to the cost of producing every ton of aluminium. The Japanese steel industry now consumes some 11 million kiloliters of heavy oil a year worth nearly 350,000 million yen. Any increase in crude oil prices would deal a severe blow to the industry which has been forced to curtail production to deal with a slump in demand both at home and abroad.

II

Market for Big Powers' Industrial Surplus Most of the countries around the Indian Ocean are politically young. They are emerging from the colonial era and moving towards self-sufficiency based on planned economic growth and industrialization. The big powers have an eye on these countries as their industrial surpluses can find an easy market there. Though the countries around the Indian Ocean produce raw materials for industries in the developed countries, they are industrially underdeveloped and far behind the developed countries in the production of industrial goods. The poor production of energy in the countries around the Indian Ocean was due to lack of resources to produce electricity and exploit mineral wealth.

Due to industrial backwardness and lack of scientific knowhow the countries around the Indian Ocean import sophisticated machinery and manufactured goods from the developed countries at high prices and their trade balances generally show large deficits except in cases of oil exporting countries (Annexure-III). It may be mentioned here that U.S.A., France and the United Kingdom are trying to recycle-petro dollars by selling large quantities of arms to Arab countries. None of the oil producing countries in the Middle East has any significant capacity to manufacture modern weapons. The following sales, both cash and credit deliveries, during 1965-73, to Iran, Saudi Arabia, Iraq an Kuwait are set out in Table 3 which was included in testimony given to the U.S.House appropriations sub-committee on foreign operations and related agencies by Defence and Pentagon Department Officials in June 1974.

TABLE 3 – *U.S. arms sales (Thousand U.S. $)*

Country	1965	1966	1967	1968	1969	1970	1971	1972	1973
Iran	12,896	52,188	36,866	56,717	94,894	127,717	79,352	200,737	234,309
Saudi Arabia	5,839	9,200	38,854	23,356	5,586	4,937	28,774	24,354	91,030
Iraq	949	4,256	5,564	205	29	-	255	-	-
Kuwait	-	-	-	-	-	-	Less than $ 500	Less than $ 500	

Source: *Airospace Daily,* 16.9.1974.

American arms sales to Saudi Arabia which totalled $ 600 million between 1950-53 soared to $ 14.6 billion in just 20 months after October 1973.[26] On 18 February 1976, the U.S. agreed to sell Saudi Arabia $ 1.2 billion worth of military services and hardware. This deal brings the value of military equipment and services contracted for by Saudi Arabia from U.S. within a period of six months to a staggering total of $ 5 billion. According to U.S. sources, the $ 144 billion Saudi Five-Year development plan provides $24 billion for arms.

Similarly, U.S. arms sales to Iran in the 1972-76 period totalled $ 10.4 billion.[27] In 1976 two major 'oil for arms' deals were entered into by Iran. The oil money will be used to purchase upto 300 General Dynamics F-16 fighter aircraft, some *Spruance* class destroyers and possibly a limited number of Boeing reconnaissance aircraft. The number of U.S. official and private citizens in Iran, a large percentage of whom are involved in military programmes, has increased from approximately 15,000-16,000 in 1972 to 24,000 in 1976; and it could, according to U.S. officials, easily reach 50,000-60,000 or higher by 1980.[28]

Among other Western countries, France, U.K., Federal Republic of Germany and Italy also supply a substantial quantity of arms to the oil producing countries of the Middle East, as well as other countries around the Indian Ocean. In January 1974, France concluded a deal with Saudi Arabia for supplying Mirage fighter bombers, low level SAMs, air-to-air missiles, AMX-30 tanks other advanced military equipment in exchange for an undertaking by Saudi-Arabia to supply France with 800,000 barrels of petroleum crude a day for the next 20 years.

Similarly, West Germany has signed an 'arms for oil' deal by which it will supply Saudi Arabia with technological expertise in exchange for oil. In late 1976 Iraq concluded an agreement with France for delivery of 60 to 80 F-1 Mirage jet fighters. If the recycling of petro-dollars continues in this manner it will exacerbate the arms race in the Middle East and turn the region into a zone of permanent conflict.

Table 4 shows U.S.S.R.'s pattern of trade with the countries of the Indian Ocean region. Its major trading partners are Afghanistan, Bangladesh, India, Iraq and Iran. The U.S.S.R's trade deficit with these countries totalled 175.7 and 159.7 million roubles in 1972 and 1973, respectively. Indo-Soviet trade has been rising rapidly in recent years, from about $460 million in 1973 to about $830 million in 1975. Indian officials estimate it will pass the $900 million mark in 1976.[29] Though the U.S.S.R. does not import oil from the Middle East, she has supplied arms to Egypt, Iraq, South Yemen, Kuwait and other countries. From 1955 to 1973, the Arab States, particularly Syria and Libya, have received Soviet arms worth about $6 billion.[30] Between 1955 and 1967 Cairo received military aid worth $ 1.5 billion and the cost of the new equipment delivered from June 1967 to mid-1971 was estimated at $ 3 billion. During January 1975 to March 1976 Egypt received an estimated $1,500

TABLE 4 – *U.S.S.R. and its principal trading countries in Indian Ocean region (Million Roubles)*

Country	1972			1973		
	Imports	Exports	Balance	Imports	Exports	Balance
Afghanistan	30.8	38.1	+ 7.3	34.4	33.7	- 0.7
Bangladesh	7.8	8.8	+ 1.0,	9.8	43.5	+ 33.7
India	312.5	138.5	-174.0	366.0	222.8	-143.2
Iran	134.0	95.5	- 38.5	127.7	127.8	- 0.4
Iraq	61.6	90.1	+ 28.5	190.6	141.5	- 49.1

Source: The Europa Year Book 1976, vol. 1, p. 1106

million worth of Soviet arms. However, after the diplomatic debacle in 1974,

Egypt turned to the West for arms supplies. In August 1976, France clinched a multi-million dollar deal to sell Egypt a modified version of *Crotale* anti-aircraft guided missile.

On 3 April 1977 the Kuwaiti Defence Minister said his country signed its first arms deal with the Soviet Union. He confirmed that Kuwait bought "certain sophisticated weapons," but gave no details about the types of arms or their price.

France's imports from its major trade partners in the Indian Ocean region in 1973 totalled 11,962.3 million francs, or 7.2% of its total imports of 166,123 million frances. However, in 1974, france's imports from this region (excluding South Africa) reached 35,963 million francs or 14.12% of its total imports of 254,651.0 million francs. In 1974 France's exports to South Africa totalled 1,678.4 million francs.

France's economic stake in South Africa concerns mainly the supply of military hardware. It is not only collaborating with Pruma in producing Panhard armoured cars in South Africa but has also been regularly supplying *Alouette* and *Super-Frelon* helicopters to the latter since 1963. It supplied Mirage-III fighter-bombers and *Mysteres* and agreed to supply three *Daphne* submarines and a number of corvettes (delivery of these ships, however, is now at stake due to the recent UN arms embargo on South Africa. By an earlier agreement France agreed to supply radar equipment worth over five million Francs to South Africa. France also cooperated with South Africa to develop an all-weather surface-to-air missile system, known as "Cartus". During the last decades, major Western countries (U.S.A., Great Britain and Italy) have helped South Africa to build a secret three billion-dollar arsenal of sophisticated weapons ranging from tanks and radar to self-propelled guns.[31] The French contribution in making South Africa a nuclear power is quite significant. France which had agreed to sell South Africa two nuclear power reactors played a crucial role in preventing the latter to head off an atomic test in August 1977. It was only after the French warning of "serious consequences" if South Africa touched off a nuclear explosion, that the U.S., Britain, Germany and the Soviet Union could persuade South Africa to give an assurance of desisting from developing nuclear explosive devices now and in the future.[32]

France has also been a regular supplier of conventional arms to Pakistan. France has agreed to supply a nuclear reprocessing plant to Pakistan. Iran has asked France to build a series of four nuclear power plants. Iran and France, on 16 October 1977, signed a $3 billion contract for the continuation of two French 900 megawatt nuclear power plants on the Karun River, north of Abadan.

In 1973 U.K's imports from the major countries of this region were £ 1,789.6 million or 13.06% of its total imports of £ 15,815.8 million. In 1974 U.K.'s imports from these countries jumped to £ 3,860.3 million which is

16.72% of its total imports of £ 23,092.8 million. Britain still is and always has been South Africa's main trading partner taking 32% of her exports and supplying 17% of her imports in 1974.

China's principal trading partners in the Indian Ocean region are Australia, Egypt, Malaysia, Pakistan, Singapore and Sri Lanka. China's trade surplus with these countries totalled $ 117.5 and $ 151.5 million in 1972 and 1973 respectively.

In 1972 Japan's imports from her principal trading partners around the Indian Ocean reached $ 8,142.8 million which is 34.69% of Japan's total imports in that year. In 1973 Japan obtained 33% of its total imports from these countries. But in 1974 Japan's imports from these countries reached $ 26,754.3 million or 43.08% of its total imports. Japan's exports to these countries, in 1972, totalled $ 4,025.9 million or 14.08% of its total exports and suffered a trade deficit of $ 4,116.9 million. In 1973 Japan's exports to these countries reached $ 6,049.5 million or 16.38% of its total exports and again suffered a trade deficit of $ 6,594.7 million. In 1974 Japan's exports to these countries jumped to $ 10,213.8 million or 18.38% of its total exports and incurred a trade deficit of $ 16,540.5 million. Two - way trade between Japan and ASEAN countries totalled more than $ 12 billion in 1975. In 1976 Japan was the largest trading partner of ASEAN countries accounting for about 18% of ASEAN's two-way global trade.

Japan depends on oil for 77% of its energy requirements and 78% of this amount comes from the Middle East. To ensure the continued flow of oil for her industries, Japan is now trying to strengthen her economic relations with the OPEC nations. Iran has become Japan's second best foreign customer, while Japan has taken first palace among the importers of Iranian oil. The two-way trade between the two nations reached a value of $ 7 billion in 1976.

During 1966-70 Japan's trade with South Africa increased at the rate of 26% a year and replaced the United States as South Africa's second best consumer.[33] This explosion of trade was the result of Japan's massive industrial expansion that made her to import huge industrial raw materials from South Africa. Moreover, South Africa also proved to be an encouraging consuming centre for Japan's light industrial products such as textiles and consumer goods. It is this economic stake that compelled Japan to swing away from solidarity with the Afro-Asian countries and moved closer to the West. (Despite a U.N. ban on permanent investment in South Africa, Japanese firms sell and assemble cars in South Africa.) Japan's imports from South Africa include vast quantities of coking coal, manganese, uranium, ferrous alloys, pig iron, maize, wool and sugar. Japan's exports to South Africa consist of goods relating to industrial infrastructure, such as micro-wave communication equipment, hydro-electric turbines, transformers and oxygen plants.

III

Seabed Resources

With the rapid development of deep-sea research in the developed countries of the world for the exploitation of the sea-bed resources of the different oceans, the floor of the Indian Ocean will become in the near future a centre of economic exploitations by the big powers. The major share of the oceanic wealth, excluding offshore oil and gas deposits, is obtained in 'nodules' of manganese oxide containing varying amounts of manganese (35% or more), nickel (1.0% to 1.6%), copper (0.75% to 1.5%), cobalt (0.2% to 0.3%) and molybdenum (0.05%) plus lesser traces of some 30 other metals.[34] Marine scientists believe that one square mile of the sea-bed where these nodules are found may be covered with 70,000 tons of minerals which would yield about 30,000 tons of manganese, 3,600 tons of aluminium, 2,300 tons of cadmium, 17,000 tons of iron, 400 tons of cobalt, 1,2000 tons of nickel and 650 tons of copper. The value of the manganese nodules can be as high as £ 400,000 a square mile.

The technology for recovery and processing of the nodules—which range in size from a grain of sand to a small boulder—is extremely sophisticated, and extremely expensive. Most major nodule deposits are in the bottom most depths of the ocean, 15,000 feet or more straight down, and hundreds of miles from the nearest land mass. Merely locating nodules, much less mining and separating the various minerals, is a major scientific challenge. In these circumstances, it is not likely that the developing countries around the Indian Ocean will have any share in the economic exploitation of marine minerals at least in the twentieth century, as these countries are yet to develop research on depriving under-sea vehicles, and marine technology.

Scientists are yet to locate the nodule zones in the Indian Ocean but the discovery of deep-sea mineral deposits in the Red Sea between Port Sudan and Jeddah, by American scientists, has increased the strategic importance of this area. The deposits are at a depth of about 2,000 metres and rich in several minerals, including zinc (4.6%), copper (1.8%), lead (0.14), and smaller quantities of gold and silver. The value of these deposits has been roughly estimated to be about two billion dollars.[35] Like the Americans, French scientists are also taking interest in exploring mineral deposits in the Red Sea, and it is expected that the U.S.S.R. and Japan will follow suit in other parts of the Indian Ocean. The big powers have already shown interest in diamonds mined from the sea-bed off South Africa, Sri Lanka, India, Indonesia and Burma.

Surveys conducted recently along the coast of the Indian Ocean have confirmed that there are rich deposits of natural gas, oil, and heavy minerals which are of great economic and strategic importance to the big powers.

World production of oil during 1970 was 45 plus million barrels per day.

Of this 17% came from offshore oil. By the end of 1974, the daily production of oil climbed to 60 million barrels per day of which 23% came from offshore oil. By 1980, resource requirements are forecast at 85 million barrels per day, of which 33% will come from offshore. It has been estimated that by 1985 approximately 50% of all oil produced will come from offshore. Over the same period, gas production in 1970 was 18 million barrels per day equivalent, of which 15% come from oil production efforts. By 1974, this figure had climbed to 23.4 million barrels per day equivalent, with 35% to 40% coming from offshore operations. By 1980, 50% to 55% (of the total barrels a day equivalent) will be coming from offshore. Market research indicates that 600 to 625 mobile offshore drillings will be available by 1982.[36]

Extensive explorations for oil and gas have been taking place in the territorial waters of the littoral states of the Indian Ocean and it is expected that, with the discovery of more oil gas wells in the region, the production of offshore oil and gas will increase substantially.

Fish

The Indian Ocean abounds in fish. Marine food will become humanity's major food item as food resources on lands are not keeping pace with the rising population.

The Indian Ocean has estimated to have a potential for 11 to 20 million tonnes of marine fish.[37] Almost all the countries around the Indian Ocean catch a good quantity of fish from the Indian Ocean. In 1973 major countries in this region caught 8.8 million metric tons of fish which is 13.46% of 65.7 million tons— the total fish catch of the world.

At present big power fishing export is not very significant in the Indian Ocean. Only the USSR and Japan fish in this ocean. Other countries, due to lack of proper knowledge of fish-beds in the Indian Ocean and long distances from their coasts, could not take interest so far.

Transport

The Indian Ocean (except its southern part) because of its location in the tropical region provides navigation throughout the year, thus allowing the transport of trade goods, military hardware, and oil in a most economical way. 80% of Western Europe's oil supplies, and a growing proportion of United States' oil supplies pass through the waters of the Indian Ocean and around the Cape of Good Hope. 70% of the strategic materials used by the NATO powers pass though the same route. When the Suez Canal was closed 26,000 ships rounded the cape every year. Over 90% of Japan's petroleum and significant quantity of its mineral imports have to be shipped across the Indian Ocean via the Malacca and the Lombac Straits.

Traffic through the Suez Canal now averages 54 ships a day and the monthly tonnage is twice that before closing of the canal in 1967. During 1977, earnings form the Suez Canal were at $ 1.25 million a day. At present ships of upto 70,000 dwt can pass through the canal but after completion of the first stage of Egypt's ambitious Suez Canal Development Project which will widen and deepen the Canal to permit fully loaded vessels of upto 150,000 dwt to transit and at a later stage to widen and deepen it further to allow supertankers of upto 260,000 dwt to pass through, importance of the Suez Canal will grow further. After completion of the second stage of development of the Suez Canal scheduled after 1980 not only Egypt's annual income from the canal will jump upto $ 1,000 million, seaborne trade between east of Suez and west is also expected to increase at an annual rate of eight to ten per cent. Supertankers carrying oil from the Middle East and now passing through the Cape route would pass through the Suez Canal after completion of its second phase of development and bring significant economic relief to the big powers and other industrialised countries of the West.

Foreign Investments

From the above survey it becomes evident that the littoral countries of the Indian Ocean possess huge quantities of industrial raw materials. Encouraged by this and also the availability of cheap labour the Western powers invested capital in different industrial sectors like petroleum, mining, manufacturing and distribution, telephones, etc. The flow of direct investment in oil production in the Middle East took place in a massive way.

Foreign investment in sea has been mainly concentrated in the following sectors: (1) large scale manufacturing for domestic markets in the Indian sub-continent and other populous countries, and (2) export-oriented manufacturing and extractive industries in the south-east Asian countries. Table 5 shows that maximum direct investment took place in India followed by Malaysia, Pakistan and Indonesia but, maximum per capita investment took place in Singapore, followed by Malaysia, Ceylon and Thailand. The simple reason for this was that the new governments in the Indian sub-continent after gaining independence did encourage direct foreign investment where it was essential and connected with sophisticated technologies.

TABLE 5 – *Foreign investments in South and Southeast Asia.*

Countries	GNP Per capita 1968 ($)	Direct investment end-1967 ($ million)	Direct investment per capita end-1967 ($)
India	100	1,290	3
Pakistan	100	340	3
Indonesia	100	250	2
Thailand	150	200	6
Burma	70	10	1
Afghanistan	80	10	1
Ceylon	180	140	12
Nepal	80	5	1
Malaysia	330	680	66
Singapore	700	180	88

Source: Grant L. Reuber etc., *Private foreign Investment in Development* (London, 1973), p. 264.

In Afghanistan, Nepal and Burma foreign investment is very little or non-existent except in hotels and petroleum manufacturing sectors.

The South-east Asian countries are attracting foreign investment in the extractive industries such as tin, rubber, oil and forestry in Malaysia; oil, several non-ferrous metals and forests in Indonesia. Foreign investment in manufacturing industries, mainly for import substitution is quite significant in Malaysia, Thailand and Indonesia. Export oriented manufacturing investment remains a speciality in Singapore. Petroleum refining, petroleum exploration services, shipping and financial services are all involving substantial foreign direct investment in Singapore.

TABLE 6 – *Foreign investments in West Asia*

Countries	GNP Per capita 1968 $	Direct investment end- 1967 ($ million)	Direct investment per capita end- 1967 ($)
Iran	310	710	26
Iraq	260	190	22
Saudi Arabia	360	870	122
Yemen	70		-
Kuwait	3,540	620	147
Muscat	250	70	124
S. Yemen	120	70	57
Bahrein	490	120	610
Qutar	3,490	90	113
Trucial States	1,920	110	820

Source: Grant L. Reuber etc., op. cit.,p. 279.

During the pre-independence period, dominance of foreign investment in different sectors was common in the countries of the western coast of the Indian Ocean. But after becoming independent these countries' intervention, in foreign dominated sectors, was limited to the acquisition by national development banks of small minority share holdings. On the other hand, Malagasy, Mozambique, Sudan and Tanzania encouraged foreign investments in petroleum refineries for local market needs. The main mining resources in Ethiopia (Potash) and Malagasy (chromite and mica), are mainly controlled by foreign direct investors. Major foreign controlled plantation interests remain in tea and tobacco in some of the east African countries. In the manufacturing sector foreign money is still dominating in few simple import-oriented industries only such as, textiles, footwear, cigarettes, matches, cement, elementary electrical and plastic goods. Some foreign investment is also involved in more advanced manufacturing industries e.g., automobile accessory manufacturing, assembly of light vehicles, fertilizer plants, food-processing industries etc. But now-a-days frequency of nationalization is increasing in these sectors. Most of the banking and insurance services are still in the foreign hands though some nationalization has taken place in Tanzania and Sudan.

TABLE 7 – *Direct foreign investments in the Eastern Coast of Africa*

Countries	GNP per capita 1968 ($)	Direct investment end-1967 ($ million)	Direct investment per capita end-1967 ($)
United Arab Republic	170	60	2
Ethiopia	70	50	2
Sudan	100	40	3
Tanzania	80	50	4
Kenya	180	170	16
Mozambique	200	100	14
Malagasy	100	70	11
Somalia	60	10	5

Due to geographical, historical and economic factors US's maximum investment lies in Latin America and not in the Indian Ocean region. Annexure-IV shows that by 1967 U.S. investment in Latin America stood at $ 9,160 millions which is 52.49% of the U.S. total investment of $ 17,450 million in the developing countries of the world, whereas U.S. investments in Asia and the Middle East reached $ 3,560 million or 20.40% of the total investment. U.S. investment in the Middle East has been concentrated in the petroleum industries. U.S. manufacturing investments in copper and aluminium and other metal industries are found in India and Pakistan, U.S. investments are quite significant in export- oriented manufacturing plants in Singapore and petroleum industries in Indonesia. U.S. investments in oil companies and manufacturing industries in Singapore now stand at about one billion dollars. However, U.S. investment in the east African countries is not very significant.

U.K.'s investment in the developing countries of Asia is higher than any other developing regions of the world. In 1967, U.K.'s investment in Asia reached $ 2,070 million which is 30.71% of, total investments of $ 6,740 million in the developing countries of the world. This is due to U.K.'s historical relation with the countries of the region. Next attractive centres for U.K.'s investments are in Africa followed by the Middle East, U.K.'s investment in the Middle East is mainly concentrated in the extraction of crude oil in Kuwait, Iran, Iraq, Abu Dhabi and Qatar; refining and marketing in many other countries in the region. Major British mining interest is found in Malaysia (tin). British interests have been very significant in banking, insurance, import trading and light manufacturing industries in many countries around the Indian Ocean. Major agricultural plantation investments are found in Malaysia (rubber), Ceylon, India, and Kenya (tea). Yet, the largest concentration of British manufacturing investments extending across a wide range of industries including heavy engineering and the chemicals lie in India followed by Pakistan, Ceylon, Malaysia and Kenya. Significant British investment is involved in telecommunications satellite stations in Bahrein also.

Significant French investment in Petroleum industry is found in Abu Dhabi, Iran, Iraq, Muscat and Oman and Qatar. Investment in mining (chromite) is particularly important in Malagasy.

Japan's investment in the area is closely connected to project that would secure essential raw materials for its manufacturing industries or markets for exports. It is also interested in manufacturing and assembly plants. In petroleum production major Japanese investments are in the Kuwait-Saudi Arabia Neutral Zone, Abu Dhabi, Qatar and Indonesia. Manufacturing investment in heavy engineering projects are important in Malaysia and shipbuilding in Singapore. Though Japan is an Asian country its investment in Asia is far less than in Latin America. Up to 1967, Japan's investment in Asia stood at $ 200 million or 28.52% of its total investments in the developing countries of the world as against $ 390 million or 55.71%, in Latin America and $ 10 million in Africa and $ 90 million in the Middle East. After the 1973 oil embargo, Japan took special care to improve her trade and investment interests in the Middle East countries. Japan is building the World's largest petro-chemical complex in Iran at a cost of Yen 550,000 million. She is also constructing a 500,000 barrel per day oil refinery in Iran. Japan is assisting Saudi Arabia in establishing major petro-chemical and steel mills. For Kuwait, Japan is not only supplying oil tankers but it is also cooperating extensively in development of telecommunication facilities. In Iraq, Japanese are assisting in projects involving thermal power generation and an agricultural irrigation system.

Table 8 indicates that Japanese investment in South-east Asia totalled $ 1,860 million by the year 1974. In 1975 Japanese investment in this region totalled $ 2,373 million. Of this amount $ 1,620 billion was invested in Indonesia, $ 276 million in Malaysia, $ 271 million in Singapore, and $ 206 million in Thailand.

TABLE 8 – *Japanese investments in south-east Asia in $ million.*

Country	1970	1971	1972	1973	1974	Total (1951-74)
Malaysia	14	12	13	126	48	250
Singapore	9	15	42	84	51	226
Thailand	13	9	30	34	31	194
Indonesia	49	112	119	341	376	1,190
Total	85	148	204	585	506	1,860

Source: *Japan Quarterly* Vol. XXIII, No. 2, April-June 1976, p. 134.

The South African connection is important to Indian Ocean politics. After the discovery of diamonds and gold in South Africa in the 19th century there was huge inflow of foreign investment in the country. By 1936 foreign investment in South Africa reached 43% of the total foreign investment in Africa. By 1960 the inflow of capital from all external sources had reached £ 1,300 million of which £ 900 million came from the U.K., £ 212 million form the U.S.A., £ 68 million from France and £ 67 million from the World Bank.[38]

TABLE 9 – *Total private foreign investment in million pounds.*

	1956	%	1961	%	1966	%
Mining	454	37	435	33	497	28
Manufacturing	303	25	249	26	515	29
Wholesale and retail trade	149	12	179	14	240	13
Insurance	30	2	39	3	47	3
Other finance	197	16	233	18	365	20
Other industries	102	8	89	7	125	7
Total	1,235		1,324		1,789	

Source: Ruth First, Honathan Steele and Chaistable Gurney. *The South African Connection* (London, 1972), p. 82.

Investment in mining got a boost in South Africa in 1977 when Canada, Australia and a consortium of 12 big European and American banks decided to put over $ 1,000 million for the mining exploration in the country and development of Natal.[39] According to reports, investments in prospecting, mining and refining the so-called 'heavy ores', titanium, manganese, zircon, iron and magnesium are involved in the Natal province of South Africa.

One of the central objectives of U.S. policy in South Africa is to protect American economic interests there. In 1971, the U.S. state Department in a private memorandum pointed out that over 300 U.S. firms in South Africa have a total of about 1% of U.S. direct private foreign investment abroad and collect about 1.6% of the total world wide profits.[40] At present there is approximately $ 20 billion in foreign investment in South Africa and no more than a fifth is in American hands.

The U.S.A. gets about 65% of titanium, so vital to its space industries, missile construction, and supersonic aircraft, from Australia. Similarly South Africa is the major supplier of seven out of 20 minerals which the U.S.A. imports for more than 50% of it requirements. Table 10 shows key minerals imported from South Africa as a percentage of total U.S. imports.[41]

TABLE 10: U.S. *Imports of key minerals from South Africa as percentage of total.*

Antimony Ore	43
Chrome	30
Ferrochrome	35
Copper	6
Chromite	21
Ferromanganese	36
Platinum group Metals	48
Vanadium	57
Vermiculite	100

Source: Conflict Studies, No. 77, November 1976, p. 16.

On the whole, U.S.A. depends largely on developing countries for the supply of 69 strategic raw materials for its industries. And a major part of these minerals are being supplied by countries around the Indian Ocean. West European countries and Japan also depend a great deal on this region and have negotiated long term contracts to ensure regular supply of minerals for their industries.

The U.S.S.R. and China are new aspirants for economic exploitation of this region. Though their present volume of trade is much lower than those of the Western powers, the latter consider the growing economic influence of the Soviet Union as a great challenge. The Western powers fear that their important industries are dependent on the supply of raw materials from the Indian Ocean region, and their economy has a direct bearing on trade with this region. If countries of the Indian Ocean region, under the influence of the Soviet Union, stop supply of vital raw materials and imports of commercial and industrial goods, it will affect the progress of their countries. Here, Angola may be cited as an example. Western powers, particularly the United States, fear that a repetition of the Angola War of 1975-76 would have disastrous results for Western interests and their influence around the Indian Ocean region. They speculate that following the establishment of pro-Soviet regimes in Angola and Mozambque, and now Ethiopia, the demise of the present administration in Rhodesia is imminent.

This will open the way for the final armed struggle in Africa by eliminating the present rulers in south Africa. In this way, the immense mineral wealth of Africa may be lost to the Western economy. Even the loss of one African country, South Africa, containing 6% of Africa's land area, producing over 40% of Africa's entire industrial production, 25% of the continent's agricultural output, will be an irreparable blow to the economies of the Western countries. It is this fear of economic deprivation that has accelerated rivalry for the exploitation of economic resources in the Indian Ocean region. However, big powers' economic stake in the Indian Ocean region is so deep that

ANNEXURE - I

Minerals production (thousand metric tons) 1973.

Region	*Aluminium*	*Antimony*	*Bauxite*	*Chromium*	*Coal*	*Copper ore*	*Gold (K.g.)*	*Iron ore*	*Lead ore*	*Magnesite*	*Manganese*	*Nickel ore*	*Tin*	*Tungsten*	*Uranium*	*Zinc*
World	12,020	69,400	67,400	2,390	2,206,800	7,140	1,135,000	482,600	3,410	940	9,200	692,400	185,000	48,230	19,480	5,670
Countries around the Indian Ocean	561.8	21,003	3,657	1,029.8	199,477	370.4	877,117	77,214	425.2	303.8	2,578	76,330	129,545	5,459	2,695	583.7
World percentage	4.67	30.26	5.43	31.30	9.04	5.19	77.28	16.00	12.47	3.36	28.02	11.02	70.02	11.32	18.83	10.29

Source: UN *Statistical Year Book,* 1974.

Annexure - II

Main importers of crude petroleum from the Middle East (Mn $)

Imports from	*Japan Jan-July*		*France Jan-Dec*		*Italy Jan-April*		*U.K. Jan-Aug*		*W. Germany Jan-June*		*U.S.A. Jan-May*	
	1975	*1976*	*1974*	*1975*	*1975*	*1976*	*1975*	*1976*	*1975*	*1976*	*1975*	*1976*
Middle East	1,190,17	1,233.42	640.90	616.90	514.50	559.47	483.05	490.55	430.14	479.11	431.11	724.14
Total from all sources	1,595.99	1,644.87	841.34	788.01	604.77	685.79*	675.35	692.53	636.18	704.95	1,589.81	1,926.08
Percentage of total from Middle East	74.6	75.0	76.2	78.3	85.1	81.6	73.5	70.8	67.6	68.0	27.1	37.6

Source: *quarterly Economic Review of Oil in the Middle East* (London), 4th quarter 1976, p. 14.

Annexure - III

U.S.A. and its trading partners (Million dollars)

Country	1972			1973			1974		
	Imp	*Exp*	*Balance*	*Imp*	*Exp*	*Balance*	*Imp*	*Exp*	*Balance*
Iran	199	559	+ 360	343	772	+ 429	3,132	1,734	– 1,398
Kuwait	49	111	+ 62	65	120	+ 55	13	209	+ 196
Saudi Arabia	194	314	+ 120	515	442	– 73	1,671	835	– 836
India	427	350	– 77	437	526	+ 89	561	760	+ 199
Pakistan	40	183	+ 143	40	239	+ 199	61	398	+ 185
Thailand	116	170	+ 54	141	256	+ 115	184	369	+ 185
Malaysia	301	128	– 173	440	157	– 283	773	377	– 396
Singapore	265	385	+ 120	467	684	+ 217	553	988	+ 435
Indonesia	278	308	+ 30	505	442	– 63	1,688	531	– 1,157
Australia	807	842	+ 35	1,067	1,439	+ 372	1.042	2,157	+ 1,115
South Africa	325	602	+ 277	377	746	+ 369	609	1,160	+ 551
Total	3,001	3,952	+ 951	4,397	5,823	+ 1,426	10,287	9,518	– 769

Source: *The Europa Year Book*, 1976, Vol. II, pp. 1640-41.

Annexure IV

Stock of direct investment is developing countries by capital exporting country, region and sector, End-1967 ($)

	Latin America	%	*Europe*	%	*Africa*	%	*Middle East*	%	*Asia*	%	*Total*	*Petroleum*	%	*Mining*	%	*Manufacturing*	%	*Other*	%
United States	9,160	52.49	750	4.30	1,370	7.85	1,780	10.20	1,780	10.20	17,450	6,600	37.82	2,050	11.75	4,730	27.11	4,060	23.27
United Kingdom	1,710	25.27	230	3.41	1,950	28.93	840	12.46	2,070	30.71	6,740	2,090	31.01	580	8.61	1,580	2344	2,490	36.94
France	470	16.10	290	9.93	1,730	59.25	160	5.48	280	9.59	2,920	850	29.11	420	14.38	920	31.51	730	25.00
Netherlands	930	51.96	100	5.59	320	17.88	170	9.50	260	14.53	1,790	1,340	74.86	20	1.12	320	17.88	160	8.94
Germany	710	59.17	180	15.00	140	11.67	20	1.67	70	5.83	1,200	40	3.33	50	4.77	1,020	85.00	90	7.50
Canada	810	72.97	20	1.80	60	5.14	10	0.90	70	6.31	1,110	20	1,80	210	18.92	200	18.02	680	61.26
Italy	400	45.45	180	20.45	250	28.41	30	3.41	30	3.41	880	190	21.59	20	2.27	590	67.05	80	9.09
Japan	390	55.71	X	-	10	1.43	90	12.86	200	28.57	700	80	11.43	60	8.57	430	61.43	140	20.00
Belgium	120	17.39	80	11.59	480	69.57	X	-	20	2.90	690	20	2.90	100	14.49	370	53.62	210	30.43
Switzerland	270	65.85	80	19.51	20	4.88	10	2.44	20	4.88	410	-	-	10	2.44	210	51.22	200	48.78
Sweden	70	35.00	10	5.00	80	40.00	-	-	20	10.00	200	-	-	70	35.00	110	55.00	10	5.00
Portugal	10	8.33	30	25.00	90	75.00	-	-	-	-	120	20	16.67	30	25.00	50	41.67	30	25.00
Australia	-	X	X	-	-	-	-	-	100	100.00	100	10	10.00	20	20.00	30	30.00	50	50.00
Denmark	20	66.67	10	33.33	X	-	-	-	X	-	30	-	-	-	-	20	66.62	10	33.33
Norway	10	100.00	X	-	X	-	-	-	-	-	10	-	-	-	-	10	100.00	X	X
Austria	10	100.00	X	-	-	-	-	-	-	-	10	-	-	X	-	10	100.00	X	—
Total	15,160		1,950		6,510		3,110		4,900		34,350	11,260		3,640		10,570		8,880	

Note: Totals may not add up because all figures have been rounded to the nearest $ 10 million, amounts less than $ 5 million being shown as 'X'.
Sourece: Grant L. Reuber op. cit p. 283.

their rivalry in the Indian Ocean will not come to an end unless all the countries of the Indian Ocean region including the hinterland states form an organization of the Indian Ocean community and put combined pressure on the big powers to keep the Indian Ocean as a zone of peace.

REFERENCES

1. See J.P. Anand, "Big Powers and the Indian Ocean", *Institute for Defence Studies and Analyses* (IDSA) *Journal,* Vol. VIII, No. 4, April-June 1976.
2. *Indonesia Times,* (Jakarta), 9 February, 1977.
3. Ibid., 4 February, 1977 and 12 May 1977.
4. *Asia Research Bulletin,* (Singapore), 31 January 1977, p. 285.
5. Bangkok Post, (Bangkok) 27 December, 1976.
6. Bangkok Post, *26 August 1977.*
7. Ibid, 12 February 1977.
8. *Sea Power,* (Washington, D.C. 20006), Vol 17, No. 2, February 1974, p. 18
9. Ibid., P. 16.
10. Ibid, p. 18.
11. *Quarterly Economic Review, Oil in the Middle East,* (London), 4th Quarter, 1976, p. 14.
12. *Indonesia Times,* 9 February, 1977.
13. *Petroleum Economist,* (London), January, 1976, Vol. XLIII, No. 1. p. 14
14. *Mainnichi Daily News,* (Tokyo), 6 July, 1976.
15. *Indonesia Times,* 5 July, 1976.
16. *Straits Times,* 5 July, 1976.
17. *Business Recorder,* (Karachi), 20 August, 1960.
18. *Sea Power,* n 8, December 1973, January 1974, p. 32.
19. Ibid, p. 33.
20. *Straits Times,* 17 December, 1976.
21. Ibid.
22. *See Europe Year Book,* 1976, vols. I and II (London, 1976).
23. See. K. Subrhamanyam, "The Energy Crisis-I", *Foreign News and Features,* Vol. V(7) (New Delhi). 16 February 1974, p. 4; *Working People's Daily,* (Rangoon) 10 January.
24. *Indian Express* (New Delhi), 7 January 1974, *East African Standard* (Nairobi), 8 January 1974.
25. *Indonesia Times,* 10 February 1977.
26. Paul Martin, "Getting Stronger: 'The Best Base that America has ever had' in Saudi Arabia", *Bangkok Post,* 28 February.
27. US Committee on Foreign Relations, US *Military Sales to Iran: A Staff Report to the Sub-committee on foreign Assitance of the committee on Foreign Relations, United States* (Washington, 1976), p. vii.
28. Ibid.
29. *Indonesia Times,* 30 December 1976.
30. See H.M.L. Beri, "Soviet Arms in West Asia", *Institute for Defence Studies and Analyses (IDSA), News Review on West Asia* (New Delhi), January 1977 p. 6.
31. *Ceylon Daily News,* (colombo), 15 July 1971.

32. *Newsweek* (New York), 5 September 1977, p. 19.
33. Ruth First, Jonathan Steele and Christable Gurney. *The South African Connection: Western Investment in* Apartheid (London, 1972), p. 142.
34. *Sea Power,* N. 8, August 1976, p. 13.
35. E.A. Gallion (ed.) *Use of the Seas* (Prentice Hall, 1968). p. 49, cited by K.R. Singh "The Indian Ocean—The Need for a Community Approach". *IDSA Journal,* Vol. 4, April 1972, p. 536.
36. E.D. Hoofnagle, "The possible Impact of the shore oil Production on World Trade," *Sea Technology* (Arlington), September 1976, p. 18.
37. *Hindustan Times*, (New Delhi), 28 February 1977.
38. Colin Legum (ed.) *Africa: A Handbook* (London, 1967), p. 346.
39. *Ceylon Daily News,* 2 July 1977.
40. n.•33. p. 85
41. David Rees, "Soviet Strategic Penetration of Africa", *Conflict Studies,* No. 77, (Lodon), November, 1976, p. 16.

XX The Indian Ocean as a Zone of Peace: Problems and Prospects

B. VIVEKANANDAN

Ten Years ago, on December 16, 1971, the UN General Assembly adopted a resolution declaring the Indian Ocean "for all time as a zone of peace. "Earlier, in September 1970, the Lusaka Conference of the nonaligned countries had adopted a declaration "calling upon all states to consider and respect the Indian Ocean as a zone of peace from which great power rivalries and competition as well as bases conceived in the context of such rivalries and competition, either army, navy or air force bases, are excluded. The area should be free of nuclear weapons."

When the UN General Assembly passed the resolution in 1971, the strategic environment in the Indian Ocean and in the adjoining region was not so complicated as it is today, although following British Prime Minister Harold Wilson's announcement on January 16, 1968 that his country would withdraw its forces from all areas east of Suez except Hong Kong, there were indications that other naval powers, especially the Soviet Union, had already made plans to move into the Indian Ocean and establish their presence there. The Soviet Union had launched a major programme of naval expansion with a view primarily to increasing its naval presence in the Indian Ocean on the thin argument that the Indian Ocean constituted a vital domestic communication route for the USSR.

> In view of its geographical position, the USSR needs a large fleet in order to maintain the necessary internal contacts with remote parts of the country. Vitally important communication routes linking the European with Asianatic and Far Eastern parts of the USSR pass through the Mediterranean and across the Indian Ocean. It is quite clear that the existing land routes cannot be any substitute for these more economic and convenient sea routes.[1]

Reprinted from *Asian Survey*, Vol. 24 No. 12, 1981, pp. 1237-1249, with permission of the Regents of the University of California

This was a new element that had the potential to complicate the situation in the Indian Ocean According to S.Koslov, the Novosti Press military correspondent:

It is obvious from this argument that this line had been carefully drawn an at independent strategy for justifying a permanent naval presence in the Indian Ocean whatever the objective situation in this area or the opinion of the countries concerned, including the littoral countries. Indeed the Soviets had commenced their drive as early as 1967, when their strategic publication, *Military Strategy*, for the first time included South and Southeast Asia in the Soviet sphere of interests and in its search for new facilities for Russia naval vessels along the shores of the Indian Ocean. In February 1968 the Soviet naval chief, Admiral Gorshkov, visited the Indian naval establishments in Bombay and Visakhapatnam. In March 1968 a Soviet flotilla called at a number of ports in India, Sri Lanka, Iraq, Somalia, and South Yemen. The Soviet naval presence in the Indian Ocean became quite spectacular when, in January 1971, a number of Soviet warships passed through Singapore as if they meant to parade their strength before the Commonwealth Prime Ministers then in conclave there.

These determined effort by the Soviet Union to expand its naval operations in the Indian Ocean do not seem to have been anticipated by Britain when it decided to withdraw from all areas east of Suez. However, as declining power with reduced resources, Britain could do little by itself to counter the Soviet move in any meaningful manner. Taking the growth potential of the Soviet Navy into account, it realized that, as in Western Europe, the only effective answer to the growing Soviet naval activity in the Indian Ocean was to ensure the firm involvement of the U.S. navy in the Indian Ocean. As a first step it decided to develop Diego Garcia jointly with the U.S. into a naval communication centre. Before agreeing to a further expansion of facilities in Diego Garcia, however, it made an effort to see if a viable security arrangement involving littoral countries like India could be worked out for the Indian Ocean. Sir Alec Doughlas-Home, the Foreign Secretary, visited India in February 1972 with a view to ascertaining if the littoral countries, and especially India, would be willing to participate in such an arrangement. When he found that the response was negative, Britain agreed to a U.S. proposal for a further expansion of facilities in Diego Garcia.

Both superpowers now enjoy naval facilities at different points of the Indian Ocean. In the western section, which is of special strategic importance, the United States has naval facilities in Diego Garcia, Berbera (Somalia), and Mombasa (Kenya), while the Soviet Union has facilities in Aden and the Socotra Island (South Yemen), Unim Qasr (Iraq), and the Dahlak Island in the Red Sea. The Soviet Union also maintains a large network of "mooring buoys" all over the Indian Ocean as rendezvous points for its warships. It is estimated that the two superpowers keep about thirty naval vessels each in the Indian

Ocean. Other countries maintaining a significant naval presence in the Indian Ocean are Britain, France and more recently, West Germany.

For different reasons the Indian Ocean region is an area of vital interest for many important countries. While some of these interests are a legacy of traditional connections, others are a fallout of the major changes that have occurred in the strategic perspective of the world in recent years. Britain, for example, has wide-ranging economic interests in the region. About one- third of its total foreign private investments (worth 28,870 million pounds) is located in the areas adjoining the Indian Ocean.[2] The Indian Ocean constitutes a major junction of sea communication. For countries like Australia, Britain, China, France, India, Japan, the United States, West Germany, etc., all their major supply lines of oil, foodstuffs, and raw materials pass through the Indian Ocean. Indeed, for economic, political, and military reasons, the Indian Ocean assumes vital importance for the littoral countries—Australia, Burma, Bangladesh, India, Indonesia, Iran, Pakistan, Sri Lanka, etc.— and also for Britain, China, France, Japan, the United States, and the Soviet Union. In fact, each one of them considers the Indian Ocean region as an area vital to its security and economy.

In a strategic sense, the Indian Ocean region first assumed exceptional importance in the 1970s. The oil crisis of 1973 and its ramifications sent shock waves around the world. The industrial nations of the West saw how, in an economic war, the oil in the Middle East and the Gulf could be used as a most potent weapon against all the oil-dependent countries, whether in the East or in the West, and reduce them to thralldom. Indeed the developments of 1973 and after revealed how vital the oil supplies from the Indian Ocean region were for the industrial West and how vulnerable the West is to oil pressures. This realization encouraged the Soviet Union to make a determined effort to gain control over the oil sources in the Middle East and the Gulf and thereby, if possible, to throttle the West slowly to submission.

There is no doubt that the Indian Ocean is a major source of petroleum today.The Persian Gulf, the Middle East, Northeast Africa, and Indonesia together hold about three-fourths of the known oil deposits of the world. The shipment of oil through the Indian Ocean and access to oil in the region on equitable terms has, therefore, become a very vital strategic question for all Western countries and for many other important oil-importing countries. And the major threat to these countries seems to emanate from the Soviet Union. According to Lord Home: "The eye of the Soviet Union is always on oil, for the Russians know that for the next 30 years the United States will be dependent on the Gulf area for a large part of its supplies; while Germany will be even more so, and all Western Europe is very vulnerable."[3] Similarly, most of the industrially developed countries of the West also depend on the region for their supplies of important, even vital, raw materials and minerals like bauxite, copper, iron, manganese, phosphates, thorium, tin, uranium, etc. The region also holds substantial deposits of gold. It would appear that the inroads the Soviet Union has made in Africa in the guise of extending support to liberation struggles there by using surrogates such as the Cubans are aimed mainly at control of the mineral wealth of Africa. The Soviet Union is fully aware of the vulnerability of the West on these counts and seems to be engaged in a resolute effort to have a voice in the supplies of all items from the region on which it is not itself dependent. This accounts for the overwhelming interest of the West in forestalling any Soviet attempt to gain control over the supplies of any of these items from Asia or Africa.These factors have added a new dimension to the strategic environment in the Indian Ocean region.

The politico-economic and military advances that the Soviet Union has made in the Indian Ocean region during the past decade have introduced new complications in the security perspective of the region. Since 1971 the Soviet Union has conduced "friendship" treaties with a number of nonaligned countries or the Indian Ocean littoral and in the hinterland—Afghanistan, Ethiopia, India, Iraq, Mauritius, Mozambique. Tanzania, etc. In the Middle East alone it has "friendship" treaties with the five countries. Except in a few cases, these "friendship" treaties have afforded the Soviet Union a convenient means by which to control the domestic and foreign policies of those littoral countries. The treaties have reduced some of these countries to the status of client states, and pushed them irrevocably into the Soviet orbit. Indeed, in most cases the treaties have put direct and indirect limits on the freedom of the littoral countries, so much so that in spite of being nonaligned, they are no longer able to take any position in world politics against the Soviet Union even if the merits of a particular case incline them to do so. Besides, these "friendship" treaties reflect the pattern of relationships that the Soviet Union wants to develop in Asia and Africa. The Soviet Union has significantly increased its economic and military aid to a number of countries of the region with a view to tightening its grip over them and weakening or destroying the influence of the West. The Soviet navy seems to have been designed as an instrument in this task. According to Admiral A.K. Chatterji, a former Chief-of-Staff of the Indian navy: "The Soviets are using their navy in a classic way for applying the necessary military, political and economic pressures to the countries of the Indian Ocean littoral."[4] And the Soviet Union's use of surrogates such as the Cubans in the guise of "liberation forces" has served to tighten its grip over the new nations of Africa.

How effective these "friendship" treaties could be as trip-wires of domination and intervention and how they could be used to overrun sovereign nations is clear from the manner in which the Soviet Union intervened in Afghanistan in December 1979, exactly a year after it had signed a "friendship" treaty with that country. What is noteworthy in the context of the present study is that when the Soviets found that President Hafizullah Amin was thinking in terms of terminating the Soviet-Afghan "friendship" treaty and asking the Soviets to leave Afghanistan, the Kremlin unilaterally invoked a convenient provision of the "friendship' treaty and occupied Afghanistan, spreading the spurious story that the Soviet army had entered Afghanistan at the invitation of President Amin. To ensure that the truth never came out, the army of occupation stilled the voice of Amin at the earliest opportunity. Even then a closer scrutiny of the available evidence clearly shows that there is no truth in the Soviet claim that its army was invited by President Amin. The story seems to have been planted as part of a futile Soviet attempt to mislead the world into believing that there was some legal basis for the Soviet intervention in Afghanistan.[5]

The Soviet intervention in Afghanistan has tremendous implications for the security of the Indian Ocean region. It has shown above all that the Soviet scheme to expand its frontiers by the use of force is still in operation. In the history of the modern world, Russia/the Soviet Union is the only country with an unbroken record of continuous territorial expansion over three-and-a-half centuries. The Revolution of 1971 does not seem to have made the slightest difference to Russian expansionism. Unlike other imperial powers, which have gone through the process of expansion and contraction, Russia has expanded its frontiers all along and in all directions and has systematically consolidated its gains. From Europe it moved first towards the east until it annexed all territories up to the shores of the Pacific Ocean in 1638. Then it moved northwards in Siberia in the nineteenth century. By 1884 the Russian Empire extended as far as the northern borders of Afghanistan. World War II helped the Soviet Union in expanding further and in consolidating its position in Eastern Europe.

The Soviet Union started making inroads into the Arab world from 1956 onwards. It has sought to entrench itself deeply in West Asia, South Asia, Southeast Asia, and Africa. In 1979 it occupied Afghanistan, the vitally important buffer state between the Soviet Union and the Indian subcontinent. And it looks as though it was deeply interested in destabilizing Pakistan and in opening a land corridor to the Indian Ocean through Baluchistan. Certainly such a design was clear from the warning Soviet Foreign Minister Andrei Gromyko administered to Pakistan in February 1980 during a visit to New Delhi. From the Soviet point of view the opening of a land corridor to the Indian Ocean is of great strategic importance since otherwise, in the event of a war, the West can effectively check Soviet naval operations by controlling various choke points such as Helsingor, Sklagerrak, the Bosporus, the Dardanelles, Gibraltar, Suez, the Malacca Straits, the Panama Canal, and the Tsushima Straits, and bring the operations of the Soviet navy to a grinding halt. A direct opening to the Indian Ocean through Baluchistan, however, can more than make up for such action, and in fact is the only corridor through which the Soviet Union can hope to reach the India Ocean quickly and circumvent the effects of choke-points under the control of the West. The scheme, therefore, clearly exists, and it would be foolish to underestimate the Soviet determination in the matter. What is not clear, however, is the time-frame drawn by the Kremlin to achieve this objective. However, as Asoka Mehta rightly put it: "Russia's moves across the Asian chess-board have been shrewd, expertly executed, and effective. It commands influence in regions—West, South, and South-East Asia—that the Czars had never dared to dream.:"[6]

The Soviet advances in the Indian Ocean region have created a new situation in Indian Ocean politics in recent years. With the Soviet occupation of Afghanistan, the Soviet forces are only about four hundred miles away from the vital oil region of the Gulf, posing a major threat to access to oil from the Gulf and oil shipments through the Straits of Hormuz. There is a similar Soviet threat to the supplies of raw materials and minerals from Africa to theWest. From the point of view of the countries of the West, therefore, the most crucial thing for the safeguarding of their interests in the Indian Ocean region is to frustrate the Soviet effort to gain control over the region or over any strategic part of it such as the Gulf. Undoubtedly maintenance of free access to the region's petroleum supplies is of paramount importance to all oil dependent nations. The interests of the industrialized nations of the West would be damaged if the Soviets were either to gain control over oil supplies from the Middle East or to be in a position to disrupt the communication channels through the Indian Ocean in times of crisis and the West would be ruined if an economic war were fought with the Gulf under Soviet control.

It may be noted that the Soviet Union as the largest producer of oil (18% of world production) has encouraged the price hikes ordered by the Organization of Petroleum-Exporting Countries (OPEC), and has been one of the highest beneficiaries of those hikes. It does not need to import oil, but it did obtain oil from Iraq in exchange for arms and sold it to Europe at high prices. Indeed oil has been the principal medium of repayment for Soviet assistance to industrial projects in Iraq and for Soviet arms supply to that country. However, it is said that in the 1990s the Soviet Union too may have to depend upon imported oil, and it is certainly not the kind of nation to hesitate to apply pressure and bully the oil-producing countries of the Middle-East if they hesitate to provide oil on acceptable terms. There is, therefore, a distinct possibility of intense Soviet involvement in the politics of the Gulf in the future.

An important preoccupation of the Soviet Union seems to be to gain recognition as a dominant Asian power with a decisive voice in the management of affairs in Asia in general and in the Gulf in particular, and to hasten the decline of Western influence in the region. One of the dimensions of the Soviet plan for system of collective security in Asia as outlined by Brezhnev in his address to the International meeting of Communist and Worker Parties in Moscow on June 7, 1969 is to bind Soviet relations with the various countries of Asia with a security cord so that eventually, as the most powerful state within Asian face, the Soviet Union may establish a hegemonistic relationship with them. Of course, the Soviet Union has sought to demonstrate everywhere its own status as a dominant global power with an ever-widening sphere of interests. If, somehow, the Western presence is simultaneously eliminated, the local rivalries in the strategically important Gulf region would give the Soviet Union ample opportunities ~~to~~ establish a firm foothold in the region, especially given the additional advantages of the geographical proximity of the Soviet Union to the Gulf. While it is true that the Soviet Union has yet to establish a firm foothold in the region despite its new ties with Iraq and Kuwait, it may be able to do so once its naval buildup in the Indian Ocean becomes formidable or once it gains a land-based access to the Indian Ocean near the Gulf. However, the Soviet Union has, by now, acquired a status in the politics of the Gulf that makes it a factor in the decision- making process of the countries of the Gulf and the Middle East.[7] The apprehension is that if the Soviet domination becomes a fact of life, it may lead to a new process of self-Finlandization of the weak Gulf states, one after another. Perhaps by its intervention in Afghanistan, the Soviet Union has made a very significant advance in the process of throwing its military might around and applying real pressure on the oil-producing Gulf states.

It is to be noted that during the past decade considerable change has taken place in the power position of the Soviet navy in the Indian Ocean and its use as an instrument of diplomacy in the Indian Ocean region. The Soviets had hardly rated as a naval power in the 1950s or used their navy as coercive instrument of diplomacy in the region because of the lack of credible capability. That, however, is no longer the position today. Not only has the Soviet navy increased its capabilities by means of a massive, long-term naval development strategy, but the Soviet Union has also used these capabilities more actively in support of its foreign policy. Ever since the Soviet warships moved into the Indian Ocean in 1968, there has been a steady increase in the size of the Soviet naval presence, and there is every reason to believe that a sizeable portion of the newly built Soviet warships would be deployed mainly in the Indian Ocean. It is estimated that the Soviet Union produces about ten oceangoing surface warships every year. In 1980 the Soviet Union launched thirteen new submarines, nine of which were nuclear powered, including one of the colossal *Oscar* class and another of the *Typhoon* class.[8] That the Soviet Union spends as much as 14% of its GNP on defence shows that militarism is rampant in the Soviet Union.

The massive Soviet naval programme and the ever-increasing politico-military activity of the Soviet Union in the India Ocean region have made it abundantly clear that the aim of the Soviet Union in the region is to dominate it. This is a new situation, one which is radically different from the situation that obtained in 1971 when the UN General Assembly first passed the resolution declaring the Indian Ocean a zone of peace. Although the littoral countries are broadly in agreement with the concept of a peace zone in the Indian Ocean region, their approaches to the presence of the big powers differ. This is clear from the attitudes of two important littoral countries—Sri Lanka and India. The two have taken a leading role in the effort to make the Indian Ocean a zone of peace. However, while Sri Lanka sees danger in the increasing Soviet presence in the Indian Ocean, India under Indira Gandhi thinks that Colombo's fears are exaggerated. While Sri Lanka views both the superpowers with apprehension, the Indian attitude is, or has tended to be, soft towards the Soviet Union[9]. However even in India there was a shift in the attitude towards the question between 1977 and 1979 under the Janata Government headed by Morarji Desai since this government seems to have shown a greater awareness of the dangers posed by the Soviet naval activity in the Indian Ocean. In the Indian Parliament on July 14, 1977, Desai said that "there is the race between two powerful nations. It is from that we have to save Asia and that is what we are engaged in."[10] This balanced approach, however, has since been modified with the return of Indira Gandhi to power in 1980. The "tilt" towards the Soviet Union now seems to have been restored.

There are also differences between India and Sri Lanka over the basic peace zone concept itself. While Sri Lanka wants the Indian Ocean region to be a nuclear-free zone, India, for obvious security reasons, disapproves of this proposal. The fact that China possesses atomic weapons is a major constraint militating against any unilateral abandonment of nuclear options by India. At the UN Special Session on Disarmament, Prime Minister Desai said on June 9, 1978 that "it is idle to talk of regional nuclear free zones when there would still be zones which could continue to be endangered by nuclear weapons. Those who have such weapons lose nothing if some distant area is declared non- nuclear. The nations without nuclear capacity who imagine that their inclusion in such zones affords them security are suffering from delusion." And he added: " We are convinced that there cannot be a limited approach to the question of freedom from nuclear threats and dangers, but the whole world should be declared a nuclear-free zone."[12]

One of the most unrealistic elements in the perception of those campaigning to get the Indian Ocean region recognized as a peace zone is their attempt to detach the Indian Ocean for the global strategic map, unmindful of the Indian Ocean's strategic importance to global security, and accord it separate treatment. It is something like a repeat performance of President Tito's attempt to demilitarize the Mediterranean when the navies of the big powers began to shadow each other there. The recent revolutionary changes in weapons technology have rendered the globe so small that it is highly impracticable to seek to insulate any particular region as a permanent zone of peace. Moreover, in view of the Soviet claim that it has a right to use the Indian Ocean as an economical communication route between its Asian and

European parts, one cannot foresee a situation in which the Soviet navy would be absent from the Indian Ocean altogether. The Soviet Union's plans for the future indicate that its growing navy will find an easy outlet into the warm waters of the India Ocean. Once in the Indian Ocean the Soviet Union would play more politics with the relatively weak Indian Ocean littoral states than would other countries with a view to undermining the interests of the West in the region.

Of course, in the strategic perceptions of the littoral states and those of the superpowers themselves there exists a hiatus. Despite detente and SALT, the superpowers have not abandoned their plans to encircle each other. Undoubtedly, from a global strategic point of view the Indian ocean offers exclusive advantages to the Western powers over the Soviet Union since the Trident-I submarine-launched missiles from the Indian Ocean can hit the Soviet industrial areas of southern Russia. From that angle Western naval presence, especially that of the United States and Britain, in the Indian Ocean can deter the Soviet Union with increasing efficiency from undertaking any adventure and simultaneously reduce Soviet naval pressures in other waters also. Significantly the Indian Ocean provides no similar advantages for the Soviet Union since it cannot operate from there against any U.S. or Western European targets. In a realistic analysis of the situation one cannot overlook this important factor either. The demand, therefore, to keep the Indian Ocean free from the British or U.S. military presence, although it has emanated from some littoral countries from an entirely different perspective, could in effect be of help in meeting the strategic needs of the Soviet Union.

It should be conceded that a demilitarization of the Indian Ocean is impossible without the willing cooperation of the two superpowers. However, the overall attitude of the superpowers indicates that in any real sense their cooperation in this respect would not be forthcoming. Both of them abstained from the vote on the question in the UN General Assembly. Their talks in Moscow, Washington, and Berne on the proposal to make the Indian Ocean a zone of peace reveal that they have no intention of keeping their navies out of the Indian Ocean. Nor are they going to keep the Indian Ocean a nuclear-free zone. Indeed, during those talks they indicated that their immediate objective was not to dismantle the existing facilities (including those on Diego Garcia) but to limit their military activity in the Indian Ocean.[12] While the United States wanted to freeze the deployment of the military might of the superpowers at the present level, the Soviet Union wanted U.S. nuclear vessels not to enter the Indian Ocean region.[13] The Soviet Union and the U.S. then rejected each other's demands. However, it looks as though the talks between the superpowers not only formalized the existing situation but also bilateralized the question of their naval activity in the Indian Ocean at a time when the littoral countries are trying hard to multilateralize it through the United Nations.

Perhaps it would have been possible to free the Indian Ocean from the large-scale naval deployments of the powers if the littoral states had, working as a moral force also, acquired reasonable naval capabilities and expressed their readiness to shoulder the responsibility of ensuring unfettered freedom of navigation through the Indian Ocean in war and in peace. However, they seem to possess neither the will nor the capacity to do so. The *Military Balance*, 1980-81, shows, that Iran has three destroyers and four frigates; Pakistan: six submarines, one light cruiser, six destroyers, and one frigate; Bangladesh: two frigates; Burma: two frigates and 36 gunboats; Indonesia: four submarines, seven frigates and three corvettes; and India: eight submarines, one aircraft carrier with a capacity of 22 aircraft, one cruiser, 29 frigates, and four corvettes.[14] The navies of other littoral states consist mainly of patrol and coastal craft. In power terms, therefore, they hardly count. There is little chance of all these navies acting in unison, and even if they did so, they would hardly be able to guarantee freedom of navigation through the Indian Ocean in times of crisis. Nor can they hope to match the presence of any single superpower.

The Soviet occupation of Afghanistan in December 1979, which, according to President Jimmy Carter, "could pose the most serious threat to world peace since the second world war," changed the strategic situation in the Indian Ocean region overnight.[13] It is true that since world War II the Soviet Union had never spared any opportunity to make full use of its military strength to achieve its political and ideological objective throughout the world. Although it had been trying in a surreptitious manner to widen its influence among the Asian and African nations since 1956, the Soviet occupation of Afghanistan is the first example in the post-war years of its use of military and airforce to occupy a strategically important, nonaligned Asian country — a country it had served for centuries as an effective buffer between Russia, the Soviet Union and the Indian subcontinent. According to the *Strategic Survey, 1979*:" The Soviet Union's demonstrated readiness to use her forces to create political *fait accomplis* beyond her borders would overshadow relations with other countries as well. Repeated unsubstantiated accusations against Pakistan, China and Iran of providing the Afghan rebels with weapons could foreshadow a Soviet attempt to use force against these countries.[16]

Eight years ago, in 1973, in a ponderous article on the security environment in the Indian subcontinent, Robert Jackson wrote in *International Affairs:*

> The Soviet Union is still pent up behind the geographical barriers erected around the sub-continent against the Tzars by the British Empire. Russia and Pakistan are still separated by that carefully negotiated strip of Afghan territory that gives Afghanistan a common frontier with China. And beyond the Wakhan Strip, Pakistan and Pakistan occupied Kashmir interpose another barrier between India and Russia."

What the Soviet Union sought to accomplish in December 1979 through its occupation of Afghanistan was to destroy this geographical barrier, which according to Jackson had insulated the Indian subcontinent effectively from Soviet incursions, and to make the possibility of the Soviet Union establishing a "continental" contact with India more distinct and immediate than ever before.

The Soviet occupation of Afghanistan has grave implications for the security of the Indian Ocean region and the interests of the various countries that heavily depend on the region for their vital supplies. However, strategically speaking, the Soviet intervention in Afghanistan is a clear indication of the ascendancy that the Soviet Union has established in Asia by its growing militarism.

It also points to the Soviet ambition to dominate the Gulf and the Indian Ocean region. U.S. determination to vigorously counter the Soviet's moves was made explicit by President Carter when he said: " Any attempt by any outside force to gain control of the Persian Gulf region will be regarded as an assault on the vital interests of the United States. It will be repelled by use of any means necessary, including military force."[18] The same determination is manifest in the outlook of President Reagan. In view of these new developments it is inconceivable that the U.S. navy or the navies of the Western European countries would stay out of the Indian Ocean region under any circumstances as long as the Soviet threat to this region, so vital to their security, looms large. In view of the unrelenting determination of the Soviet Union to gain a firm foothold in the Gulf one feels that the Soviet effort ultimately is to grab the Middle East and its oil at some stage and throttle the West into submission. In achieving this aim the next Soviet victim would probably be Baluchistan, a province in Pakistan where a secessionist movement, led mainly by pro-Moscow communists, is quite active.[19]

If the Soviets succeed in stirring up the secessionist movement in Baluchistan to a high degree from their base in Afghanistan and in getting a Karmal-type of government established there, it would inevitably enable them to gain direct access by land to the Indian Ocean, upset all the present Western naval strategies against the Soviet Union, and pose a formidable threat to the supply of vital resources—oil, raw materials, and minerals—from the region to the West. As the Soviets are just about 250 miles away from Karachi and close to the Straits of Hormuz through which the oil tankers pass, there is every reason to believe that the West would not allow Moscow to establish a direct opening to the Indian Ocean through Baluchistan and would do everything to fortify Pakistan heavily against any further Soviet drive towards the South. Whatever the perceptions of the other neighbouring countries such as India, it is in the overwheming security interests of the West to continue its strategy. Indeed this strategy may strengthen India's security also inasmuch it will ensure that the Soviet connection with India would remain a safe "oceanic" connection rather than a more dangerous "continental" tie.

A realistic approach to the Indian Ocean problem will have to take into account all these factors, including the vital interests of both the littoral and non-littoral states in the region. In fact, today the Soviet factor is the principal source of apprehension for the Indian Ocean countries, littoral and non- littoral alike, as it was in Europe in the late 1940s and the 1950s. In view of the latest moves of the Soviet Union in Afghanistan and its threats to Pakistan, its intentions, contrary to the simplistic views of some of the littoral states in the region, seem to be inimical both to the littoral countries and to the Western countries. Its efforts seem aimed at having more and more client states in the region and thwarting all attempts to bring about basic changes in the security environment in Asia through such measures as the normalization of relations between India and China or between India and Pakistan, since such normalization would substantially reduce India's dependence on the diplomatic support of the Soviet Union. However, there is no doubt that its aim is to establish the presence of its navy permanently in the Indian Ocean. This likely permanent presence of a steadily growing Soviet navy, in the absence of any other matching presence, would inevitably result in Soviet dominance of the region and would prove a threat to the freedom and independence of the countries of the region.[20] Moscow could on the one hand tilt the global balance of power against the Western powers, and on the other deny petroleum and other vital minerals and raw materials of the region to the West. With this new ascendancy in Asia, coupled with its surreptitious efforts to penetrate into Africa as deeply as possible, the Kremlin hopes to achieve its objective of dominating the Gulf and the Indian Ocean region. The silver lining, however is that there is an increasing realization of this fact among the littoral countries of the Indian Ocean, especially because of the Soviet intervention in Afghanistan. And one of the overwhelming security interests of the West in the region since 1980 has been to prevent the Soviet Union from gaining access to the Indian Ocean by land through Pakistan.

As the possibility of keeping the Indian Ocean free from the activities of the big powers seems to be remote, it would perhaps be more fruitful to forestall the possibility of any single big power dominating the India Ocean. It is quite obvious that the attempt being made by certain countries to insulate the Indian Ocean artificially from the world strategic map seems to be an exercise in futility. It makes more sense to strive for a balancing presence of the various navies at a minimum level in the Indian Ocean so that no single navy would be in a position to pose a real danger to the stability and security of any littoral state in the region while the countries concerned are reassured that their vital interests there are not jeopardized by the dominance of any adversary in the region.

REFERENCES

1. Svyatoslav Kozlou, "The Truth about the Soviet Presence on World Oceans," *Soviet News* (London), April 17, 1973, p. 175.
2. UK Central Office of Information, *Britain 1980: An Official Hand book* (London, 1980), p. 370.
3. Lord Home, "The Scope and Limits of Detente," *The Year Book of World Affairs 1980* (London, 1980), p. 34.
4. A.K Chatterji, "Soviet Strategy in the Indian Ocean," in T.T. Poulose, ed., *Indian Ocean Power Rivalry* (New Delhi, 1974) p. 11.
5. For details see V. Vivekanandan, "Afghanistan Invasion Viewed from India," *Asia Pacific Community* (Tokyo), Fall 1980, no. 9, pp. 60-99.
6. Asoka Mehta, *Changing Alignments in Asia* (Bombay: Asia, forthcoming).
7. Shahram Chubin, *Soviet Policy Towards Iran and Gulf* (London, 1980) Adelphi Papers No. 157.
8. An Oscar-class submarine is 24,000 tons and has the capacity to carry out at least 20 long-range anti-ship cruise missiles capable of declining a nuclear warhead about 315 miles. A. 30, 000 tone *Typhoon class* is over three times the size of a Polaris submarine. *The Times* (London), December 17, 1980.
9. B. Vivekanandan, "Naval Power as in the Indian Ocean: A Problem in Indo-British Relations, "*The Round Table* (London), January 1975, pp. 39-72).
10. *The Hindu* (Madras), July 15, 1977.
11. Quoted in the *Indian Express* (New Delhi), November 30, 1978.
12. *Christina Science Monitor* (Boston, Mass.), December 26, 1977.
13. *The Hindu,* December 21, 1977.
14. International Institute of Strategic Studies. *Military Balance, 1980-81* (London, 1980), pp. 42 and 66-73.
15. *The Times,* January 25, 1980.
16. International Institute of Strategic Studies, *Strategic Survey, 1979* (London, 1980), pp. 54-55.
17. Robert Jackson, "Great Powers and the Indian Sub-continent," *International Affairs* (London), January 1973, p. 41.
18. *The Times,* January 25, 1980.
19. For an analysis of the separatist movement in Baluchistan, see Selig S. Harrison, "Baluch Nationalism and Superpower Rivalry," *International Security* (Cambridge, Mass.), Winter 1980/81, 5:3, pp. 152-163.
20. In India the Jammu and Kashmir Chief Minister, Sheikh Abdullah, said in March 1981 that the Indian subcontinent is faced with "The menace of Soviet Russia" at its doorstep. After Afghanistan, he said, it would be the turn of Pakistan, and finally Russia would be on India's borders. See the *Indian Express*, March 1, 1981.

XXI India at the End of the Century: Transformation into an Asian Regional Power

M.V. BRATERSKY & S.I. LUNYOV

In the 1970's, India became a centre of force in the South Asian subcontinent; in the 1980s it became a regional naval power. This article will argue that the 1990s will be the decade in which India emerges as a major regional power of Asia.

The analysis of the prospects for the evolution of India's strategy in the subcontinent and the Indian Ocean is based on several premises. First, in the present multipolar world, with the Cold War coming to an end, the countries of a region themselves play a decisive role in the formation of the system of international relations in the region. Second, in regions that are not vital to superpowers, the involvement of the latter in regional affairs becomes less pronounced. Third, supranational structures—international and regional organizations—in the Third World are as yet unable to play an effective part in the regulation of a region in, for instance, preventing conflict and restraining the ambition of various powers. Strategic games within regions are to a great extent free from the limiting factors(nuclear parity, awareness of a common threat, regional integration,developed political consciousness)that operate in West-East relations or those among the developed countries.

Fourth, the regional strategy of the new centres of force proceeds from the size of their military and their general economic and external economic potential, and also from the latter's quality,i.e., the technological level of production.These parameters also determine priorities in the search for raw materials and power sources, as well as opportunities and technological advantages of entering markets of various countries. In this way, the main directions of external trade and military strategy(strategic interests) may be defined. Finally, the measure of the "youth of the nations" will determine the balance of economic and military components of a country's regional strategy. This

Reprinted from *Asian Survey*, Vol. 30, No. 7, 1991, pp. 862-874 with permission of the Regents of the University of California.

applies both to the country employing the strategy and to the countries that are its object. The measure of their "youth"is general development—social, political, cultural- psychological, and economic. The further their development level is from the world's criteria of development, the greater is the role of violence in their foreign policy mentality and the importance of the military component in their strategy.

India and South Asia: Political and Military Dominance

To appreciate the geostrategic situation of the South Asian countries, some essential features must be recognized. Pakistan, Bangladesh, Nepal,Sri Lanka and the Maldives are situated at the periphery of the subcontinent in which India is the central power. Only India had a common border with the other states. India accounts for 73.4% of the territory, 76.5% of the population, and 97.2% of the gross national product. 'A similar situation exists in the military field. India's natural and complete dominance predetermines the characteristic features of the regional situation.'

The main characteristic of a power is its ability to establish a system of alliances and to project its power (military or economic, the proportion depending on the development level of the country, its environment and its foreign policy strategy) over a certain area. The purely qualitative difference between a world and a regional power lies in the area of its influence. India, according to all characteristics, has no doubt begun to play the role of a regional power. The period of the emergence of a power is sometimes marked by somewhat chaotic policy due to the necessity to adapt its new interests to the existing strategic environment. Only foreign policy practice can show what "niches" are relatively free and which of them may be filled by the new military-economic power. It was in exactly this way, by means of tests and tryouts, that India's regional strategy was evolved.

After winning independence in 1947, Indian ruling circles regarded the cardinal strengthening of India's position in South Asia as a major goal of its foreign policy course. That objective was considered as the main one throughout the 1960s, although the 1962 China-India conflict made the latter revise some aspects of its foreign policy. The early 1970s was marked by an event that shaped developments in the subcontinent and the adjoining waters for many years to come. The results of the 1971 India-Pakistani war and the emergence of Bangladesh created a situation in which there remained no country in South Asia able to seriously challenge India's predominance.

The task of ensuring security is regarded in India in terms of the entire subcontinent. Therefore, it is natural that ruling circles strove to keep neighbouring countries within India's sphere of influence, to consolidate its commanding positions" in South Asia and to oppose resolutely the participation of external forces—the superpowers in the first place—in the settlement of regional problems.[2] India realized that its tough stand in the regional field led the weaker countries to strive to develop relations with global powers in order to bring controversial issues in their bilateral relations with India's participation. From the early 1980s on therefore, New Delhi began to strengthen its relations with the states of South Asia.

A big stake was placed on the South Asian Association for Regional Cooperation(SAARC) set up in 1985. Formally, the initiative to establish a regional organization for economic and cultural cooperation of the countries of the region came from the president of Bangladesh, but such ideas had wide currency among the Indian political elite. According to the Indian side, such an organization would provide a unique platform for the revision of the status of the region in world politics by strengthening its independence from external factors (i.e., superpowers and world economic bodies), and for politically and organizationally reaffirming the zone of direct Indian interests and legitimizing India's leadership in the region.[3] Of course, not all of the member countries agree with that interpretation of the nature of the organization; therefore, it is still rather ineffective. It could be supposed, however, that with the increase of India's economic involvement with its neighbours, the time may come for SAARC to become if not a military, then an economic and political bloc of the ASEAN type.

The clash of India's foreign policy goals with those of other South Asian states at times leads to complications; periodically, its relations with Pakistan, Bangladesh, Nepal, and Sri Lanka become aggravated. Such things happen due both to the real situation on the subcontinent and to the "elder brother" complex, as well as the quite natural psychological approach of Indians to South Asia as a single entity. India's neighbours are especially comprehensive of its military might. The growth of Indian military expenditure became manifest in the early 1980s, but especially pronounced in the second half of the decade. In the 1986- 1987 fiscal year the increase was 27%, the following year 18%, and in 1988-89 it was 8%. While prior to 1985 the military's share of the GNP was 3.5%, in 1986-87 it amounted to 3.9% and in 1988-89 it rose to 4.2%(in Pakistan it is 6.4%). In 1981 defence expenditures made up 18.1% of India's budget (Pakistan's was 33.6%), in absolute terms Rs 132 billion (over $9.6 billion).[4] The strength of the Indian army is at 1.362 million, the fourth largest in the world. It has over 3,000 tanks, including 50 T-72s, and over 700 combat planes. In 1989 India successfully launched the Agni missile with a range of 2,500 kilometers. Soon, tactical surfact-to-surface Prithvi missiles (150 km range), Akash surface-to-air, and Trishul air-to-air missiles will be added to its arsenal.

The purchase of foreign-made weapons is of no small importance to India. About 75% of its imported arms are from the USSR, but India has been trying to diversify its sources of arms supply to avoid dependence on a single country. It has obtained VTO Sea Harriers, ship-based missiles and aircraft carriers from Great Britain; Mirage 200 fighters from France; and SSK-2500-U submarines from Germany. Recent years have witnessed the development of American-Indian military-technological ties, the pivotal event in that field being the U.S. consent to help India develop a light combat aircraft to serve as the basis of the Indian Air Force. The Indian government's usual policy has been to buy technology, not the weapon itself, which is also prompted by the desire to retain independence in the field. Since 1986, India has been the biggest Third Word importer of arms, purchasing $5.2 billion worth of weapons in that year alone.[5] The defence R&D expenditure also increased sharply—from Rs 950 million in 1981-82 to Rs 4.648 billion in 1987-88, to Rs 5.616 billion in 1988-89, 32% of appropriations for research were earmarked for military purposes.[6]

India's position vis-a-vis nuclear weapons is quite ambivalent. It alredy has become a "threshold power". According to its leaders, it needs only to "turn a screwdriver" to start producing nuclear arms.[7] India did not sign the Nuclear Nonproliferation Treaty on the ground that the People's Republic of China possesses nuclear weapons and therefore India should be "ensured" against them. Like some other countries, India believes that the aim of the treaty is to preserve the nuclear weapons monopoly by the developed countries.[8] At the same time, it is difficult to regard as justified the stand supported by many political and public groups in India that world power status involves the possession of nuclear arms.

The Indian government wanted to avoid international control of its nuclear activities which, it claimed, would undermine the prestige and sovereignty of the country. New Delhi rejected all accords of that kind, and the Indian mass media published opinions of various experts maintaining that the possession of nuclear arms is the most reliable and economical means of ensuring defence capability. According to their calculations, a modest programme of nuclear deterrence (50 warheads) would cost approximately 5700 million, which would amount to 8% of defence spending by the year 2000.[9] The appeals for India to enter the nuclear club are to a great extent prompted by the apprehensions caused by Pakistan's nuclear programme. The appearance there of nuclear weapons will, in the opinion of the Indian military, devalue India's superiority in conventional arms.[10] The pro-nuclear stance in India is also fed by the existence of nuclear arms in the PRC and the nuclear presence of great powers in the Indian Ocean.

One should not try to describe the rapid buildup of Indian military power in a unilinear way. The increased military might of China and Pakistan, the strengthening of military and political links between them, and Pakistan's military collaboration with Saudi Arabia and a number of other Islamic states could not but cause anxiety in India. Since the late 1970s, the Indian government has also had to take into account the negative changes in the Indian Ocean area and Southwest Asia casued by the militarization of that region, the conflict in Afghanistan, and the Iran-Iraq war.

A comparison of Indian and Pakistani military power shows India's superiority in all major components: Indian army strength is 2.5 times greater than that of Pakistan; India has twice as many planes, 1.7 times as many tanks, and 1.6 times as much artillery. Moreover, the military technology at the disposal of the Indian Air Force is much more sophisticated than that of Pakistan. The strained relations between the two cause India to keep some two- thirds of its land forces on their common border. However, in analyzing the defence efforts of India, the Chinese factor also has to be taken into consideration. We agree with Jasjit Singh, director of the Indian Institute of Defence Studies and Analyses, that the treatment of the military and political situation by many experts exclusively in terms of the subcontinent provides a distorted picture.[11] India's military doctrine envisages not only its complete superiority over Pakistan, but also the maintenance of sufficient forces to contain China. Indian politicians think that a military conflict with the PRC is improbable at present, but the military elite exercises no small pressure on the government in demanding a military buildup in the Chinese direction. Another thing to be considered is that India regards American links with Pakistan as a part of Pakistan's general military structure aimed against India. Public opinion polls show that the bulk of the Indian population believes in the existence of a conspiracy against India's unity and regards the military threat as a serious factor.

One more consideration should be noted. The developed countries consider that the main, perhaps the only, threat to their security comes from external forces, while in the developing countries a prominent role is also played by internal threats. The threat to security"from inside" is no less important than the threat "from outside," and several army divisions are deployed to fight separatism. This explains the highly negative Indian response to any information about attempts of external forces to use India's internal contradictions.

Change in the Superpower Approaches: Late 1980s-1990s

The end of the Cold War era became a reality not only in East- West relations but also in the Third World. It is not yet possible to speak about the final end of the superpowers' rivalry in the Third World, and particularly in the Indian Ocean area, but the beginning of a reassessment of their roles in the region has become apparent.[12] The approach envisaging an ideologically neutral assessment of the conformity of foreign policy results to inputs has acquired a greater weight in the formulation of policy decisions. It turned out that the U.S.-USSR rivalry in obtaining zone of influence in the Third World often proved counterproductive for several reasons: the superpowers developed overly heavy commitments vis-a-vis their allies; the price of alliance relations proved to be too high; the depth of the involvement of the superpowers in regional conflicts involving their allies heightened as allies began to manipulate their patrons. And what was most significant, it turned out, the "zero- sum game" in which the U.S. success was regarded as a defeat for the USSR, and vice-versa was often played in regions of no great strategic importance to either the U.S. or the USSR.[13]

As to the Indian Ocean, the U.S. seems to have concluded that it is mainly interested in two strategic positions—the Straits of Hormuz and Malacca. It is significant that it was able to convince India of that view, which has smoothed down a littlle the American-Indian frictions over the Diego Garcia base and has set apart the U.S.and Indian"spheres of interest" in the Indian Ocean. The echoes of the Iran-Iraq war, paradoxical as it is, have somewhat calmed India regarding the role allotted by the U.S. to its navy in the Indian Ocean. India, of course, did not give up its demand that the navies of the non-littoral states be pulled out of the region, but in the opinion of real politicians—-and Indians are real politicians—the U.S. proved that it needs the 7th Fleet and the Diego Garcia base to protect its interests in the Persian Gulf.[14] At any rate, the talk of the American threat to India has become more muffled.

An important change in American-Indian relations took place in the mid-1980s. Judging by all evidence, the U.S. has got accustomed to the fact that India, with its help or without it, is steadily becoming a promiment centre of power, not only in South Asia but in the Asian region as a whole. Today, the U.S. has no prominent interests in India, but it would perhaps prefer to have normal relations with New Delhi in the future. This is seen in the tremendous expansion of technology transfer to India, including the results of purely military research,[15] as well as from the de facto recognition of India's right to ensure stability in the region(all the more since neigther the U.S. nor the USSR have recently expressed a wish to perform that function). In fact, the super-powers have even informally encouraged India to act as a controlling force, as was seen from their broad support of the India-Sri Lanka agreement.

In the new strategic context, India established itself as the regional power in the surrounding strategic water area. In 1987 the India-Sri Lanka Accord was signed, in accordance with which India endeavored to settle by military means the internal ethnic conflict in Sri Lanka. Another event, which evaded the attention of most analysts, occurred a little later: Indian paratroopers and navy units carried out a blitz operation to suppress the anti-government coup attempted in the Maldives Republic, demonstrating India's capabilities in its new status.

As the global tension has been easing, the great powers have been showing more willingness to recognize the natural dominance of India in the region. India's regional strategy is not limited to its dominance in South Asia; it also attaches greater attention to the notion of "region" the major part of the Indian Ocean and a part of West Asia, as well as South Asia.Actually, a single strategy became characteristic of India both vis-a-vis the subregion and the Indian Ocean basin.

Political, Cultural and Psychological Factors shaping India's Strategy

There are no serious differences within India regarding the main goals of its foreign policy. Its permanent objectives have been: to utilize contradictions between the two sociopolitical systems, and the superpowers symbolizing these systems, to provide the most favourable conditions for economic development; to meet internal social challenges; to guarantee the integrity of the country and keep a high level of defence capability; to achieve informal leadership in the Nonaliged Movement; and to gain dominating positions in the region. Up to now, the major foreign policy actions of India have been based on a broad consensus of various groups in the ruling circles. In the past, the coming to power of parties opposed to the Indian National Congress (I) did not lead to cardinal changes in the substance and priorities of Indian foreign policy. There are grounds to believe that the Congress (I)'s defeat in the 1989 general elections also has not resulted in cardinal foreign policy changes.

There is no doubt a difference in the foreign policy platforms of the Congress and the opposition National Front coalition that came to power in 1989. Regarding regional policy, the new V.P. Singh government is on the whole less ambitious in its foreign policy strategy, and because its social base is a little different from that of the Congress, it is less interested in an all-around consolidation of the position of Indian big business in other countries. The nuclear lobby is less active in the National Front than in the Congress. But the differences among India's various political forces cover not so much the macro objectives as the means of their achievement. In this, the present government prefers a broader recourse to compromise and concession in dealing with countries of the region, while the Congress at times in the past resorted to forceful methods. Since V.P. Singh came to power, greater priority has been given to India's "immediate" region policy as compared with its broader relations with global powers. During his first months in office, Singh did not visit any of the global powers unusual for Indian politics. Though today Indo-Pakistani relations remain strained because of the Kashmir issue, India's relations with its other neighbours have notably improved. Of course, this does not mean that the National Front will reject force in case of grave changes at the regional level that threaten India's national interests; nor should one think that the new government is against India becoming a mighty regional power.

In contrast to the vast majority of other developing countries, India's foreign policy adequately reflects not onlly the state but also national interests and the internal alignment of forces. This explains in certain measure the absence of sharp zigzags in the country's foreign policy in the past. The middle class (about 100 million) adds great stability to foreign as well as domestic policy, for its approaches to the foreign policy values are always taken into consideration. A certain role is played by India's 40-year long adherence to parliamentary democracy. India's standing in the Nonaligned Movement and the Third World is quite high; at present it enjoys the status of spokesman of the developing countries. Therefore, its example as a country promoting democratic values while developing economically may favourably affect the processes underway in the Afro-Asian world.

The cultural and civilizational features of Indian society facilitate the strengthening of the Republic's positions. On the one hand, India has always possessed a powerful reservoir of intellectual and spiritual traditions.Neither self-isolation and contempt for the surrounding world, nor excessive aggressiveness and intolerance—features of most other Asian civilizations—have been characteristic of Indian society as a whole. Thus,there exists a "culturological rationale" for the expansion of influence of that country in the world. On the other hand, it should be noted that, compared with the vast majority of Third World states, it is qualitatively easier for India to embrace and to make use of the achievements of sophisticated technology. This enables it to implement economic modernization and enter a new stage of the scientific and technological revolution, which creates new opportunities to expand economic ties with other states including those of the Indian Ocean area.

India's Ocean Policy

India is situated in the very heart of the northern part of the Indian Ocean and has a tremendously long coastline. Many reasons, including geographic, arouse India's interest in exercising control over the surrounding waters: because of its vulnerability from the sea, India has to pay close attention to the security of its sea borders; the bulk of India's foreign trade goes through its ports (the value of goods shipped by sea is up to 15% of its GNP [17]); India has an oil-rich shelf, other mineral resources and marine products. Therefore, the problems of control over adjacent waters and creation of the necessary military infrastructure have had a high priority in the country's policy since the very moment of independence.

India has been building up its navy quite rapidly; the naval budget increased 200 times in 40 years and forms 13% of the entire defence outlay.[18] The country's new role entailed the creation of the potential for projecting its power at sea. The large-scale naval programme launched in the 1970s had two aims: to protect the Indian coast and marine economic zones and to set up the potential for controlling a broader water area, comprising at the time the South Asian region. It seems that the main accent was on the second task, and as the Indo-Pakistani conflict demonstrated in the early 1970s, India's naval power was sufficient for the security of the country's coast. The development of naval carrier- and land-based avisation, the development of the submarine fleet(the biggest in the region) and the induction of ships for protracted endurance cruises into the navy suggested broader aims.

The 1970s was the decade of India's emergence as a regional power; in the 1980s it became a regional naval power. The naval programme, called upon to make India's presence felt in the entire Indian Ocean, had also to preempt the development of the Chinese Navy and guarantee the future security of India's territory from the sea. The sophistication level of the Indian Navy changed along with its strategic character: while in the 1950s and 1960s, small patrol and escort ships served as the nucleus of the naval force, in the 1970s and 1980s this role belonged to frigates. In the 1990s nuclear attack submarines and aircraft carriers will provide the main striking power of the Indian Navy.

Real signs of these plans being implemented became apparent recently. On February 3, 1988, the Indian flag was hoisted on the nuclear submarine-hunting submarine, Chakra, leased from the Soviet Union, which is to be used to train Indian crews. According to available reports, by the end of the century the USSR might sell to India five to seven nuclear submarines of the Victor 3 class (a 1979 project) and Charley 2 (able to carry cruise missiles). Both ship types are in the inventory of the Soviet Navy. Another line in the buildup of the strategic potential of the Indian Navy is the development of the aircraft carrier fleet. The British-built carrier Viraat was commissioned in 1951, and the second Indian carrier, Vikrant the former British Hermes, was handed over to the Indian Navy in 1955. Indian naval authorities plan to have five aircraft carriers: two in the eastern and two in the western marine zones-i.e., in the eastern and western parts of the Indian Ocean—and one under repair or reconstruction.

The political issue is whether the level of the developed navy and the ocean strategy conform to the country's external economic power. Military power, projected on economic positions already achieved, plays an important stabilizing role, while "bare" military power, deprived of an external economic basis, becomes a rather destabilizing factor because, in the first instance, it is intended to protect the status quo, whatever it is, an in the second, it may be used to break it up, and violently.

In this case it is difficult to produce precise criteria, but it is clear that the probability of the use of military force to solve foreign policy problems is higher in Africa and Asia than, let us say, in Western Europe. In great measure this may be accounted for by the strategic balance existing in Europe and the presence there of nuclear arms, but not an unimportant role is also played by the level of development of the countries—the state of their civil societies, high prosperity level, and absence of violence in internal politics. If such criteria are applied to India, it will be seen that although it has reached a relatively high level of social, political, and economic development, violence still remains prominent in its internal political life. This cannot but affect its foreign policy mentality. Furthermore, the politically and socially unstable external environment, where the level of violence is even higher, does not facilitate the rejection of military means in foreign policy. As we see it, all this prompts India to rely to a greater extent on its military and naval power in the conduct of foreign policy than it would have done in the more developed regions of the world.

The military machine created up to now is sufficient to control the adjacent strategic water area and to protect India's national security. Its further buildup within certain limits will create a reserve of defence strength, but if the military buildup is going to be expanded further, that may produce a double negative effect. First, it will provoke other countries of the area, and second, excessive defence spending may undermine the economic power of the country in the future. Its economic potential makes India a great power only when compared with its poor environment. It is highly important at the present stage for India to maintain an optimum balance between the buildup of economic and military power, to see to it that political ambitions do not outrun economic possibilities.

Real politics, maintaining a balance of all elements of foreign policy and economic strategies, presupposes the projection of military power in the same direction in which the external economic ties develop. India's naval strategy seems to follow that principle. Of course, considerations of national security make corrections in the picture, but on the whole the Indian Navy develops approximately according to that scheme. As has been already pointed out, the western part of the Indian Ocean lies within the zone of vital U.S. interests, which inhibits India's progress in that direction. Strategic vistas open in the South and East. Today, there are no reasons or possibilities for India to expand its strategic zone, but some attempts are being made: Indian ships pay friendship visits to Australian ports and sail in the South of the Indian Ocean; Indian scientists have opened up a second base in the Antarctic continent; and India keeps advocating a pullout of the fleets of non-littoral states from the Indian Ocean area. This is a quite legitimate demand, but those who criticize this policy see behind it India's wish to make its rather powerful navy the only real force in the water area.

Economic Measurements of India's Foreign Policy Strategy

Some countries of the region and the area adjacent to it, the NICs(newly industrialized countries) in particular, have apparently made a qualitative technological leap and entered the "post-industrial era". They have developed in the most sophisticated forms and have already taken a solid place in the world economy. Perhaps the tremendous difference in levels of technological development will prevent India from competing with them as an equal in the near future. What India can and does offer is cheap, but not highly skilled labor and the products of the traditional industries metallurgy, heavy engineering, construction. Japan has developed the most sophisticated science- intensive industries, yielding to the NICs the markets for household technology, consumer goods, and industrial products. Gradually, the NICs have also mastered a higher level of production, and perhaps India will be next in line for markets to supply products that have already become traditional. But this prospect cannot be regarded as a strategic objective. The economic power of a country is determined in an ever greater measure by the speed at which the results of the scientific and technological revolution are introduced into production and the flexibility and capacity for constant innovation. This was realized by Indian leaders, who set the task of entering the twenty-first century as as industrially developed power and who pay great attention to the modernzation of the Indian economy. But all this as yet is but plans that may serve as the foundation of real economic expansion only in the future. The present technological state of the Indian economy, with the exception mentioned, allows us to expect a serious strengthening of India's economic position only in the most underdeveloped countries where there is demand for its industrial products because of their low price. Among such countries are the South Asian states, countries of East Africa, insular states of the Pacific, and some countries in Southeast Asia — Cambodia, Laos, Vietnam, Burma.

Another important feature of the economic situation in the Indian Ocean area to be taken into account in an analysis is the economic policies of the big industrial powers in the region and the processes of regional economic integration. Abstracting oneself from the traditional geographic division and taking as the basis the economic characteristics—integration, the level of economic and technological development, and the place in world production and trade—one finds that the Indian Ocean area and the adjacent regions comprise (if the Middle East is regarded as a separate region) at least two clear-cut zones: the ASEAN countries, including the NICs that are economically gravitating to Japan and the U.S., and groups of less developed countries in the area. Main commercial streams, the movement of capital, and the transfer of technology are shaped accordingly. The ASEAN countries occupy a more and more prominent place in the foreign trade turnover of the U.S. and Japan, which seriously inhibits India's economic advance in that subregion.

While India's entrance into the market of a number of the countries of the area is made difficult due to its lagging behind in technological development and the emergence of the U.S.-Japan ASEAN triangle, its economic advance into the less developed countries of the area may also face strong competition from China. The latter is not only the principal strategic rival of India in the region, but also an important economic force not to be ignored by India. At present there is a race betwen the econonmies, which soon will turn into a sharp competition. Naturally, India's ocean strategy takes that prospect into account.

This very general analysis helps to delineate roughly the zone relatively open in the region for India's economic advance in foreign trade, investments, and joint enterprise. There are, to be sure, no grounds to believe that India will not try (it does it even today) to develop economic ties with the most advanced countries as well. India is engaged in quite broad-scale construction work in the developed countries, and some of its research and goods compete successfully in the most glutted markets of the U.S. and western Europe.[19] But the main vector of its economic offensive may be aimed today only at a rather limited number of countries- the poor in the first place, which rim the western and southern parts of the Indian Ocean. In 1987 India's GNP was one-quarter of the total GNP of all countries in the Indian Ocean area. If the countries lying in the eastern part of the ocean and those who gravitate toward the Pacific are excluded, India's share will grow by almost two fifths, and without the Arab countries (i.e. if only the western and southern parts of the ocean are taken into consideration) it will amount to 57%.

One should also keep in mind the rapid rate of development of the Indian economy. In the second half of the 1980s, the real increase in GNP exceeded 6%(fourth among the 25 economically most developed countries.) The forecasts of IMF experts made in 1985 to the effect that by 2010 India's GNP will be about $1 trillion (at 1980 prices) and by 2020 over $2 trillion, have up to now proved true(in 1987 it was $241.3 billion).

India's economic links with the Indian Ocean countries are developing slowly, but in the line of ascent. Suffice it to point out that India is on the list of principal importers of 17 out of 27 countries of the area.[20] It actively exports capital, both public and private. Old cultural ties with many countries, including the presence of many Indian residents abroad, should also be taken into account. All this facilitates the expansion of economic links. Lately, Indians who reside in other countries have been playing a prominent part in the carrying out of India's technological policy and in attracting foreign capital. A vivid example is the U.S. Indian community. In the conditions of the rapidly developing Indian economy, the shift of the center of world economic development to the Pacific, and a decline in the interest of developed countries in the developing states (which seems to be a rather stable trend), India can certainly fill the vacant low technology niches in the region and be instrumental in the economic and political integration of the Indian Ocean area countries.

India's Role in the Twenty-first Century and the Postures of the Superpowers

The relative weakening of the superpowers' economic, military, ideological, and political roles is but a tendency, though evidently a leading tendency. The world moves farther and farther away from the bipolar model and the USSR and the U.S. must define their attitude to these processes. The system of international relations has become much more complex in the multipolar world and is no longer determined largely by Soviet-American relations. An important role will be played by prospective centers of force of the India or Brazil type, which are rapidly increasing their political and economic weight.

As we see it, there exists even now no real confrontation of Soviet and

American interests in South Asia. The onward movement of democracy in India and its striving to stabilize the situation in the region-supporting bourgeois-democratic movements in the neighboring countries, on the one hand, and making attempts at evolutionary democratization of the regimes in these countries, on the other—conform to the contemporary foreign policy interests of the superpowers.

It should be admitted, without absolving the two superpowers of their share of guilt and taking into consideration other factors, that the main source of pressure on the stability of the system of international relations in the Third World, of elements of tension introduced into it, and the flareup of regional conflicts is underdevelopment. In this context, India's efforts to develop its own economic potential, to overcome the backwardness of the countries of the region, and its approach to the creation of a mode of universal economic security deserve to be supported. Perhaps, some may be irritated by the prospect of India becoming a world power, but this process can only be slowed down a little, not stopped. It is in the U.S. and Soviet interests to support the emergence of a new world center with ethical and spiritual values close to European civilization—in contrast to many other countries-which may become a staunch bulwark in the defence of democratic and human ideals.

Conclusions

In the 1970s India became a centre of force of the South Asian subcontinent, and in the 1980s a regional naval power. The 1990s are going to be the time when India emerges as a major regional power of Asia. Such development to a great extent has been facilitated by the weakening of the East-West confrontation, the revision by the United States and Soviet Union of their regional priorities, particularly in South Asia, and the deescalation of their rivalry in the zone of developing countries.

India's growing military and naval power is called upon to ensure security, stabilize the situation in the subregion, and support India's advance into the markets of some countries of the Indian Ocean area (East Africa, Indochina, and the southern insular states). The level of India's economic and technological development and its international political and military strategic limitations make it possible for India to move into markets mainly of the less developed countries of the southern and eastern parts of the Indian Ocean. India's economic dominance in these countries is not to be expected; it is hindered both by world integration processes and the relatively low technological level of Indian exports. In the foreseeable future, India will advance mostly in the sectors of traditional industrial and low technology production (consumer goods).

The expansion of India's "Zone of responsibility" is not on the whole inconsistent with the interests of the world community, the U.S., and the USSR. This is due to the following reasons: (1) India will stabilize the situation and facilitate economic development and democratization processes in an area where the leading world centers are not prepared to fulfil these functions; (2) economically, India will not dominate totally as its military power is not to be used for the purposes of aggression or suppression of positive processes taking place in the region; this is hardly probable because of the peculiar nature of the internal political, cultural, and psychological foundations of its foreign policy; and (3) India is "inscribed" in the humanistic and civilizational characteristics of the contemporary world and will help move the transformation of the underdeveloped countries in that direction.

REFERENCES

1. Calculated on the basis of the *Europa World Year book,* Vo. 1-2 (London, European Publications, 1989).
2. Robert Hardgrave, *India Under Pressure: Propsects for Political Stability* (Boulder, Westview Press, 1984), pp. 144-46.
3. Leo E Rose, "A Regional System on South Asia: Problems and Prospects, in Leo E. ed. *Asian Security Issues: Regional and Global* (Berkeley: University of California).
4. *The Hindu* (Madras), 5 March 1988.
5. *Time,* 3 April 1989.
6. *The Hindu,* 5 March 1988, and *Telegraph* (Calcutta) 3 October 1983.
7. *Telegraph,* 28 March 1989.
8. K.N.Singh, Deputy foreign minister of India, called the policy "Nuclear apartheid", *The Hindu,* 21 September 1987.
9. V.G. Kulkarni, "Indo-Pakistan Rivalry Boosts Nuclear Option" *Far Eastern Review,* 9 April 1987, p. 43.
10. *Telegraph,* 7 March 1987.
11. *Illustrated Weekly of India,* 23 July 1989.
12. See, for instance, R.W. Jones and S.A. Hildreth, "Regional Powers and Defence Introduction," in R.W. Jones and S.A. Hildreth, eds. *Emerging Powers: Defence and Security in the Third World* (New York, Praeger, 1986), p. 16.
13. On the USSR, See E.A. Koldoziej and R.I. Kanet, *The Limits of Soviet Power in the Developing World* (London, Macmillan Press, 1988).
14. D. Braun, *The Indian Ocean: Region of Conflict or "Zone of Peace?* (London: Goom Helm, 1983), p. 138.
15. See, S.I. Lunyov, *India i SSha, Mezhgo udarstvennyye otnosheniya* (Moscow Nauka, 1989), pp. 159-60.
16. B.S. Gupta, "Hegemony in South Asia" *World Focus,* 8:11, 12 (Delhi) 1987, p. 63.
17. *Telegraph,* 26 February 1988.
18. M.G. Gupta, *Rajiv Gandhi's Foreign Policy: A Study in Continuity and Change"* (M.G. Publishers 1987), p. 62.
19. See J. W. Mellor, ed. *India: A Rising Middle Power* (New Delhi: Henemann, 1981), pp. 357-61.
20. There are different assessments of the number of countries of the Indian Ocean area. The figure 27 comprises the states directly adjacent to the Indian Ocean.

XXII India and the Post-Cold War World : Opportunities and Challenges

S.D. MUNI

For nearly 30 years, India has been on the receiving end of international strategic, political, and economic transformation. Earlier, during the first decade of its independence, India tried to mobilize Afro-Asian countries in order to influence the then emerging global political and strategic order. Its objective was to secure greater democratization in international relations and free access to available economic resources and opportunities for the politically weak and economically poor countries of the Third World. Accordingly, Afro-Asian countries became the principal arena for India's policy and diplomacy during the 1950s. But such Indian endeavor did not go very far and had to be almost abandoned by the beginning of the 1960s. India was unable to sustain its efforts not only because it suffered military humiliation at the hands of China in the 1962 border conflict, but also because Jawaharlal Nehru, the architect of India's foreign policy and an authentic spokesman for the Third World's aspirations and hopes for a just world order, died in 1964.

But even if these two developments had not taken place, India's role would have remained marginal to the emerging world order of that time and perhaps subsequently. The forces at work at the global level—the superpowers and the intensifying cold war between them—were all too powerful and determined to shape the world in a way that suited their interests. Against them, countries such as India, which stood for an alternative order responsive to the aspirations of the weak and the underprivileged, were too weak themselves and too disorganized to make their impact felt. The Cold war, which was brought to India's doorstep as a result of the U.S.-Pakistan military alliance forged in 1954, to a very large extent blunted India's initiatives and innovations in shaping the new world. The better part of its efforts thus remained preoccupied with reacting to the changes taking place so as to contain their damage

Reprinted from *Asian Survey*, Vl. 31, No. 9, 1991 pp. 956-975. with permission of the Regents, University of California.

potential and learn to live with them accordingly.

As in the late 1940s and early 1950s, the world has once again been undergoing major transformations. The Cold War structures are now being dismantled and dissolved in their military- strategic as well as politico-ideological dimensions. This process of dissolution and dismantling, however, is not uniform or consistent but gradual, staggered, and in some significant respects, selective.

In a way, the voices of euphoria in the West that claim the Cold War has been won by the West, both militarily and ideologically, make the process look one-sided and highly asymmetrical. The economic and political mess in which the Soviet Union finds itself gives credence to these voices. But it is possible that as the varied and diverse ramifications unfold, some of the euphoric voices may start to fade. The end of the Cold War process leaves serious uncertainties, as one cannot say for certain how much of its strategic and ideological structures will really be dismantled and what in their place will constitute an alternative nexus in great power interaction. The rivalry and competition among the great powers and their coalitions, which up to now have acted as the engine of change and dynamism in the world, may not completely disappear, although the forms, extents, and manifestations are bound to change.[1] Also uncertain are the parameters of the emerging global order that will replace the disappearing or eroding structures and stigmas of the Cold War. At best, one can only discern some broad directions of the transformations taking place and make projections on their basis to see what is in store in the coming months and years.

Indian foreign policy, as that of several other Third World countries that are generally at the receiving end of global changes and transformations, has to grasp the thrust of emerging directions in order to evolve adequate and sustainable responses to them. The challenges and opportunities inherent in them have to be identified and assessed so that respective national interests, in the immediate as well as the long term, may be advanced. From the point of view of India's foreign policy, a number of specific developments in the strategic, political, and economic fields in the current situation deserve careful attention. Let us look at them one by one.

Strategic Implications

All countries of the world have welcomed the dismantling of the Cold War and its concomitant processes of relaxation of tensions between the two superpowers and their respective allies and their strategic and conventional disarmament. However; when these processes are looked at from the vantage point of India and the Third World, they appear to be—as the Cold War essentially was—Eurocentric developments. In terms of fallout on the Third World security situation, the overall global relaxation of tensions has no doubt had a benign effect, as can be seen in the lowering of regional conflict levels, for exmple, in Afghanistan and Indochina. But this is not enough. These persisting regional conflicts are still to be resolved fully. Besides, there are no signs of the naval disarmament question being taken up on the agenda of superpower talks.

As a result of the end of the Cold War, many dormant and simmering regional conflicts and bilateral rivalries have suddenly become live again creating prospects of armed flareups and internal turmoil. The crisis in the Gulf reflected a strong element of this newly worsened security situation for Third World countries, and it also had several other complexities built into it, including aspects of North-South resource and economic conflicts. Many of the still potential and simmering regional conflicts in the Third World may receive an impetus from the flow of surplus arms from Europe as a result of new arms control agreements. The changing dynamics of the "arms bazaar" in North America and Europe as a consequence of the so-called "peace dividend" is adding to the prospects of an ever growing diffusion of weapons and military technologies in the Third World.[2] All this has to be viewed with considerable concern by a country such as India which has been forced into an expensive and unproductive arms race to match the introduction of sophisticated military technologies supplied to its neighbours under the exigencies of the Cold War. Mention may be made of continuing reports of China-Pakistan nuclear collaboration and the recent disclosures of China's supply of M- 11 missiles (capable of carrying nuclear and chemical weapons) to Pakistan.[3]

The withdrawal of superpower military presence has been uneven. While Moscow's doctrines of "reasonable" and "sufficient" defence resulted in a considerable drawdown of a Soviet presence—from Afghanistan and Vietnam on the one hand, and in relation to China, Mongolia, and Eastern Europe on the other—NATO remains intact. In particular, the U.S. presence in Asia is still unaffeoted. The U.S. President has promised the Congress to reduce U.S. defence expenditures by 25% by the year 1995, but if the U.S. behavior in relation to the Gulf is any indication, there are serious doubts that the promise will be fully kept.[4] In any case, the U.S. presence under this promise of overall 25% reduction will not be more than 10% to 12% in Asia, and that too would be confined largely to the presence in Japan, South Korea, and the Philippines. A good deal of this reduction may, in fact, be made up by encouragement to allies such as Japan and Australia to increase their share of the Western defence burden.

If we add the factor of internal turmoil, political uncertainties, economic crisis, and theratened disintegration in the Soviet Union, the global strategic, balance seems heavily tilted in favour of the United States and its Western allies. Notwithstanding the much talked about emergence of Western Europe—particularly the 1992 Europe with a united Germany—and Japan as economic giants, the global balance in military-strategic terms may be described as a one-superpower-situation.[5] In many ways, despite vital qualitative differences, this is akin to the situation that obtained during the late 1940s and early 1950s that characterized the beginning of the Cold War.

Accordingly, for Indian foreign policy the dilemma of the early 1950s has qualitatively reemerged. The choice now, as it was then, is difficult and tricky, i.e., choosing between joining the Pax-Americana and the charting of an independent Nehruvian course.[6] In some respects Indian policy seems to be groping in the direction of the former. The Soviet Union, under the thrust of its new dynamism in foreign policy and specific strategic and economic initiatives toward China and the Asia-Pacific countries, has subtly distanced itself from India(and other countries such as Vietnam), its former strategic ally and close friend. Against this, the U.S. relationship has been warming up since the beginning of the 1980s. Even former presidents Carter and Reagan appreciated India's discriminating approach toward getting the Soviet military intervention form Afghanistan, contrary to the imnpression created by diplomatic rhetoric and media images. Thus, the U.S. became India's major trading partner in 1983-84, surpassing the Soviet Union. In the fall of 1984, New Delhi and Washington signed a Memorandum of Understanding (MOU) that opened the way for further cooperation between the two countries, covering greater economic exchanges, transfers of sophisticated U.S. technology (supercomputer, etc.) to India, and cooperation in the defence field. The prospects of cooperation under the MOU grew much brighter during the Rajiv Gandhi regime (1985-89), getting an additional boost from the beginning of the end of the Cold War.

Many Indian analysts, policy makers, and interest groups have argued for strengthening this cooperation with the United States. Several possible advantages have been identified including the easy flow of much needed technology and investment. The two-way trade between India and the U.S. stood at nearly US$7 billion per year by 1990. In the strategic field, the increasingly visible shift in the U.S. position in Indo-Pakistani issues in favor of India is being cited to underline the positive aspects. This shift is particularly notable in the U.S. urging of Pakistan not to encourage the Punjab and Kashmir secessionists and to seek instead a political and negotiated solution to the Kashmir dispute on the basis of the 1972 Simla Agreement. Many Indians are also delighted at the prospect of the U.S. discontinuing security assistance to Pakistan due to the latter's presistence in developing nuclear capabilities for military purposes. The U.S. endorsement of an assertive regional role for India, as witnessed in relation to sri Lanka and the Maldives during the late 1980s is also being mentioned to buttress the arguments in favor of joining the Pax Americana.

A final Indian policy decision to line up with the Pax Americana in the post-Cold War global stragegic balance would, however, not be without its constraints and costs. India would have to identify itself closely with the broad thrust of emerging American strategic interests from, say, Guatemala, to the Gulf, to Guam. The U.S. expectation that India would join the so-called multinational force in the Gulf to push Iraq out of Kuwait may be recalled in this regard. Washington's unhappiness over the political fuss created by the opposition parties in India on the question of refusing U.S. war planes enroute to the Gulf war was clearly evident.

In the post-Gulf war strategic order envisaged by the United States for Asia, India may be expected to endorse and collaborate with U.S. moves in the Indian Ocean region. Such moves may cover India's sensitive neighbourhood, as in the recent examples of landing U.S. marines in unusually large numbers in Bangladesh in June 1991 under the pretext of the cyclone relief and of revival of the proposal to expand the Voice of Americas transmission station in Sri Lanka notwithstanding provisions against and expansion in the Indo-Sri Lankan Agreement of 1987. Indian policy of closer identification with, and reliance upon the United States will also make it vulnerable to restraints on India's nuclear and missile programme. The application of the Pressler Amendment (no U.S. aid to countries pursuing nuclear-weapons programme) to India in a U.S. House of Representatives resolution may be seen to point in this direction.[7]

On the nuclear question, of course, much would depend upon the outcome of the U.S. endeavor to secure from Pakistan a credible non-proliferation assurance that could be presented to the U.S. Congress to facilitate the presidential certification necessary to restart security assistance.[8] It is pertinent for advocates of the Pax Americana line to properly understand that the U.S. has been pursuing a policy of parallelism toward India and Pakistan in order to evolve a regional balance in South Asia that would serve its perceived interests. India, which has consistently resisted such a parity approach, will have to learn to live with it if it seeks greater strategic proximity to the United States. In the field of economic cooperation, India will have to open up its economy through heavy liberalization measures as demanded by the U.S., which may even mean accepting distortions in some sectors of its industrial, trade, and domestic capital market policies.[9] The overall impact of this on India's standing and image in the Third World cannot but be nagative and undesirable.

It would, therefore, be more advantageous for Indian foreign policy to rediscover Nehruvian elan in its global strategic posture and adjust it to the emerging post-Cold War order. This does not mean any attempt to dilute India's economic and technological areas of cooperation with the united States in the bilateral context, but rather preserving its global economic and strategic options. It may be recalled here that Nehru in 1949 had proposed to do exactly this, to forge closer economic and scientific cooperation ("somewhat ally") with the U.S. without surrendering his independence on political and strategic initiatives on global issues.[10] He was then spurned by the United States. One hopes that India, if it adopts such a stance now, may be better understood by

U.S. policy makers, for in this policy option there will be no Indian desire to hurt U.S. economic or strategic interests in any way. To ensure a proper understanding by the U.S. on the revival of Nehruvian elan in foreign policy, Indian diplomacy may have to explain carefully the Indian stance to Washington decision makers and American public opinion at various levels.

The prospects of pursuing a Nehruvian option in the post-cold war order are indeed promising. This takes into account the reality of new emerging power centres, such as Europe, China, Japan, and a number of "regional influential" actors in the Third World. It may be necessary to keep in mind that China's emergence as a major political force in south east Asia and a principal arms supplier in the whole of Asia, along with its desire to exchange sensitivities and concerns (such an in Tibet) with India will need a far more dynamic Indian response to preserve and promote perceived interests in Asia. Nehruvian elan also gives greater resilience to Indian foreign policy in coping with an emerging unstable security situation in the Third World.

Above all, the Nehruvian option seeks to maintain a continuing beneficial (mutual) interaition with the Soviet Union. It is necessary to reiterate that a number of uncertainties are hanging over the Soviet situation and that India has deep stakes in Soviet stability and strength; both the radical and the conservative forces in the prevailing flux of Soviet political dynamics have expectations and affection for India. The Soviets are deeply committed to expanding economic cooperation with India, which may tide them over the difficulties of the present phase of perestroika.

The Soviet leadership is also as interested as ever in continuing to operate within the framework of the peace and friendship treaty concluded with India in 1971. An understanding to extend the treaty by another twenty years has already been worked out. In this respect, the significant change that has taken place in the Soviet stance as a result of Gorbachov's foreign policy initiatives is that the USSR may not like to be dragged into the Indian side to any military or political disputes with Pakistan and China. This is a vital and somewhat unhappy change from India's point of view, but it may be possible for India to adjust. After all, Delhi, on its part, has also been cautious in the past in getting too closely identified with the Soviet position on some sensitive issues such as the collective security arrangement in Asia (1967-70) and Soviet intervention in Afghanistan (1980- 87). There have also been obvious and successful attempts on India's part to diversify its dependence on Moscow in the area of arms supplies, particularly since the early 1980s. It would, therefore, be ideal if a mature relationship with the Soviet Union can evolve at the levels of growing economic cooperation and lower strategic expectations within a Nehruvian policy framework.

The Emerging Political Context

When one talks of reviving Nehruvian elan in India's foreign policy in the changed context, its political projection in the form of nonalignment comes in for intensive questioning. The basic argument advanced is that nonalignment, being a product of the Cold War, has no relevance in the post-Cold War global order. Nonalignment between and against whom, one asks? In the specific context of India, where "nonalignment" came to be perceived as a synonym for an anti-American and pro-Soviet stance, the call for the irrelevance of nonalignment smells of preference for Pax Americana. A close look at the nonaligned framework and also the manner in which India pursued this policy suggests that the relationship between the Cold War and nonalignment, if one existed at all, was superficial and only theoretically tenable. On the ground, nonalignment flourished and prospered during the 1960s and 1970s within the international strategic space created by detente. Since the 1970s, several of the nonaligned countries have entered into close military and strategic relationships with one or the other superpower but still merrily pursue their nonalignment as a foreign policy posture.

An important aspect of nonalignment since the 1970s has been the assumption of a greater role in international economic issues. This is an area where the post-Cold War world will witness much greater dynamism, characterized not only by North-South tensions but also competition among the developed countries for resources and markets in countries of the Third World. The latter, then will have to coordinate their strategies and approaches to cope with the pressures emanating from such competition. Some of the developing nations in Africa and Southern Asia may experience yet another form of economic pressure resulting from their increasing marginalization in the post-Cold War order. The greater competition for aid and concessional financing from the Soviet Union and East European countries and the emergence of new trading blocs in North America, Western Europe, and Asia-Pacific will result in such marginalization of poorer countries.

The Third World will have to meet this challenge through concerted diplomatic efforts in international economic fora, but it does not have many instrumentalities to do this except the G-77 (of which G-15 is a streamlined but still ineffective component) which works within the United Nations system, and the nonaligned movement, which operates at a much broader level. The post-Cold War order, thus, may call for strengthning of the economic dimension of nonalignment.

Another significant aspect of nonalignment has been in the field of decolonization. This objective stands nearly achieved, except in some pockets like South Africa where the struggle for independence, antiracialism, and democratic political order is also on the verge of being accomplished. But this in no way exhausts the concerns of Third World countries over protecting their effective freedom of action in world affairs and their fight against the legacies and persisting tendencies of domination and exploitation. And added to them are the new security concerns resulting from the fall-out of the end-of-Cold-War processes.These include the emergence of regional dominants, a greater flow of arms, unaccomplished and partial disarmament measures, continued occupation of territories and domination over people, and continued threats of intervention and interference (e.g., Panama, Nicaragua, Kuwait).

Even the U.S. action against Panamaian President Noriega, the allied military action against Iraq, and the post-Desert Storm move to create a "safe haven" for Kurdish refugees, however justified on ethical grounds, set a precedent of violation of international law that many Third World countries do not relish. Such developments in future can be avoided only if the nonaligned countries intervene diplomatically in time to diffuse such crises before they explode into violent conflicts. The Nonaligned Movement's (NAM) inaction and lethargy was particularly appalling during the Iraqi invasion of Kuwait and later during the allied attack on Iraq. There is need to check the tendency of the great powers to steamroll the U.N. into endorsing actions initiated by them, and nonalignment has to be geared up to meet some of these challenges. India, therefore, would do well to work not only to preserve the nonaligned forum but also to adjust it to the demands of the post-Cold War and post-Gulf War world order so as to serve its own interests and those of similarly placed Third World countries.

While discussing the political context of the post-Cold War situation and India's response to it, one has to take note of two burgeoning ideologcal forces that are seriously affecting the regional situation in around South Asia. They are the rise of religious and ethnic fundamentalism on the one hand and the assertion of democratic aspirations on the other.[11] These two sets of forces started gaining momentum as major political challenges even during the Cold War years, particularly in the 1980s. But they gained strength and ferociousness with the beginning of the end-of Cold War processes. At present, they threaten peace and stability not only on Western and South Asian regions but also in Soviet and Chinese Central Asian provinces and Europe. Democratic forces have made impressive advances in India's immediate neighborhood, particularly in Pakistan, Nepal, Bangladesh, and Burma, but the endeavor to bring genuine and stable democratic political order in these countries is in no way over. The forces of military influence and authority are still fairly well entrenched in Pakistan and Burma and are down but not out in Bangladesh. In Nepal, while a democratic political order is being evolved under a constitutional monarchy, challenges from the vested interests of the rejected system are still to be fully met.

Indian foreign policy, while facing the internal challenge of ethnic disorder and separatism (in Punjab, Kashmir, and Assam) will also have to contend with the spillover of ethnic and democratic turmoil in neighboring countries. This will call for better management of bilateral relations and greater harmony at the regional level. The emergence of democratic regimes in Nepal and Bangladesh does provide hope for Indian policy, but the challenges in Pakistan, Sri Lanka, Bhutan, and Burma have become in no way less formidable. There are glimmers of hope that SAARC (South Asian Assoriation for Regional Cooperation) will become stronger and more meaningful as a regional organization in responding to South Asian and global challenges confronting India and other member countries, but such hopes are still very fragile and feeble.[12]

Economic Challenges and Opportunities

As mentioned earlier, an important characteristic of the post- Cold War order is going to be greater economic competition among the global economic giants—Japan, the United States, and Europe (particularly Germany)—for resources and markets is the Third World. This competition may lead to a greater regionalization of the world economy, most probably under the influence of one or more of the global giants. Some of the giants may even organize regional groupings as areas of their respective economic influences. The integration of European markets (1992) and attempts to set up effective economic groupings in the Asia-Pacific region(Asia-Pacific community), in addition to the existing ASEAN, may be mentioned in this regard. The Third World stands disunited and confused about how to evolve a proper response to economic activism in the post - Cold War order, as the prospects of the more equitable world economic order, either through North-South negotiations or through fora like the GATT dominated by the developed world, have receded. Even the prospects of aid and assistance to the Third World from the North have declined. The economic and political confusion in Eastern Europe has further added to the Third World's worries as the Soviet Union and some of its erstwhile allies are emerging as new claimants for political attention and available development assistance.

Indian policy needs to take initiatives at two levels to face the emerging economic challenge. On the regional level, it must act to avoid a situation where India is left on the margin of newly emerging economic grouping. Efforts have to be made to impart economic content and meaning to the SAARC, which of course is easier said than done India's efforts to introduce core economic areas such as trade, industry, and currency into regional cooperation under SAARC have not succeeded mainly because of the hesitation on the part of some of the member countries, particulary Pakistan.[13] There are deep political reasons for this economic reluctance but all the countries of the region have to realize that there is no other way out. India would also do well to take some unilateral measures, say, in the field of tariff reduction for imports from its neighbors, to impress upon them that it means well and that greater regional cooperation will not be the cost of any one nation and will benefit all. In addition to this, India could also make use of the circumstances created by the end of the Cold War and changing regional strategic relationships of the super-powers to build bridges of confidence and understanding with Pakistan on political and stragegic issues, which would have a benign impact on regional cooperation.

The other level at which firm action is needed in Indian policy is the domestic one. India is endowed with impressive market potential and reserves of skilled manpower. This is being recognized all over the world and is reflected in the increasing attention paid to India by the global economic giants. The problem, as is again being universally realized, is one of turning this potential into real market and economic strength. The other major hurdles mentioned in this respect are avoidable state controls including administrative inefficiency, and inadequate infrastructural support. Referring to these hurdles, the Indian economy is being described as a tiger caged in ideology," and the prescription to release this "tiger" is to liberalize controls and improve the infrastructure.[14] Forecasting that Europe 1992 will offer a high income market for India, a German deplomat in Delhi said recently that in spite of European integration and "German unification, India's trade with Germany will not decline, provided India "modernizes its production procedures and offers a competitive package in price, quality, and timely delivery."[15] Similarly, the assistant administrator in the Asia and Private Enterprise Bureau of the U.S. Agency for International Development, said that the USAID was on the lookout for opportunities in India and wanted fast bureaucratic clearance of projects. The complaint was that while such clearances take three weeks to three months in the United States, they take as long as three years in India.[16] U.S. pressures on India to liberalize its economy have been exercised through several means and fora and are well known.

The fact that the Indian economy needs liberalization was realized late in the 1970s and considerable action toward this end has already been taken. There is also a broad political consensus in India that further liberalization is in order and should be gradually and selectively carried out. However, the Indian economy, while badly in need of technology and investment—which can be attracted only through liberalization—cannot afford to have the Western multinationals dictate to it. There are genuine apprehensions that unrestrained liberalization may flood the Indian markets with consumer items like potato chips and Pepsi Cola, without helping to harness India's potential economic strengths and capabilities. Thus India will have to chart the course of economic liberalization judiciously and carefully.[17] It should aim at not becoming a preserve of one country or the other, but skillfully exploiting the competitive drives between the U.S. the E.C. (united Germany in particular), Japan, and the Newly Industrialized Economies to build up its own inherent strength and resilience. Accordingly, while seeking new opportunities and economic ventures, India cannot afford to neglect its traditional markets in the

Soviet Union and East Europe, where it will face growing competition, and in Africa and the Gulf, where its full potential has not yet been exploited. All of this provides an economic rationale for the Nehruvian strategic option in India's foreign policy for the post-Cold War world.

Conclusion

All the projections of India's foreign policy in the emerging world order will mean nothing if India fails to prepare itself internally to face the external challenges. Two areas are of particular concern in this regard. One is the question of political stability and confidence. The situation created by political uncertainty since the 1989 elections has not been inspiring form this point of view. The two previous governments could not last in office even for one full year each, and the new Congress government also has a monority character and may not complete its full term. There are also still no reassuring indications about the on-going process of realignment of sociopolitical forces across the Indian political spectrum. With these uncertainties and inherent weakness of the domestic political order, India or any other country can take no meaningful initiatives to face the challenges of global transformations.

The other area of concern in India arises from the prevailing turn created by ethnic and separatist forces. The upsurge of Hindu fundamentalism as a fallout of the Mandir-Masjid (temple/mosque) controversy, if allowed to go uncontrolled, may tear the secular fabric of Indian society apart and result in its political collapse. India as a society and a polity simply cannot survive the persistent onslaught of the majority community's fundamentalism. The Sri Lankan situation is there for any one to draw lessons from. Therefore, Indian leadership will have to tackle this internal question sooner rather than later. Fortunately for India, its basic democratic political framework, based on federalism and secularism, has the required strength and capability to weather the storm, provided political parties, enlightened interests, and the leadership can rise above their parochial and short-term concerns.

REFERENCES

1. This point is discussed in S.D. Muni, "Global Developments: Implications for South Asia," in Air Commodore Jasjit Singh, ed., *India and Pakistan: Crisis of Relationship* (New Delhi: Lancer, 1990).
2. SIRPRI Year Book, 1990: also, Geoffrey S. Kemp, ''Regional Security, Arms Control and the End of the Cold War," *Washington Quarterly* 13:4 (Autumn 1990); and John "Controlling Arms Sales to the Third World," *Washington Quarterly* 14:2 ("Spring 1981-89).
3. *Washington Post,* June 17,1991; *Times of India,* June 18, 20, 1991.
4. Such doubts were expressed by U.S. Assistant Secretary of Defence for International Security Affairs Henry Rowen and Admiral Huntington Handistry, C-in-C, U.S.

Pacific Command, at a seminar in Pune in December 1990.

5. Charles Krauthammer, "The Unipolar Moment," *Foreign Affairs* (America and the World 1990-91) 70:1 (1991). pp. 23 - 33.
6. D.S. Kamtekar, in an insightful article, reflected this dilemma while discussing India's approch to the Gulf crisis, in the *Telegraph* (Calcutta), January, 8, 1991.
7. Angry reactions in India against this Resolution are in *Times of India,* June 15, 1991, *Indian Express,* June 16, 1991, and *Business and Political Observer,* June 15, 1991.
8. It was perhaps with the intent to soften the U.S. stance on economic aid to Pakistan that Prime Minister Nawaz Sharif in June 1991 proposed a five-power (India, Pakistan, U.S., USSR, and China) conference to resolve the "issue of nuclear nonproliferation in South Asia". Pakistan, however, did not succeed in securing a repeal of the Pressler amendment in the US House of Representatives on 13 June 1991. *Hindustan Times* (New Delhi). June 14, 1991.
9. For a brief outline of U.S. expectations towards India, see U.S. Deputy Assistant Secretary of State Teresita Schaffer's paper presented at the Pune seminar, (December 1990). In to IMF pressures and U.S. expectations. This package was outlined in the President's address to the new Parliament. (*Times of India,* July 12, 1991.)
10. Sarvepalli Gopal, Jawaharlal Nehru n. 1, Vol. 2, pp. 58-59.
11. The dimensions of ethnic challenge in South Asia are discussed in Urmila. *Ethnicity And Nation-Building in South Asia (New Delhi: Sage, 1989).*
12. The outcome of the recently held meeting of the SAARC Council of Ministers in Maldives indicated this cautious optimism. (*Times of India,* July 4, 5, 6, 1991.)
13. Reservations entertained by some SAARC members, with the exception of Pakistan to India's proposal to include core economic issues in the SAARC agenda have now been set aside or weakened. Pakistan's reluctance may be both political and due to the economic vested interests. In view of the changing options on economic issues among the SAARC members, Pakistan has agreed to let the SAARC Secreatariat study cooperation in trade. A draft of its study report was presented at the Male ministerial meeting in July 1991. Any concerted action is still far away, however.
14. *Business & Political Observer* carried a dialogue between Swraj Paul and Paul Trition, December 3, 1990. and Rahul Bajaj and Klars Schwab, January 7, 1991.
15. *Business & Political Observer,* December 3, 1990 p. 6.
16. Ibid., p. 9.
17. The debate on wholesale vs. selective liberalization has heated up in India in the wake of the new Congress Government's policy package announced on July 11, 1991. (*Times of India,* July 12, 1991.)

XXIII Concluding Remarks

CHANDRA PAL SINGH

Political geography of India as has been presented in the preceding pages goes beyond analysing relationships between politics and geography. It would be a narrow and microscopic view of reality within territorial extent of India if the search is merely for the political in geography and the geographical in politics. The scope of political geography here covers the vast man-environment system, with all its interacting subsystems, their spatial manifestations and variations within the defined territory of India, and its external relationships. The organization of human systems in society, economy and polity, their interaction with physical and biotic systems, their varying mosaics from place to place, region to region and nation of nation, lays down the overall setting in which the Indian state exists and functions. Her ranking among the countries of the world depends on the best synchronization of all the systems towards attainment of high economic standards for all her citizens, high level of participation in the world economy, and possession of high level of coercive power. Whether these goals are really desirable from the point of view of their attainability, as well as the sustainability of the environment from which they could be obtained, and whether they lead to equitable distribution of wealth by exploitation of nature, are different matters for debate.

The approach emphasizes the indivisibility of knowledge which is presently compartmentalized among different social, physical and biological sciences. The approach here also links man and nature from local to global scales. It has been demonstrated that concepts developed in social, physical and biological sciences are necessary for building a comprehensive knowledge of a place, a region or a country. When we think of nation-building then it cannot just be the interplay of social, economic and political systems of a territory, as is done in the approaches of political sociology and political economy. The multi-levels of identity are territorial and have relationships with natural systems within the territory, which shape the lives, actions and perceptions of individuals and groups of individuals that define their relationships, complex equations among themsleves, and ultimately identification and complete assimilation with territory, which exists along with similar other areal units.

The contributions in this volume are merely illustrative of the interconnections among different phenomena studied by different social and natural sciences within a country. A political geography of India conceptualizing the unity of these sciences opens up interdisciplinary exchange among them to understand her internal organization, numerous processes operating in it, giving her a particular identity and status, with which she conducts her external relationships with other states in the world.

This volume has left out several aspects of the political geography of India from discussion and illustration. For instance, electroal geography and the state of environment in India. Elections are occasions when the interplay of all factors that become input in opinion building and their expression through the ballot box are in sharp focus. One likes to know why an individual or a group of individuals at a particular place, or a region, voted the way he voted at that point of time. The placement of the individual in his total context is an exercise in understanding the complex network of social, economic, political, spatial, environmetal structures and flow of information and energy over this network, of which the individual is just one of the numerous nodes. The spatial pattern of voting and its variation from place to place, region to region, is just the result of aggregates of opinions formed by inividuals in this complex network. Electoral geography of this nature is studied in the same framework as political geography, of which the former is an essential part.

The state of the environment reflects the concerted action of man on the environment and its consequences. A political geography of India should include the state of environment and ecology as politics through social and economic pressures and processes makes demands on them and often brings about change in their state. It is a coincidence that environmental concerns have come into public focus and debate at a time when the superpower rivalry has come to an end and threat of nuclear conflict has receded. The high energy consuming behaviour with wasteful expenditure on arms race in the North and the continuously increasing demand for sustenance of the huge and fast growing human and bovine population in the South beginning from the middle of the present century had started accumulation of environmental problems. The first signs of the problem started appearing in the form of industrial and urban pollution. Whether the cold war had ended or not the environmental problems were going to burst upon the world in any case. Concentration of polluting industries, proliferation of nuclear plants, application of chemical ferilizers in agriculture, deforestation, extraction of minerals, oil and metals, all of them at a continuously accelerated rates in the human effort for increasing gross domestic product, were bound to have an adverse effect on the environment and ecology. Politics played a crucial role in its overt and covert support for exploitation of resources and marketing the products. It has given rise to the subdiscipline of political ecology in the recent years. The recent trends of privatization, aided with modern means of communication with satellite technology, actively supported by the state (in some cases even promoted by the

thereby making greater and quicker demands on the environment. Every country now has a ministry of environment which monitors the exploitation of resources and its consequences within its territorial jurisdiction. It is important to know who gets what and where in a country because that will determine the nature of its internal politics and policies to attain the national objectives. On the other hand, relations with other countries are also governed by the resources and technology to exploit them. The new world order is based on economic power of individual or group of countries which have formed trading blocks. The place of traditional diplomats is being taken by economist or business dislomats in international politics to ensure a good place in the global economy. The political discourse at international level constitutes economic interests more than anything else. The wave of privatization all over theworld has unleashed forces that are going to overstretch the limits of the earth, which will destroy the environment throughout the world. The polarization of countries into the North and the South, creation of trading blocks in the northern hemisphere, and mushrooming of pressure groups for protection of the environment and ecology in all the countries have changed the parameters of international politics. A political geography of India would be incomplete without taking into account economic development and its impact on the environment in India.

The traditional geopolitical theories have been rendered obsolete with the end of the Cold War and emergence of a unipolar world. New alignments and power centres are emerging which have globe influencing capacity. The creation of European Community (EC) and North American Free Trade Agreement (NAFTA), and the growing membership of Association of South East Asian Nations (ASEAN) have changed the players on international scene from NATO, SEATO, CENTO, Warsaw Pact and Non-Aligned countries. The place of India in this changed scenario will depend on her ability to utilize her physical and human resources and creating a sizable niche in the world economy. A political geography of India has to take into account the new geopolitical situation in the world, replacing or modifying the geopolitical theories that were customarily included earlier. Some of the papers included in the volume could be helpful in that direction.

Some of the studies included in this volume have suggested certain geographical frameworks for political studies in India. For instance, the heartland-rimland framework of Rudolph and Rudolph, and the broad cultural-linguistic regions conditions by the physical geography of India as given by Panikkar, Venkatachar, Ray and others. Even in general parlance the use of Hindi heartland is often made. These need to be investigated, analysed and refined further to be utilized by all the social scientists working on India. Indian political geography would have been richer if it had seized upon the ideas and concepts of Panikkar in 1955. Such a mistake must not be repeated now. The exchange of concepts and tools devised in other social sciences would enrich all, including political geography.

Index